ELECTRIC FIELDS
IN
VERTEBRATE
REPAIR

Natural and Applied Voltages
in Vertebrate Regeneration and Healing

ELECTRIC FIELDS IN VERTEBRATE REPAIR

Natural and Applied Voltages in Vertebrate Regeneration and Healing

Richard B. Borgens
Kenneth R. Robinson
Joseph W. Vanable, Jr.
Michael E. McGinnis

With
Colin D. McCaig

Alan R. Liss, Inc., New York

Address all Inquiries to the Publisher
Alan R. Liss, Inc., 41 East 11th Street, New York, NY 10003

Printed in the United States of America.

Library of Congress Cataloging-in-Publication Data

Electric fields in vertebrate repair : natural and applied voltages in vertebrate regeneration and healing / Richard B. Borgens . . . [et al.].

p. cm.
Includes bibliographies and indexes.
ISBN 0-8451-4274-7
1. Electrotherapeutics. 2. Nervous system—Regeneration.
I. Borgens, Richard B.
RM871.E4 1989 88-33363
615.8'45—dc19 CIP

Cover Illustration: Many scientists who study regeneration and wound healing can trace their interest to studies of regeneration of the amphibian limb. This illustration suggests the regrowth of a limb in the Eastern Redspotted Newt and was inspired by a composite photograph from the book, *Principles of Regeneration*, by Richard Goss (1969, New York: Academic Press).

Contents

Author Affiliations

Richard B. Borgens, Center for Paralysis Research, Department of Anatomy, School of Veterinary Medicine, Purdue University, West Lafayette, IN 47907

Colin D. McCaig, Department of Physiology, University of Aberdeen, Marischal College, Aberdeen AB9 1AS, Scotland

Michael E. McGinnis, Center for Paralysis Research, Department of Anatomy, School of Veterinary Medicine, Purdue University, West Lafayette, IN 47907

Kenneth R. Robinson, Department of Biological Sciences, Purdue University, West Lafayette, IN 47907

Joseph W. Vanable, Jr., Department of Biological Sciences, Purdue University, West Lafayette, IN 47907

Introduction

We offer this book to the person desiring an introduction to the biology of endogenous electrical fields and its role in the regeneration and healing of vertebrate tissues. This area of scientific investigation and the related use of applied electricity in medicine have a very long history, the latter dating to before the emergence of our understanding of electricity as a physical entity. Beal (1969) describes what may be the first use of electricity as a therapeutic agent—Antero (a former slave of the Roman Emperor Tiberius) discovered that his symptoms of gout were alleviated after stepping on an electric fish. Use of the Mediterranean torpedo fish as a therapeutic agent apparently became more common in Rome after this time. Dioscorides, an herbalist and surgeon in Nero's army, used discharges from the fish to treat not only gout but migraine and other physical problems such as prolapsed anus (Beal, 1969, 1974).

Many historical treatments chronicling the development of the physiological sciences usually separate the *scientific* appreciation of endogenous currents and voltages from the *medical applications* of electricity. Actually, such an arbitrary division is neither convenient nor instructive, since most of the great contributors to our understanding of the role electricity plays in excitable tissues were physicians as well as scientists. There was little distinction after the Enlightenment. Theologians and philosophers usually possessed some degree of medical training, which was considered a necessary component of proper education. For example, Luigi Galvani was a competent physician and therapist of his day. Not only did he investigate the electrical basis of contraction in frog muscle, he used electrotherapy as a part of his medical arsenal. John Wesley, the great theologian and Methodist reformer, was also of this ilk. He wrote several popular pamphlets on medical subjects—one dedicated to the medical use of electricity (*The Desideratum: Or Electricity Made Plain and Useful. By a Lover of Mankind and of Common Sense,* London, 1760). In another text he wrote that he was "firmly persuaded there is no remedy in nature for nervous disorders of every kind, comparable to the proper and constant use of the electrical machine . . .

the palsy [should] be electrified, daily for three months, from the places wherein the nerves spring, which are brought to the paralytic part'' (Schiller, 1982; Wesley, 1776).

Early practitioners made use of an emerging knowledge of the human body as a vehicle for electric conduction (described by Gray in 1720), as well as the concepts of polarity and insulation (discovered by Du Fay in Paris in 1733—Benjamin Franklin named Du Fay's ''vitreous'' and ''resinous'' kinds of electricity ''positive'' and ''negative''). Applications of both steady and static electricity to treat myriads of unrelated diseases and traumatic injuries became commonplace after Alessandro Volta [the discoverer of direct current (DC)] described his battery in 1800. Simultaneously a physiology based on physics was emerging—the ''unity of the organic and inorganic parts of nature, dear to the German mind'' (Schiller, 1982), found its expression in the studies of the great father of modern physiology, E. DuBois-Reymond (Fig. 1). DuBois-Reymond elegantly described the physical basis of spontaneous nervous conduction, finding it to be electrical in nature, and the controversial ''animal electricity'' (after Galvani) was vindicated. ''Electrotherapy'' then became respectable.

I have not discussed in detail the development of an understanding that ''animal spirits'' were electrical in nature, or that the laboratory use of electricity was instrumental in the understanding of how the neuromuscular system works (from Galvani to DuBois-Reymond), as such a discussion is common to most treatments of the history of physiology. Less well appreciated is the great debt that the development of certain medical specialties owes to the early practitioners of ''electrotherapeutics.'' Francis Schiller (1982) traces the roots of the great science and medical specialty neurology to all three concepts. He writes, ''Less well anchored in our historical awareness is the decisive role that electrotherapy and diagnosis have played in emancipating neurology as a specialty.''

In this regard it is interesting to trace the intimate weaving of *neuroscience* and *electrotherapy* into neurology—the application of neuroscientific principles to medicine. No better example in the history of medicine can be found than the collaborations of Jean-Martin Charcot (1825–1893) (considered by many to be the father of modern neurology) and the medical pioneer, Armand Duchenne de Boulogne. These collaborations began at a low ebb in the personal life of Duchenne and at a high point in Charcot's professional life. Duchenne moved from Boulogne to Paris after his wife died. He had fallen into a troubled time, both professionally and personally. His in-laws

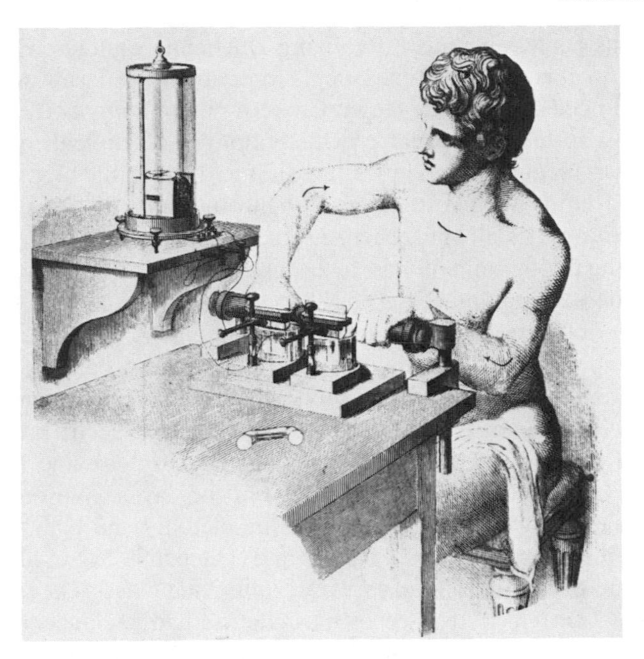

Fig. 1. Measurements of skin currents and skin wound potentials using a galvanometer. (Reproduced from DuBois-Reymond, 1843.)

refused him access to his only son, he was shunned by the medical communities in Boulogne and Paris, and his work (and notions on electrotherapy) were ridiculed. Charcot, on the other hand, was one of the most prominent physicians in Paris and director of the infamous Salpêtrière. The Salpêtrière (from ''saltpeter''—a component of early gunpowder and explosives) began as an arsenal built on the left bank of the Seine in 1565. Civil strife in France brought thousands of homeless to Paris and swelled the already significant numbers of Parisian beggars, prostitutes, and mentally and physically disabled. By royal edict, Richelieu formed the general hospitals of Paris; the Salpêtrière became a treatment and confinement ''hospital'' primarily for poor and disabled women and prostitutes, and later for beggars. By 1680, it had evolved into the largest asylum in Europe—a great proving ground for the emerging disciplines of surgery, neurology, and psychiatry (at the expense of its hapless inmates). Realizing Duchenne's genius, Charcot gave him access to this inexhaustible supply of patients and saw to all of his professional needs. Duchenne taught Charcot photography and exposed him (by example) to his own structured methodology of clinical investigation. Duchenne was an innovator. He was the first to develop the biopsy technique in medicine (during his investigations of neuromuscular disease). The biopsy needle, forerunner of all subsequent needles, was his invention. He is also considered the patron saint of muscle disease (Capildeo, 1982), credited with the first formal description of the childhood form of muscular dystrophy. He was also the first to use artificial applications of electricity as a *diagnostic* as well as therapeutic tool. Through the intelligent use of Faraday's induction coil, Duchenne stimulated precise points where nerve trunks entered muscle bundles (as Wesley had suggested). This localized electrification during therapy and diagnosis evolved into a strong methodology and reversed the "therapeutic dilettantism" (Schiller, 1982) of the day. It allowed Duchenne to draw many important conclusions on the anatomy and pathology underlying neuromuscular disease. Charcot and Duchenne were humane men working in an insane and, in many ways, inhumane place. The conventional application of current to the deeper musculature had been accomplished by the insertion of electrified blades or needles (favored by the anatomist Magendie), a painful technique. To alleviate this torture, Duchenne developed the surface electrode for neuromuscular stimulation. When Charcot appointed Duchenne to a permanent staff position in 1862, he called him his "master" (Capildeo, 1982; Schiller, 1982). Duchenne was also indirectly instrumental in legitimizing electrotherapy in Europe. Although electrical treatments had been available in hospitals and through private practitioners since the beginning of the 18th century, by the middle of the 19th century it became more respectable—in part through the scientific inquiries of DuBois-Reymond (as discussed earlier), but also through a book on "galvanotherapy" written by Robert Remak (dedicated to his mentor and protector, Alexander von Humboldt). Remak was an early proponent of the neuron theory and the role of electrochemical influences in physiology. His book (1858) was published 3 years after a text by Duchenne (1855) on the localized application of electricity for diagnostic and therapeutic use. Remak was deeply impressed by Duchenne's approach and was inspired by visits to his wards (Schiller, 1982). In America, Benjamin Franklin (1706–1790) experimented with the use of static electric discharge on paralyzed patients (Geddes, 1984). His experiments are described in detail in a letter to Dr. John Pringle, reprinted at some length by Geddes (see also Robinson, Introduction to Chapter 1, this volume).

I have described the medical application of electricity in general terms. Geddes (1984) categorizes such applications into two basic types: 1) applications of static electricity; and 2) galvanism, or the "passage of a constant electric current through the whole body." In the former, static electricity was applied to the patient via a capacitor charged to a high voltage by a static electricity machine. The static discharge was directed in three ways: electrification of the whole body, called the electric air bath; the "drawing of sparks" from such an "electrified" body; and localized capacitive (Leyden jar) discharge to the patient. Geddes reports that the "response to the positive air bath was certainly dramatic. When the static-electricity machine was activated adequately, the patient's hair on the head and exposed surfaces rose. If the patient was in the dark, a halo (corona discharge) surrounded the body. The patient felt warm, exhilarated but relaxed and perspired slightly. The face was usually flushed and the pulse rate increased. Often the patient felt drowsy and desired to sleep after the treatment, which lasted about 15–30 minutes; however, periods up to three hours were occasionally used" (Geddes, 1984) (Fig. 2). A variation of the electric air bath was the "drawing of sparks" from the electrified body. The experience of the patient during such treatment was not as enjoyable as in the electric air bath. Depending on the distance that the spark was induced to jump across the air gap from the body to the electrode, the sensation could be quite uncomfortable. Hysterical paralysis, neuralgia, rheumatism, and many other ailments were treated in this way. The localized discharge of a spark to a restricted part of the body (the third method of static electrical application) was applied to almost any portion of the body for almost any ailment (Figs. 3–6). As in the "drawing of sparks," these localized discharges could be very painful. One common malady treated in this way was marital impotence in men. To treat such impotence, sparks were discharged locally to the genitalia. Great successes, probably credible, were claimed for this therapy. (I speculate that since most marital impotence arises from disinterest rather than dysfunction, it is probable that the patient developed a renewed interest in connubial matters—certainly more acceptable than a return to the physician!)

The second type of electrotherapeutic application, galvanic therapy (the application of constant current to the body), could also be applied locally or *systemically*. Geddes reminds us that such direct current application evolved from Galvani's discovery of muscle twitching in frogs in response to contact with a bimetallic couple (an electrochemical cell) and from the popularization of this "stimulating" effect of DC by Galvani's nephew, Aldini. In this short but interesting review, Geddes describes a remarkable episode in which Aldini went to London to investigate the effects of applied current on the

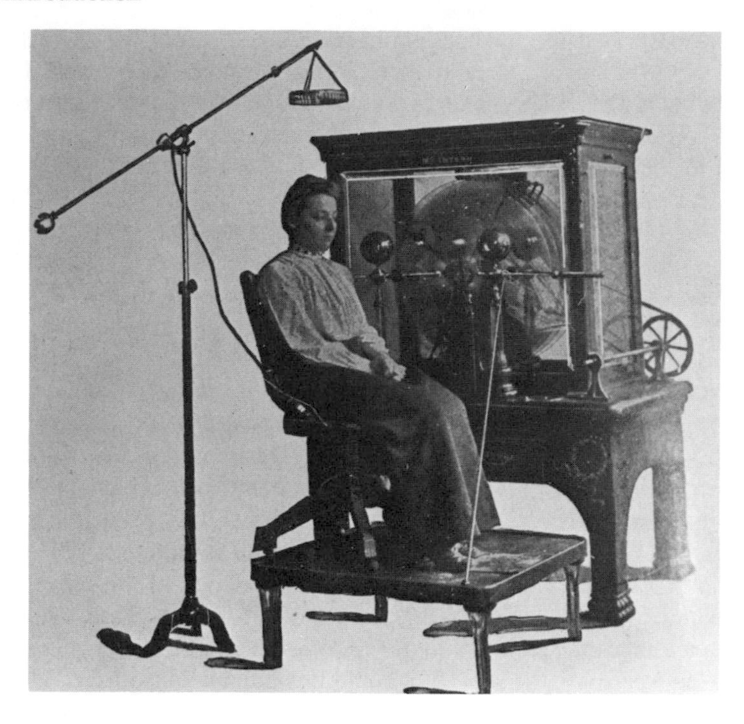

Fig. 2. A static electricity therapeutic machine circa 1860–1900; the electric air bath or static breeze.

body of a criminal who had been executed by hanging. Using a voltaic battery of 40 zinc and copper plates in dilute hydrochloric acid, electrodes were pressed to various parts of the corpse (kept at about 30°F for 1 hour after his execution and during transportation to the London College of Surgeons). Geddes recounts, "On applying the conductors to the ear and to the rectum, such violent muscular contractions were executed, as almost to give the appearance of reanimation. Recall that these events occurred more than one hour after execution and certainly did much to attribute resuscitative powers to electricity." Indeed, popularized accounts of such experiments on cadavers undoubtedly influenced Mary Wollstonecraft Shelley in the writing of the Gothic classic *Frankenstein,* published in 1818. Here electrical discharge provided the impetus for the reanimation of the dead body—reflecting the general belief that the "spark of life" was electrical in nature and also possessed curative powers (Borgens, 1983). These "curative

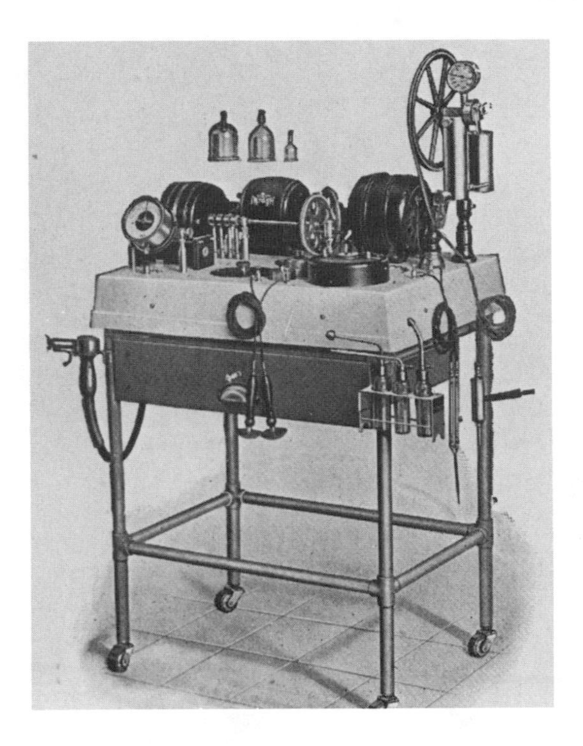

Fig. 3. "A universal electro-therapeutic apparatus affording all low tension current modalities." (Reproduced from Neiswanger, 1922.)

powers," however, became increasingly more disputed as more physicians attempted to treat patients with electricity. The rigorous scientist was more cautious. For example, Benjamin Franklin was careful to point out that his treatments to paralytics produced only temporary reversal of paralysis and that such therapy was not effective for all types of paralysis. After the hard-fought gains in credibility of the previous 100 years of investigation, by 1850 the respectability of an electrical approach to the treatment of disease began to deteriorate, in response to outright charlatanism as well as the incessant promulgating of "electrotherapy" by less careful physicians. Although electrotherapy traced its development from the minds and hands of the scientific and medical elite of 1750–1850, it became the domain of the medical proletariat after this time. An example of such a publicist was George Beard in the United States. Beard (1867) aggressively sought (and

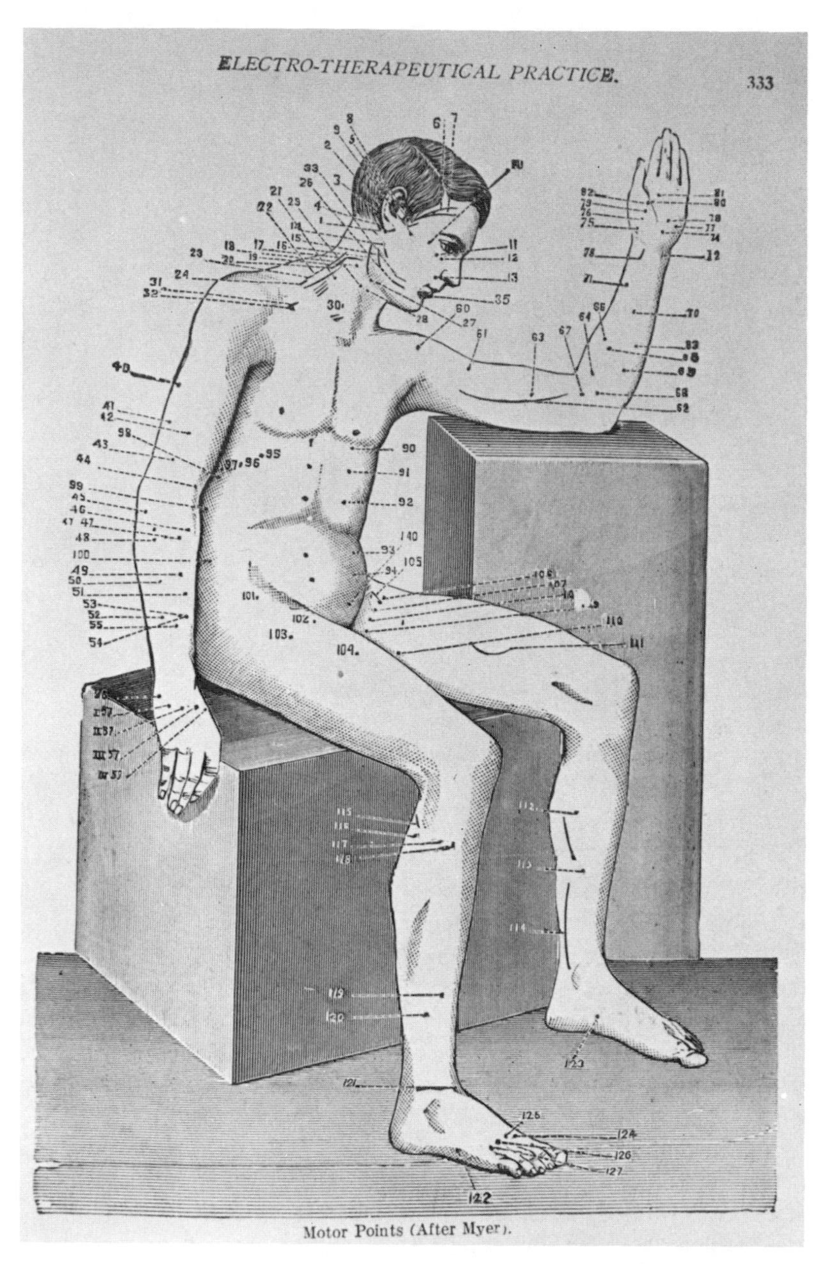

Fig. 4. Sites of electrode application for diagnosis and treatment. (Reproduced from Neiswanger, 1922.)

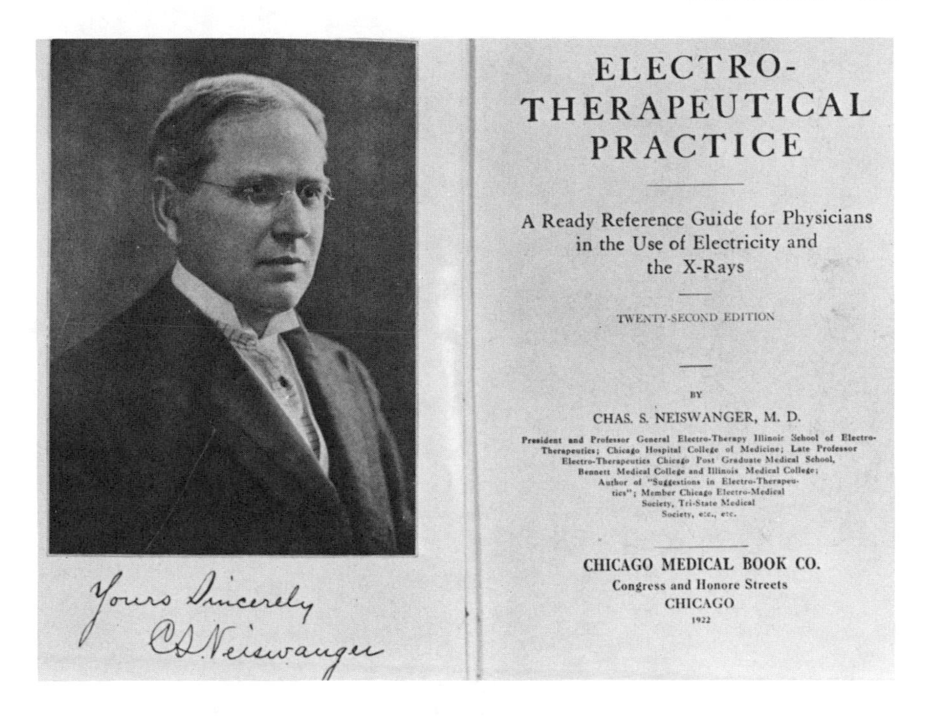

ELECTRO-
THERAPEUTICAL
PRACTICE

A Ready Reference Guide for Physicians
in the Use of Electricity and
the X-Rays

TWENTY-SECOND EDITION

BY

CHAS. S. NEISWANGER, M. D.

President and Professor General Electro-Therapy Illinoir School of Electro-
Therapeutics; Chicago Hospital College of Medicine; Late Professor
Electro-Therapeutics Chicago Post Graduate Medical School,
Bennett Medical College and Illinois Medical College;
Author of "Suggestions in Electro-Therapeu-
tics"; Member Chicago Electro-Medical
Society, Tri-State Medical
Society, etc., etc.

CHICAGO MEDICAL BOOK CO.
Congress and Honore Streets
CHICAGO
1922

Fig. 5. Title page of Neiswanger's 1922 electro-therapeutical guide.

received) great publicity and notoriety as a proponent of electrotherapy for many physical ailments. His book on the subject, coauthored by Rockwell, was printed in seven editions over a period of 17 years (Fig. 7). Although popular with the laity, he was less admired by colleagues. His ideas and claims fell before the onslaught of his critics, many of whom viewed Beard as a ''kind of P.T. Barnum of American medicine'' (Rosenberg, 1962).

I have tried to chronicle briefly the simultaneous development of the scientific and medical use of electricity. As in art, music, and other creative endeavors, periods of interest and innovation in the scientific world are subject to fad and fashion. In the last decade, we once again began to see a reversal of attitude toward the use of applied electricity to treat certain conditions in medicine (Fig. 8). I believe this renewed interest developed from a steadily growing body of literature linking endogenous electrical fields to control mechanisms of development and regeneration in animals and plants (in part reviewed in this book). Such an interest was

THE AUTHOR'S WALL CABINET.

Fig. 6. Office cabinet. (Reproduced from Neiswanger, 1922.)

almost certainly influenced by the development of the vibrating probe system for the measurement of extracellular current (Jaffe and Nuccitelli, 1974). Fundamental studies in vertebrate regeneration and healing, based in part on measurements made with this device, form the basis of this book.

In this volume we have focused on the areas of investigation that provide the modern basis for understanding the role of ionic currents and natural voltages in vertebrate regeneration and repair. These areas are: electrical control of amphibian limb regeneration (the archetype for students of regeneration); and the role of natural and applied fields in bone fracture healing, nerve regeneration, and skin wound healing. These areas of research pres-

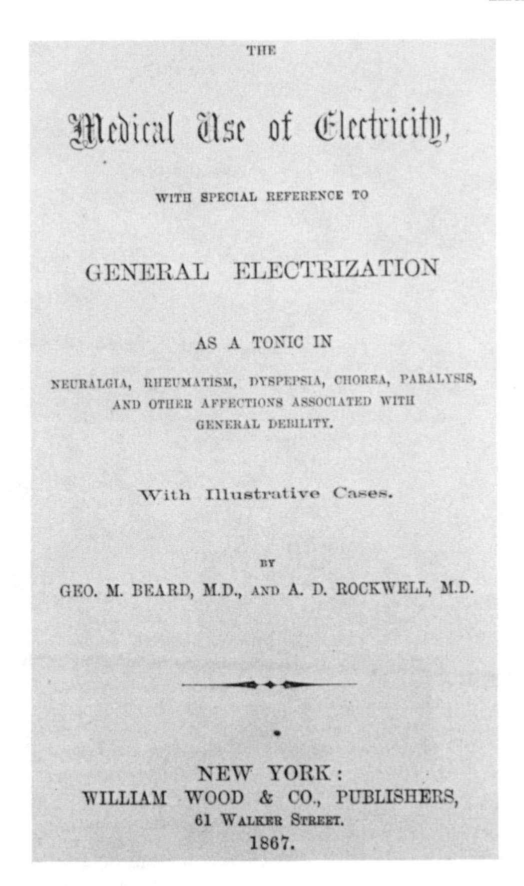
THE

Medical Use of Electricity,

WITH SPECIAL REFERENCE TO

GENERAL ELECTRIZATION

AS A TONIC IN

NEURALGIA, RHEUMATISM, DYSPEPSIA, CHOREA, PARALYSIS, AND OTHER AFFECTIONS ASSOCIATED WITH GENERAL DEBILITY.

With Illustrative Cases.

BY

GEO. M. BEARD, M.D., AND A. D. ROCKWELL, M.D.

NEW YORK:
WILLIAM WOOD & CO., PUBLISHERS,
61 WALKER STREET.
1867.

Fig. 7. A turn-of-the century medical text on electro therapy.

ently supply the most convincing direct and most convincing *circumstantial* evidence supporting the critical role of electrical controls in vertebrate developmental biology, as well as providing interesting possibilities for the development of new medical applications. The focus then, is on regeneration and repair in vertebrates and the extrapolation of this knowledge to clinical possibilities. (In the case of fracture repair, applications of electrical fields to intransigent fractures are now commonplace.)

We do not attempt to review the role of ionic currents and fields in vertebrate embryology, except in instances in which both the anatomy and

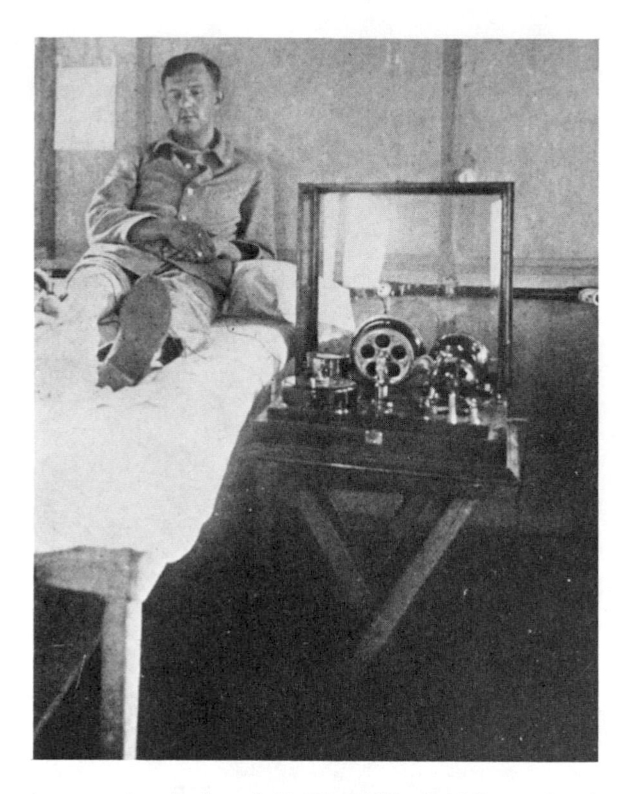

Fig. 8. A gunshot wound to the leg of this World War I soldier produced a nerve lesion causing the loss of flexion and extension of the limb. The patient was treated with "galvanism" and daily massage for 10 weeks. The attending physicians claimed a restoration of function, with the patient "walking about the hospital." (Reproduced from Neiswanger, 1922.)

physiology of development and repair closely parallel each other. This parallelism exists between the embryonic development of the vertebrate limb and its regeneration, as well as between the developing and regenerating nerve fiber.

While discussing the scope of this volume, I also wish to stress that the emphasis here will be on the role of *steady direct current* electrical fields and their exogenous application. There are several reasons for this restriction. First, with the possible exception of bone, it is becoming clear that the endogenous fields known to be relevant to development and regeneration are

DC in nature. Although we do not dispute that artificially applied radio frequency stimulation, certain electromagnetic applications, and applied alternating currents and fields may produce a variety of biological responses, we question if these modalities are related to the natural physiology of the organism or to the natural biology of the cell.

The medical overtones of most of these chapters are intended. Direct application of our developing knowledge of the electrophysiology of development, regeneration, and repair may well be one of the next great areas of expansion in applied medical science. I recall hearing Carl Brighton, one of the pioneers in the medical use of applied currents in the treatment of chronic fracture nonunions, muse that historically there are only three ''eras'' in the orthopedic management of fractures: ''*carpentry*,'' *biologically* based management, and, in modern times, *biophysics,* providing new avenues of orthopedic practice. The earliest era of fracture management, ''carpentry,'' consisted of treatments whereby broken bones were realigned and apposed by the prevailing method of the day—tethering, splinting, casting, or perhaps pinning. The second era (the biological) can be characterized by an understanding of the biological basis of fracture repair, providing novel and more successful therapies in difficult cases, such as bone grafting. Today, Brighton suggested, we are in the era of biophysics. Real advancements in our knowledge of the ''material-science'' properties of bone and its bioelectrical properties (such as its ability to behave as an electrical transducer) are translating into novel bone replacement (such as biodegradable ceramic prosthetics) or the healing of long-standing nonunions with weak, applied direct current.

Thus this book is directed to the basic scientist, the clinician, and the clinical researcher wishing a better understanding of the fundamental science and phenomenology of electric fields and regeneration and repair.

Our point of departure is Chapter 1, by Kenneth Robinson, in which he introduces the principles of electricity in biological systems and the pitfalls in measurement techniques. He then details the design and construction of the vibrating probe system for measurement of extracellular current—the most modern instrument allowing rigorous investigation of endogenous electricity.

Lastly, we wish to mention that all authors in this volume can trace their interest in this area of research to one man, Lionel Jaffe. Professor Jaffe is the originator of the modern probe system, and a vigorous and stimulating developmental biologist. We have all been, at one time or another, either his collaborator, his student, or even a student of one of his students at Purdue University. Our debt to him is inestimable.

REFERENCES

Beal JB (1969) Electricity and Disease. N. Scientist, July 24, p. 173.

Beal JB (1974) In Llaurado JG, Sances A Jr., and Battocletti JH (eds.): *Biological and Clinical Effects of Low Frequency Magnetic and Electric Fields.* C. C. Thomas, Springfield, Ill.

Beard G and Rockwell AD (1867) *Practical Treatment in Medical and Surgical Use of Electricity.* William Wood Co., New York.

Borgens RB (1983) The Electricity of Life and Limb. *Yearbook of Science and the Future.* Encyclopedia Britannica, Inc., Chicago, p. 50.

Capildeo R (1982) Charcot in the 80's. In Rose FC and Bynum WF (eds.): *Historical Aspects of Neurosciences.* Raven Press, New York.

DuBois-Reymond E (1843) Vorlaufiger abrifs einer untersuchung uber den sogenannten froschstrom und die electromotorischen fische. Ann. Phys. U. Chem. 58:1.

Duchenne de Boulogne A (1855) *De l'electrisation localisée, et de son application à la pathologie et à la therapeutique.* J.P. Bailliere, Paris.

Geddes LA (1984) A Short History of the Electrical Stimulation of Excitable Tissue Including Electrotherapeutic Applications. *The Physiologist* (supplement) 27:1. The American Physiological Society, Bethesda, MD.

Jaffe LF and Nuccitelli R (1974) An ultrasensitive vibrating probe for measuring steady extracellular currents. J. Cell. Biol. 63:614.

Neiswanger CS (1922) *Electro-Therapeutic Practice,* Chicago Medical Book Co., Chicago.

Remak R (1858) *Galvanotherapie der Nerven und Muskelkrankheiten.* Hirschwald, Berlin.

Rosenberg CE (1962) The place of George Beard in 19th century psychiatry. Bull. Histo. Med. 36:247.

Schiller F (1982) Neurology: The Electrical Root. In Rose FC and Bynum WF (eds.): *Historical Aspects of the Neurosciences.* Raven Press, New York.

Wesley J (1776) *Primitive Physick.* Reprinted as *Primitive Remedies,* Woodbridge, Santa Barbara, California.

Richard B. Borgens

Center for Paralysis Research,
Department of Anatomy, School of Veterinary Medicine, Purdue University,
West Lafayette, Indiana 47907

Acknowledgments

We acknowledge the fine original art work by David Williams and Lynn O'Kelley, graphics by Rose Beghtol-Hunt, the Purdue Research Foundation for permission to use these figures, and Mary Jo Maslin for preparation of this volume.

We wish to thank the National Science Foundation, the Spinal Cord Society, the American Heart Association, the Department of Defense, Department of Army, and National Institutes of Health for their support of our research, and the Wellcome Trust and International Spinal Research Trust for their support of Dr. McCaig's visit to Purdue, summer of 1988.

Electric Fields in Vertebrate Repair, pages 1–25

CHAPTER 1

Endogenous and Applied Electrical Currents: Their Measurement and Application

Kenneth R. Robinson

Department of Biological Sciences, Purdue University, West Lafayette, Indiana 47907

INTRODUCTION

People have been fascinated by the interaction of electricity and living organisms since the first awareness of electrical phenomena arose. Benjamin Franklin, in a letter to John Pringle dated December 21, 1757, reported on the effects of applying electrical shocks to paralyzed limbs of human subjects (Goodman, 1945). He said that sensation and voluntary motion were restored, albeit temporarily, and he concluded that "some permanent advantage might have been obtained if the electric shocks had been accompanied with proper medicine and regimen. It may be, too that a few great strokes, as given in my method, may not be so proper as many small ones. . . ." Other early scientists noted the electrical manifestations of living organisms, and over the years, especially in the first half of this century, there has been much interest in steady bioelectric currents and their roles in growth, development, and repair, as summarized by Lund (1947).

After the spectacular success of the Hodgkin-Huxley formalism in explaining electrical excitability of cells, interest in steady currents in nonexcitable systems declined and did not revive until the 1970s. The major stimulus to this renewed interest was the development of the "vibrating probe," which was capable of measuring the tiny voltage gradients produced by the currents and had adequate spatial resolution for measurements on single cells. In this chapter, I will discuss the cellular basis for steady currents, the techniques used for measuring and applying them to cells, a few examples of such measurements, and the possible roles of these currents in cell growth and polarization. My purpose here is to describe in general the biophysics and measurement technology of steady currents, not to give a comprehensive review of this expanding field.

CELLULAR BASIS OF STEADY CURRENTS

The basis of all electrical currents through cells is the electrochemical gradient of the particular ion involved. These gradients are maintained across the plasma membrane at the expense of cellular energy. In many animal cells, for example, the resting membrane potential is largely a potassium diffusion potential, and leakiness to ions other than potassium is compensated for by various ionic pumps such as the Na^+-K^+ ATPase. These pumps may be electrogenic and may or may not contribute directly to the resting potential by a significant amount. In any case, virtually all cells have a negative membrane potential (that is, the cytoplasm is electrically negative with respect to the surrounding medium), and the physiologically important ions are maintained in nonequilibrium distribution across the plasma membrane. The equilibrium potential for any ion can be calculated from the Nernst equation

$$E_x = (RT/ZF)\ln([X]_o/[X]_i)$$

where the square brackets refer to the activities of ion X inside (i) and outside (o) the cell, (RT/F is approximately 25 mV at physiological temperature), R is the gas constant, T is the absolute (Kelvin) temperature, F is the Faraday constant, and Z is the valence of the ion. A typical value for $[Na]_o/[Na]_i$ might be 15, so $E_{Na} = +68$ mV. Since resting membrane potentials are normally tens of millivolts negative, sodium is far from equilibrium and will flow into cells through any conductive pathway. All cells do have a measurable resting sodium permeability, so sodium continuously leaks into cells and is dealt with, as already noted, by the sodium pump. Calcium is generally even further from equilibrium than sodium, and specialized mechanisms exist for expelling the calcium that leaks in. Likewise, potassium, chloride, and hydrogen ions are all out of equilibrium by greater or lesser amounts in the resting state of a typical cell, although potassium and chloride may be quite close to equilibrium in many cases.

The basis of electrical excitability of many nerve cells is the functioning of voltage-gated sodium channels in their membranes. Much is known about the physiology and structure of these membrane-spanning channel proteins (Salkoff et al., 1987); indeed, the sodium channel gene has been cloned and the mRNA injected into *Xenopus* oocytes (Noda et al., 1986), with the result that functional sodium channels are expressed in the oocytes' plasma membranes. When the membrane of an excitable cell is depolarized, the probability that a given channel is in the open state is increased. When a channel opens, sodium ions will flow in, down the steep electrochemical gradient, thus depolarizing the membrane. This sodium permeability increase spreads over the surface of the cell, and inward current is generated at the active region. The sodium channels open for a short time, after which they close and become refractory to further opening, so the membrane returns to the resting state, the whole episode lasting a few milliseconds. For an exceptionally lucid, detailed account of these matters, see the monograph by Hille (1984).

The currents described above are quite different from the steady currents that are the focus of this chapter, although the underlying physiological basis is similar. By methods that will be described later, it has been shown that there are numerous instances of long-lasting, steady electrical currents whose polarity is associated with the morphological polarity of a developing or growing system. These currents last for minutes, hours, or even days, and it is thought that they may be effectors of morphological polarity. From what has been said, it is easy to see how such currents might arise. I have already pointed out that cells have a finite (although small) resting permeability to the ions that are kept far from equilibrium, and the movement of those ions will constitute one branch of an electrical current loop. If these leaks are distributed uniformly over the surface of a cell, the resultant current loops will be small, on the order of protein spacings in the membrane. However, the current paths that are measured by the vibrating probe have dimensions much larger than that, on the order of tens of microns, implying that the ion channels (or electrogenic pumps) are asymmetrically distributed in the plane of the membrane so that there is net charge entry in one region of the cell and net charge exit in another. It should be pointed out that both the entry sites and the exit sites do not have to be asymmetrically distributed in order to generate a net current; it is sufficient for one or the other to be localized.

I have outlined the well-understood basis of current flow in excitable cells, but what is the relationship of those currents to steady currents? The available evidence suggests that it is leaks rather than pumps that are localized, and I assume that these leaks are ion channels that are related (at least in their ion-selective regions) to voltage-gated channels of excitable cells. One possibility is that they are identical to voltage-gated channels, and the steady currents are the result of infrequent opening of these channels in relatively hyperpolarized membranes. It is certainly true that the magnitude of steady currents is typically a few $\mu A/cm^2$, while currents during action potentials are measured in mA/cm^2. These steady-current channels, however, are unlikely to be identical to voltage-gated channels, because the magnitude of steady currents would be expected to increase dramatically if a cell were depolarized (for example, by high external potassium), and that has never been reported. Also, there is experimental evidence that the potassium permeability of the resting membrane of the squid giant axon is not determined by excitable potassium channels (Chang, 1986). It seems more likely that these channels are not gated or are gated chemically. For example, calcium-gated potassium channels and chloride channels are known to exist in a variety of cells (see, e.g., Hille, 1984; Dascal et al., 1985; Young et al., 1984), including some of those in which steady currents have been measured (Robinson, 1979). This finding would allow for steady currents to be generated by regional differences in the concentration of the gating substance whether calcium or something else, rather than by an asymmetric distribution of channels per se. In any case, it is clear that our understanding of the underlying molecular basis of steady current is primitive and that this is an area in which research is needed.

THE VIBRATING PROBE

The major impetus to the study of steady currents has been the development of the vibrating probe, which has made the measurement of such currents possible. The requirements for such a measuring device were made clear when Lionel Jaffe (1966) polarized some 200 *Fucus* zygotes in a capillary tube so that their morphological axes all pointed in the same direction, parallel to the tube axis. He then measured the voltage difference between the two ends of the tube and from that measurement was able to estimate that each zygote was driving a current of 1 $\mu A/cm^2$ or less through itself. Jaffe and Nuccitelli (1974) then addressed the problem of how to measure such currents with a resolution of 10–30 μm. This problem requires a consideration of current flow in an extended conducting medium. The appropriate form of Ohm's law in a volume conductor is $J = (1/\rho)E$, where J is the current density at a point, E is the electric field at that point, and ρ is the volume resistivity of the medium. Convenient units for the physiologist for these quantities are, respectively, $\mu A/cm^2$, V/cm, and ohm $\times$ cm, where A is amperes, and V is volts. The unit of resistivity may seem odd at first, but if one considers that the resistance (R) of a volume of a medium will increase with the length and decrease with the area ($R = \rho \times$ length/area), this unit makes sense. The resistivities of physiologically relevant solutions range from about 25 ohm $\times$ cm for sea water to 2,000–3,000 ohm $\times$ cm for pond water. The resistivity of a typical Ringer's saline is about 100 ohm × cm. To return to the problem at hand, a quick calculation shows that a current density of 1 μA/cm^2 flowing in a medium of the resistivity of sea water (25 ohm × cm) will produce a voltage drop of about 50 nV over a distance of 20 μm. This is the core of the problem: how to measure these small voltages over short distances. Such measurements are well beyond the capacity of conventional saline-filled glass microelectrodes. The solution that emerged made use of a platinum-black ball 20 μm or so in diameter plated on the end of a metal-filled glass microelectrode. This microelectrode was mounted on the end of a piezoelectric strip. When a sinusoidal voltage was applied to the piezoelectric driver, the electrode tip (the "probe") was vibrated at the frequency of the driving voltage and (at a given frequency) at an amplitude determined by the magnitude of driving voltage. If the electrode were moving through a region of fluid in which a steady, uniform current was flowing, it would experience a voltage gradient and would report that gradient as a sinusoidal voltage difference. Modulating the DC field to an AC voltage has two major advantages. The first, and most important, is that the impedance of the electrode falls dramatically as the frequency increases to the typical operating frequency of 300–800 Hz. At these frequencies, the platinum electrode is capacitatively coupled to the medium, and its contribution to the impedance is negligible. The major contribution is from the resistance of the fluid medium near the electrode through which charge must flow. This convergence resistance sets the limiting theoretical noise of the system and is a function both of the resistivity of the medium and the diameter of the electrode, as discussed by Jaffe and Nuccitelli (1974). This limiting noise, in nanovolts, can be calculated from the formula

$$8(\rho/\tau d)^{1/2}$$

where ρ is the resistivity of the medium (ohm × cm), τ is the RC time constant of the amplifier filter (in seconds), and d is the diameter of the probe (in μm) (Scheffey, 1986a). To return to the earlier example of the expected voltages near a *Fucus* zygote, if d = 20 μm, ρ = 25 ohm × cm, and τ = 10 seconds, the noise is

$$V_{rms} \cong 3 \text{ nV}$$

In the earlier example, the peak-to-peak voltage to be measured was 50 nV, which is equal to about 17 nV rms; thus the vibrating electrode has the inherent capacity to do such measurements, since the signal is more than five times larger than the noise. In practice, the actual noise is typically twice as large as the theoretical minimum.

There are, of course, numerous other sources of unwanted signal at these low voltages, and it is there that the second advantage of AC measurements becomes apparent. Commercially available lock-in amplifiers (LIA) act as narrow-pass filters and will reject spurious signals that are not at the same frequency as that of a reference frequency. In practice, the piezoelectric driving element and the reference channel of the LIA are run from the same source, ensuring that the LIA and the electrode are at the same frequency. LIAs are also sensitive to phase, and there are several places where phase shifts can occur in the system, so the phase difference between the reference channel and a known calibrating signal must be empirically adjusted to zero. Operationally, this is done by using a point source, such as a KCl-filled microelectrode with a ~1 μm tip, and passing a known current through this electrode so that the vibrating electrode can be placed at a location of known current density and the phase adjusted appropriately. In addition to setting the phase, this operation allows a check on the performance of the system and its ability to register correctly a known signal. We have found that such a calibrating source can be constructed from a 1.5V battery, a potentiometer, a 1 MΩ current-sensing resistor, and 100 MΩ current-limiting resistor. The output of this box in connected to the KC1-filled electrode via a Ag-AgCl wire. The voltage across the 1 MΩ resistor is measured with one of the many inexpensive digital multimeters (DMM) available that have 1 GΩ or more input impedances on the 200 nA range (e.g., Soar model 530). With this arrangement, currents of 1–15 nA can be passed through the calibrating electrode, yielding current densities of about 1–15 μA/cm^2 at a distance of 90 μm from the tip of the calibrating electrode.

Since the introduction of the original design of the glass electrode by Jaffe and Nuccitelli, several simplifications and improvements have been made, many of them at the National Vibrating Probe Center, Woods Hole, MA. These are discussed by various authors in the book, *Ionic Currents in Development,* edited by R. Nuccitelli (1986), in the "Technical Advances" section. Perhaps the greatest simplification has been the introduction of wire electrodes to replace the glass electrodes. Since they were first introduced by Freeman and collaborators (see Freeman et al., 1986, for a recent discussion), several types have been used, including those made of tungsten, platinum, and stainless steel. In my laboratory, we use parylene-coated stainless steel electrodes, sold by Micro Probe, Inc., Clarksburg, MD. These electrodes are supplied with a fine point and have a 5 μm uninsulated portion at the tip. Before plating, the electrodes are cut to the desired length, and an appropriate connector is attached with silver conducting epoxy. The electrode tips are then plated, first with gold, then with platinum. This is done under microscopic observation; it can be done with the electrode in place in the bender assembly. If the Applicable Electrotechnics preamplifier (see below) is used, it has an input for plating current and a switch that permits breaking away from the measuring circuit; otherwise, the signal cable can be disconnected from the preamplifier and connected to the plating power supply. The tip of the electrode is immersed in a 0.2% AuKCN solution; 2 nA current is passed for 5 minutes, 5.5 nA current for 10 minutes, and then 20 nA current passed until the gold is about one-half of the desired final electrode diameter. After rinsing in distilled water, the electrode is immersed in 1% PtCl, and 100 nA of current is passed for 5 minutes; then 500 nA is passed until 80% of the final diameter is reached. The final 20% is achieved by applying 0.5-second bursts of 2 μA current. An inexpensive current source for plating can be constructed from a battery, appropriate current-limiting and current-sensing resistors, and a DMM, as described above in the discussion of the calibration source.

Electrodes made in the manner just described are not always good or sometimes go bad during use even though they appear to be normal. Scheffey (1986b) has published a method for measuring the capacitance of the electrodes, which is a reliable measure of the electrodes' impedances. A triangle wave is connected to the electrode through a 10 pF capacitor, and the voltage across the electrode is measured. This is done conveniently by displaying the output of the preamplifier on an oscilloscope screen. The electrode's capacitance (in pF) can then be calculated from the formula

$$C_e = 10\text{pF}(V_i/V_e)$$

where V_i is the amplitude of the input triangle wave, and V_e is the amplitude of the triangle wave across the electrode. (More exactly, V_i is the voltage across the 10 pF capacitor, but since the electrode capacitance should be much larger than 10 pF, nearly all of the applied voltage is dropped across the 10 pF.) In performing this calculation, the gain of the preamplifier must be considered, of course. Also, there will be a resistive component, which appears as a square wave and is not included in the measurement of V_e. An input triangle wave of 50 mV at 100 Hz is convenient, and the measured capacitance of a good probe should be 5 nF or larger.

These electrodes are not shielded as the older glass electrodes were, requiring the use of differential recording, with the probe going to one side and the reference (a bare platinum wire) to the other side of a differential preamplifier. Many commerical lock-in amplifiers allow for differential recording; an alternative is to use the preamplifier especially designed by Carl Scheffey at the National Vibrating Probe Center and sold by Applicable Electrotechnics.

Another advance in this area is the introduction of the two-dimensional vibrating probe, which can measure two components of the electric field vector. The original version of the two-dimensional probe was developed by John Freeman and associates (Freeman et al., 1986) and employed two loudspeakers mounted at right angles to each other and connected to a wire electrode. When the coils were driven by computer-generated sine and cosine waveforms, the electrode could be made to move in a circle or in other patterns. This version of the vibrating probe does not use an LIA; instead, the signal is analyzed digitally. More recently, progress has been made by this group toward a functional three-dimensional probe.

In my laboratory, a somewhat different two-dimensional vibrating probe is in use. This version was developed by Nuccitelli (1986) and utilizes two piezoelectric elements mounted perpendicularly to each other. A parallelogram is completed by a bent piece of titanium foil, and the electrode itself is mounted at the vertex of the foil. It is easy to show that if the two piezoelectric crystals are driven with sine waves of the same amplitude and frequency, but displaced 90° in phase with each other, the vertex will move in a circle. The signal sensed by the electrodes is sent to a two-phase LIA, which reports two orthogonal components of the electrical field. These two outputs are digitized and analyzed by a computer, and a vector is displayed on a video screen. These vectors are superimposed on an image of the biological preparation at points corresponding to the location of the electrode; this information is communicated to the system by a light pen. The vibrator assembly and the software were purchased from the Vibrating Probe Co. (Davis, CA) and the other components, such as the analog-to-digital (A-D) converter, the computer, and the light pen are the ones specified by Nuccitelli (1986).

While it is possible, in theory, to get the same information from a one-dimensional probe as from a two-dimensional probe simply by rotating the biological preparation or the probe by 90°, in practice it is quite difficult to achieve this rotation. The time involved in such repositioning, uncertainty about location, and movement of the preparation during the maneuver all contribute to the difficulty. If the geometry of the system under study is complex (for example, like that of a limb bud of an embryo), the two-dimensional probe offers major advantages over a one-dimensional probe in determining the pattern of endogenous current.

There are still many cases in which a one-dimensional probe is valuable, and M. E. McGinnis and I have simplified the original design so that we can have several such instruments, customized for specific applications requiring vertical or horizontal vibration, inverted or stereo microscopy, etc. Figure 1 shows the details of our design. The choice of microscope depends on the size, shape, and optical properties of the biological preparation. Large, opaque preparations are best viewed with a stereo microscope, while small, transparent preparations may require an inverted microscope. It is rare for vibrating probe work to demand high-performance optics; the inexpensive, fixed-stage, tissue culture miscroscopes sold by Nikon (model TMS) or Olympus (model CK) are adequate for many purposes. Evans Electronics (Berkeley, CA) sells printed circuit modules that can be used to construct an inexpensive LIA; the two required modules (a phase-control unit and a phase-sensitive detector) cost about $375. We find that an amplifier constructed from these modules, when used with an appropriate preamplifier, such as the one from Applicable Electrotechnics, performs quite well. We use an inexpensive (~ $200) function generator to drive the probe and supply a reference signal to the LIA; specifically, we use a Tenma Digital Function Generator/Frequency Counter, sold by MCM Electronics, Centerville, OH, but there are many others available. A solid, vibration-free support, such as a marble table or an air table, is also necessary, as for any electrophysiological set-up. The total cost of a driver unit, home-made LIA, preamp, and function generator is about $2,500, including some electronics and machine shop time. One of our rigs is shown in Figure 2.

Regardless of the version of the vibrating probe that is used, the user should be aware of the various artifacts that can arise in such measurements. Two important controls that must be repeated often are: 1) to verify that the system responds correctly to a known current source; and 2) to determine that it does not give a signal when a measurement is carried out on an object known to be inert, such as a dead cell or a glass bead. If these two controls are used, a host of pitfalls can be avoided (or at least recognized and dealt with). We have found that it is important not to drive the electrode at resonant frequencies that give maximum vibrational amplitudes. Such frequencies tend to be system resonances, and, at these frequencies, the vibrational amplitude is sensitive to many factors, including damping caused by the experimenter touching the manipulator and the depth of the probe in the medium. These problems can be avoided by detuning the system, that is, using a driving frequency that is 10 Hz or so off resonance, and achieving the desired amplitude by increasing the driving voltage.

Changes in fluid level because of evaporation can be a source of noise at

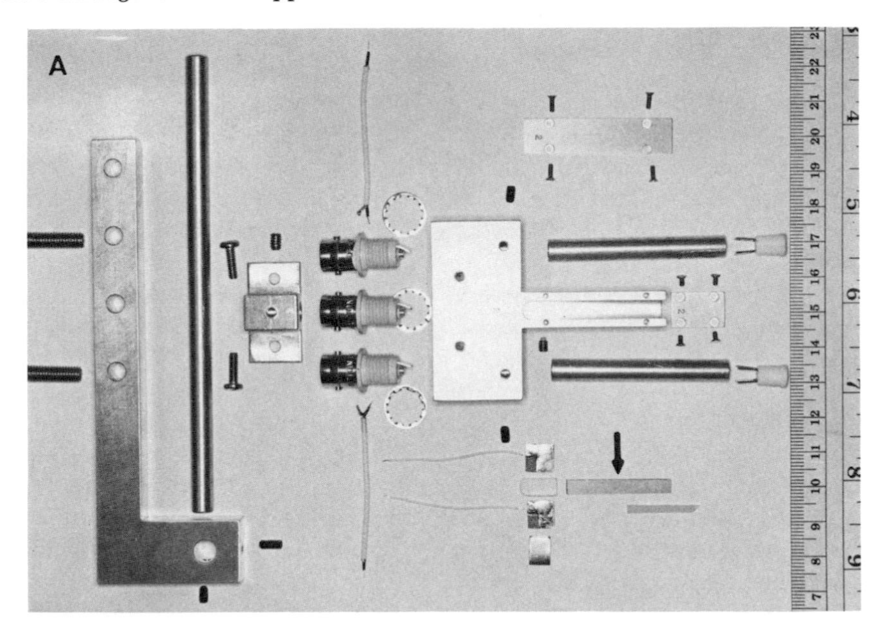

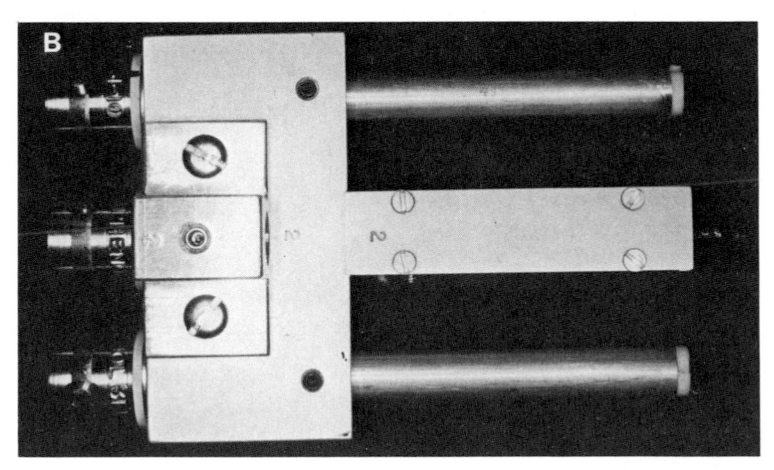

Fig. 1. A simplified one-dimensional vibrating probe. **A:** Exploded view of the assembly. In the lower right is the piezoelectric element (arrow), which is to be clamped between two small pieces of printed circuit board, both for mechanical support and electrical contact. **B:** The assembled unit. The middle extension from the main block houses the piezoelectric element; the electrode is attached at the end. The two cylindrical arms house the signal cable and reference cable; they are attached to the two imputs of the preamplifier via the BNCs. The middle BNC connects the piezoelectric element to the signal generator.

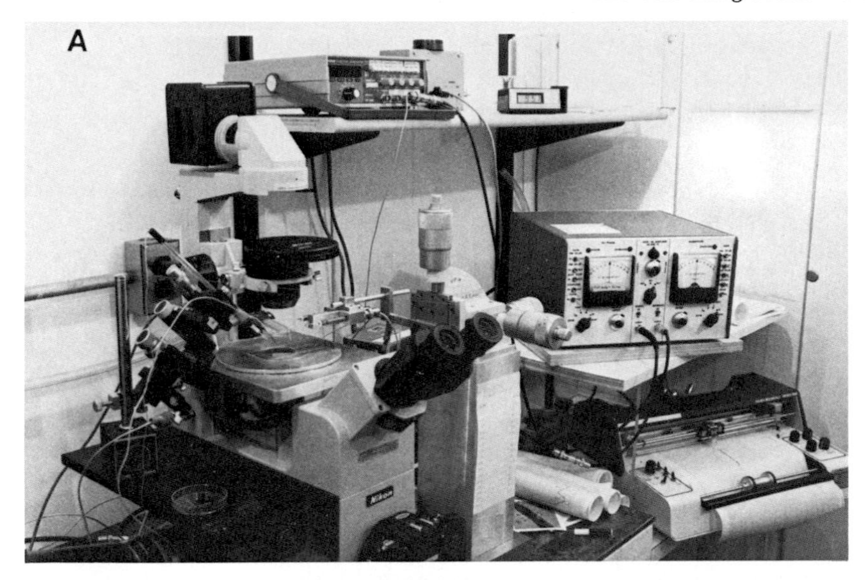

Fig. 2. **A:** Overview of our one-dimensional vibrating probe set-up. The custom-built lock-in amplifier is on the right, and the signal generator for driving the bender is on the shelf above the microscope. **B:** Closer view of the probe assembly. The electrode (the small, straight wire) and the platinum reference electrode (the thicker, curved wire) can be seen at the left end of the assembly.

the most sensitive scales, especially if the medium is warmed to 37°C; if such a change occurs, a cover can be devised (with appropriate slots for the electrodes) to reduce evaporation, or the surface of the medium can be covered with mineral oil. Other pitfalls and remedies are discussed by Scheffey (1986b).

EXAMPLES OF VIBRATING PROBE MEASUREMENTS

There is now a voluminous literature on the measurement of endogenous electrical currents in biological systems; a complete review of this subject is far beyond the scope of this chapter. Instead, I will discuss a few examples that illustrate the range of such measurements.

Tip-Growing Plants and Fungi

The first use of the vibrating probe was to measure the current pattern around developing embryos of marine algae. These embryos grow by extending a rhizoid, which involves the continuous addition of new material at the growing tip. Jaffe and Nuccitelli (1974) found that current (taken as the movement of positive charge) enters this growing tip and leaves the other portions of the embryo. This pattern has been seen in a number of other tip-growing organisms. One example from the recent biological literature is that of the water mold *Achlya.* Currents enter the tips of the growing hyphae and leave the regions behind the tip. The switch from inward to outward current occurs about 300 μm from the hyphal end (Kropf et al., 1983). The maximum current density measured by the probe was about 0.5 $\mu A/cm^2$; when extrapolated to the surface of the hypha, it was about 2.5 $\mu A/cm^2$ (Kropf et al., 1984). Protons seem to be the major current carriers in these cells. This conclusion was based on the observation that direction of the current reversed when the extracellular pH was changed from 6.5 to 8.5. Furthermore, pH microelectrodes showed that the growing hyphae alkalinized the medium near the tip (where the current was inward) and acidified it near the regions of outward current (Gow et al., 1984). Another feature of the current was its dependence on extracellular methionine, suggesting that the inward proton current might be part of a symport mechanism for supplying amino acids to the growing tip. The locus of inward current at points other than the growing tip indicated positions of branches; the inward currents there predicted the branches before any morphological changes could be seen. More recently, however, the role of the current in localizing growth has been questioned (Schreurs and Harold, 1988), since growth and the current could be decoupled.

Kropf (1986) has shown that there is a cytoplasmic voltage gradient along the growing hyphae. This electrical field is 0.2 V/cm or larger in the tip region, and is dependent on the endogenous current flow. Since the magnitude of the cytoplasmic field is about ten times as large as calculated from Ohm's law, it was suggested that this field results from a gradient of relatively immobile ions in the cytoplasm, as predicted by Jaffe et al. (1974).

Armbruster and Weisenseel (1983) also measured currents around *Achlya* hyphae. They report inward current at the tip (and up to 600 μm behind the tip) and, in addition to a steady component, they found a pulsatile component of current. They also report that the direction of the current reverses transiently during spore formation, although this may be an artifact (Thiel et al., 1988).

In general, current entry in the growth zone seems to be a feature of tip-growing organisms. In addition to *Achlya* and seaweed embryos, such currents have been measured in pollen tubes (Weisenseel et al., 1975), barley root (Weisenseel et al., 1976), and several other fungal hyphae (Gow, 1984).

Rat Skeletal Muscles

A quite different biological system that has recently been studied with the vibrating probe is the skeletal muscle of the adult rat. It is of interest in this context because it demonstrates that the vibrating probe can be used to give new information about a well-studied preparation. Using isolated, whole lumbrical muscles, Caldwell and Betz (1984) showed that current leaves the synaptic region and enters the adjacent extrajunctional region. The peak of the outward current is over the end-plate band and is typically 5–10 $\mu A/cm^2$. This current declines and becomes inward over a distance of 0.5–1.0 mm. A similar current pattern was measured around single, isolated muscle fibers by Betz et al. (1980). In response to carbachol, an acetylcholine analog not hydrolyzed by cholinesterases, the outward current in the synaptic region reverses and becomes a very large inward current of nearly 100 $\mu A/cm^2$. This finding is expected, of course, since the existence of large inward current through the open acetylcholine channel is well known, but it is satisfying to observe that the vibrating probe can easily detect it; indeed, this current is some three orders of magnitude above the detection limit of the probe.

The physiological basis of the steady currents has been determined by Betz et al. (1984). It was shown that the "current does not depend on the activation of acetylcholine channels, voltage-gated Na^+ channels, or increased electrogenic pumping in the endplate region." However,the results of a series of experiments in which the membrane potential was modulated, external chloride was removed, and chloride transport inhibitors were applied strongly implied that chloride is accumulated in the cell to greater than equilibrium levels and that the steady current is a result of reduced chloride conductance in the end-plate region. For example, they found that 9-anthracene carboxylic acid (9-AC) rapidly and reversibly blocked the steady current; 9-AC has been shown to block chloride conductance in rat diaphragm (Palade and Barchi, 1977).

Since current, taken as positive charge movement, flows intracellularly toward the end-plate region from the flanking region in the muscle cell, it would be expected that the resting membrane potential of the end-plate region would be more negative than elsewhere; Caldwell and Betz (1984) estimate that the difference should be a few tenths of a millivolt, on the basis of Ohm's law. They were unable to demonstrate such a small difference because of the large number of measurements required; however, they point out that Yoshioka and Miyata (1983) did a sufficient number of measurements on rat soleus muscle to obtain statistically significant evidence that the end-plate resting potential is about 1 mV more negative than the extrajunctional region. These potential differences, unlike those in *Achlya* hyphae, do agree with Ohm's law predictions. This finding is expected if the current-carrying ion is freely mobile in cytoplasm; nonohmic potentials are expected only in cases in which the mobility of the current carrier is significantly impeded (Jaffe et al., 1974).

Somewhat unexpectedly, it was found by Betz et al. (1986) that the steady current persists following denervation, a treatment that is known to decrease chloride conductance. There seems to be a compensatory increase in internal chloride in response to denervation, from 12 mM to 23 mM.

Lateral Root Formation

Keerti Rathore and Kevin Hotary in this laboratory have been using the two-dimensional vibrating probe to map the endogenous currents around the emerging lateral roots of radish seedlings. Figure 3 shows a photograph of a video display of a lateral root with superimposed current vectors. Two different passes were made around the root at different distances from the surface in order to produce this display. Currents of 5–10 $\mu A/cm^2$ enter the elongation zone and leave the differentiating zone. The pathways of the currents in the medium are seen vividly in this display. In response to the plant hormone indole acetic acid (IAA), this current pattern is changed dramatically; the current in the meristematic region at the tip of the root reverses direction, becoming outward. This reversal occurs rapidly, and may be involved in

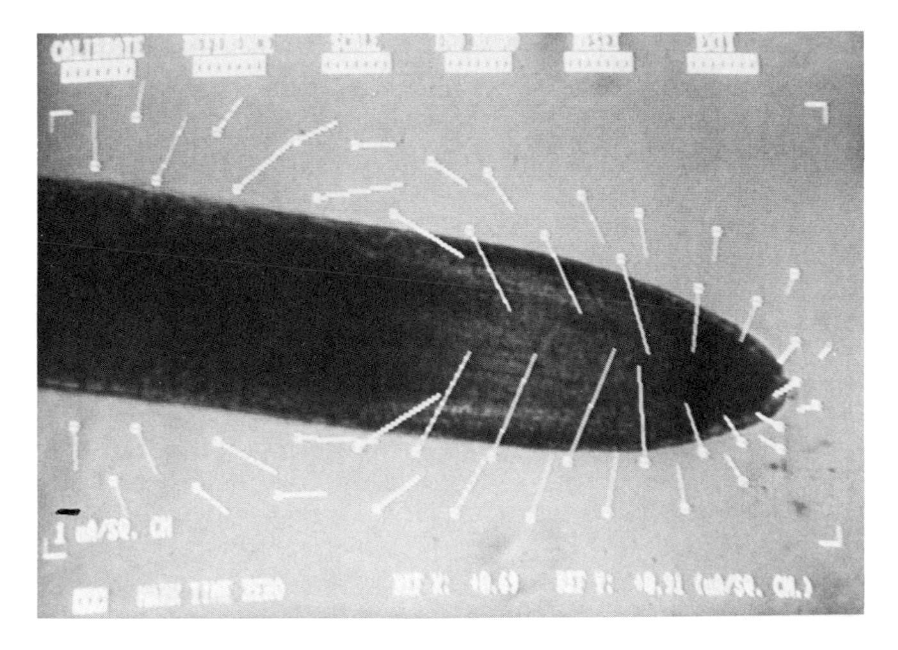

Fig. 3. The current pattern around a radish lateral root as determined with a circularly vibrating two-dimensional probe. This is a photograph of a video screen on which electric current vectors are superimposed on an image of the growing root. The dot at the end of each vector shows the location of the electrode at the time of the measurement. The lengths and directions of the vectors represent the magnitudes and directions of the currents. The short horizontal bar at the lower left represents 1 $\mu A/cm^2$; the width of the lateral root is about 200 μm.

mediating the growth response to the hormone, which in this case is a reduction in growth rate.

SIGNIFICANCE OF STEADY CURRENTS

As discussed earlier, the motivation for developing the vibrating probe and measuring steady currents was to study developmental polarity, and the main use of the technique has been on developing systems. Regardless of the implications of these measurements for developmental biology, it is clear that data from vibrating probe experiments have given much new information about the organization of the plasma membranes of cells and the long-lasting physiological asymmetries in membranes. In principle, it is possible to gain information about regional densities of ion channels in single cells by mapping them with patch electrodes; however, this technique has been used primarily to study voltage-gated channels, whose relationship to steady currents, as I noted above, is uncertain. Nevertheless, it would seem that a thorough study of the same system with both the vibrating probe and patch electrodes would yield valuable correlations and might give information about the nature of the ionic mechanism of steady currents.

Some of the cases in which the vibrating probe has given a new picture of cellular asymmetry, besides the ones discussed in detail in the previous section, include rat eye lens (Robinson and Patterson, 1983), mouse embryo blastomeres (Nuccitelli and Wiley, 1985), cleaving *Xenopus* zygotes (Kline et al., 1983), the green algae *Micrasterias* and *Closterium* (Troxell et al., 1986), and *Xenopus* oocytes (Robinson, 1979). As this list makes clear, the occurrence of steady currents is a widespread phenomenon encompassing both plant and animal cells, microorganisms, giant cells, and adult and embryonic cells. Thus there are stable, functional asymmetries in the ion channels and/or pumps associated with the morphological and developmental asymmetries of organisms. Leaving aside the role of these asymmetries in causing and maintaining polarity, the vibrating probe has made a major contribution to our understanding of cellular organization by establishing the existence of the asymmetries. In some cases, the identification of the current-carrying ions has been made with certainty; it is disappointing, however, that unequivocal identification has not been possible in most cases. There are several reasons for this failure. Normally, the investigator does not have control over (and often, has no knowledge of) the intracellular ionic composition, so if the current involves the leakage of an ion out of the cell, it is difficult to identify. Many of the pharmacological agents that have been used to block voltage-gated channels in excitable cells are of uncertain value in the study of steady currents, since we know so little about the nature of the ionic leaks involved. As other techniques, such as ion-specific microelectrodes, are used to complement the vibrating probe, a clearer picture of the ionic basis of steady currents should emerge.

What of the role of steady currents as effectors of cellular polarity? Clearly they cannot be the ultimate initiator of polarity, since their generation is dependent on an underlying asymmetry in ion transport sites. It has become apparent, however, that the currents are predictive of morphological polarity, and there is mounting evidence that steady currents are part of a feedback mechanism to convert small asymmetries (such as an extracellular gradient of a morphogen) to a gross asymmetry, such as that manifested by localized growth. There are at least three ways in which a steady current might act to polarize a cell.

The first involves the chemical action of the current-carrying ion. If this ion is relatively immobile in cytoplasm, a substantial cytoplasmic gradient of this ion may result from its local entry or exit. This situation is best typified by the developing fucoid zygote. As mentioned earlier, current enters the region of the cell where the tip-growing rhizoid emerges and elongates. There is evidence from Robinson and Jaffe (1975) and from Nuccitelli (1977) that a fraction of the inward current is carried by calcium, and a resultant calcium gradient has long been predicted. Recently, Brownlee and Wood (1986) have shown that there is a steep gradient of free calcium in the growing rhizoid tip of *Fucus serratus*. Using calcium-sensitive microelectrodes, they measured $[Ca^{2+}]$ as 2.61 μM at the tip and 0.35 μM in the subtip regions, representing a tenfold gradient over a distance of 20–30 μm. While these measurements were done on older embryos, there is evidence that local calcium entry is involved in the initial polarization of the zygote, when it chooses a direction in which to initiate a rhizoid. If zygotes are grown in the dark (to isolate them from the polarizing influence of light) in the presence of a gradient of the calcium ionophore A23187, they tend to germinate toward the higher concentration, that is, from the region where the calcium influx is greatest, as shown by Robinson and Cone (1980). This finding suggests that intracellular calcium gradients are causally important in the polarization process, perhaps by way of an interaction with the cytoskeleton, as suggested by Brawley and Robinson (1985).

Another circumstance in which steady currents are likely to produce an intracellular ionic gradient is that of the fungal hyphae. As discussed earlier, protons are the current carriers in these cells, and extracellular gradients of pH have been measured. Almost certainly there will be larger pH gradients in the cytoplasm, both because the volume is restricted (compared with the essentially infinite extracellular medium) and because the ions are likely to be less mobile in cytoplasm. Obviously, a steep gradient of pH could have profound metabolic consequences for the cell.

A second way that endogenous steady current might act to polarize cells is by the action of the transcytoplasmic electrical field that they generate. As already discussed, such fields can be quite large, as much as several millivolts in *Achlya*. One target of these fields could be plasma membrane proteins. It has been shown that mobile membrane proteins can be redistributed within the surface of the membrane by electrical fields as small as 1 mV per cell diameter (reviewed by Poo, 1981, and Robinson, 1985). The transcytoplasmic fields could also have as their target cytoplasmic macromolecules and organelles. The existence of such a mechanism has been shown in the case of the growing insect (Cecropia) follicle by Woodruff and

Telfer (1980). There is a potential difference of several millivolts between the nurse cells and the oocytes, which are connected by a cytoplasmic bridge. This electrical field blocks or augments the movement of molecules across the bridge, on the basis of the molecules' electrical charge.

Finally, the currents generated by groups of cells, such as an epithelial layer, can generate substantial electric fields at places distant from the current-generating cells (McGinnis and Vanable, 1986). These fields are easily large enough to affect other cells and may be involved in a variety of processes such as regeneration and wound healing. These matters are discussed in detail in other chapters in this volume.

APPLICATION OF ELECTRICAL FIELDS TO CELLS IN VITRO

As was discussed at the beginning of this chapter, there is a long history of interest in the effects of exogenous electrical fields on living organisms. This interest has continued for several reasons. One is that cells do respond to applied fields (see Robinson, 1985, for a review), and, regardless of any connection with natural phenomena, this is a method for modifying cellular behavior that may have clinical and other practical importance. Another reason is that the responses of cells to applied fields may give clues to role of cells' endogenous currents. The origin of Jaffe's search for endogenous currents through *Fucus* zygotes was Lund's observation that an applied field could determine the rhizoid-thallus axis of these cells. In some cases, then, an applied current can mimic and modulate an endogenous one. Electric fields can also be used as a tool to perturb the distribution of proteins on the surface of cells in order to study diffusion and the effects of molecular crowding (Ryan et al., 1988). Finally, as I noted in the last section, cells in a developing embryo or in an adult may experience electrical fields that are generated by ion-transporting epithelia during normal development or in response to injury. Studying the response of cells in vitro to applied fields may give information about cellular behavior during development and repair.

Compared with the construction and use of the vibrating probe, exposing cells to electrical fields is relatively simple. Nevertheless, there are certain problems to be considered in order to avoid the introduction of unwanted artifacts. The cells must be maintained in a way that meets their requirements for temperature, gases, substrate, nutrients, pH, etc. The introduction of electrode products into the growth chamber must be avoided, and heating of the medium from the applied current must not be great. Both the geometry and the magnitude of electrical field produced by the applied current must be known and must not vary in an uncontrolled way during an experiment.

Finally, a method for assessing the cells' responses to the field must be devised. I will describe below the strategies that have evolved in this laboratory, based on our experience and that of others, for meeting these criteria.

The requirement for minimal heating by the applied current requires that the power generated by the current be minimized. The rate of energy dissipation (power) is given by $P = V^2/R$, where V is the voltage drop across the length of the chamber, and R is the resistance of the medium in the chamber. As mentioned in the Vibrating Probe section, above, $R = \rho \times$ length/area, so $P = V^2 \times$ area/($\rho \times$ length). This equation tells us that the cross-sectional area should be minimized, meaning, in practical terms, that the depth of the chamber should be minimized.

There is another factor that must be considered in the construction of the chamber, and that is the heat exchange with the environment. The extent to which the temperature rises above ambient at steady state depends on the rate of heat generation (P) and how effectively that heat can be transferred to the surroundings, in turn depending on the thermal conductivity and thickness of the materials used to construct the chamber.

With these considerations in mind, we use chambers that are constructed from two strips of 6-cm-long coverslip sealed to the bottom of 100-mm plastic Petri dishes. (We use a thin layer of silicone grease, followed by warming to 55°C to allow air bubbles to escape.) The strips are positioned 1 cm apart, creating a channel 6 cm × 1 cm. This channel is filled with the appropriate medium, and the cells are added to it. After they have settled and attached, a third coverslip is laid over the top of the channel. The edges of this coverslip are smeared with silicone grease before it is put in place, serving to seal the lid to the chamber, whose final dimensions are 6 cm × 1 cm × ~ 0.05 cm. This basic design can be modified as required. There is no reason that the chamber cannot be shorter or narrower. The Petri dish can be of the tissue culture type and can be treated with poly-L-lysine if needed. Sterility (or its lack) may be a problem for long-term experiments; if so, the chambers can be gas-sterilized after assembly.

The dimensions of the chamber are such that heating is not usually a problem. The electrical fields that are of physiological interest are generally 100 mV/mm or less, so the voltage drop across the 6-cm-long chamber is 6 V or less. If the chamber is filled with a medium that has a resistivity of 100 ohm × cm, the total resistance is 100 ohm × cm × 6 cm/l cm × 0.05 cm = 12 K ohm. This is an approximate figure, since the depth of the chambers may vary substantially from 0.05 cm, but it is adequate for estimating the required current, 0.5 mA, to produce a field of 100 mV/mm.

The actual voltage drops across the chambers are always measured at the end of the experiment. We have measured the steady-state temperatures produced by such currents by inserting a small thermistor into the chamber, and we find that the temperature increases less than 0.5°C above ambient. Electrical contact is made to the ends of the chamber by agar bridges, which are constructed from glass tubing (about 10 cm long, 0.5 cm ID) and filled with medium gelled with 1% agarose. The agar bridges are brought into the chambers through holes in the lids (or sides) of the Petri dishes and make contact with the medium that is filling the chamber. The other ends of the bridges terminate in saline-filled beakers containing Ag-AgCl electrodes, which are in turn connected to a constant current power supply. Our power supply is based on the LM334 (National Semiconductor) three-terminal current source. This device is inexpensive and requires only an external variable resistor and a 30 V power supply in order to function as a constant current source. We operate six such devices from a single power supply, allowing us to supply current to six chambers; the six currents are independently set and monitored. It is also possible to use a high-voltage power supply and a large resistor (at least ten times as large as the resistance of the chamber) as a constant-current source. The important point is to drive a constant current through the chamber even though the resistance of the chamber may vary due to the somewhat uncertain contact of the agar bridges with the ends of the chamber. Some aspects of the physical layout of this equipment are shown in Figure 4.

The length of the bridges that I have specified (10 cm) is adequate for relatively short-term experiments (up to 12 h); for longer experiments, a longer bridge should be used or, if that becomes impractical, the bridges should be changed. In any event, products from the Ag-AgCl electrodes must be kept out of the chamber.

The chambers described above can be mounted on the stage of an inverted microscope equipped with a long-working-distance condenser. Various

---

Fig. 4. **A:** Five chambers attached to the constant current power supply. The digital multimeter (DMM) on the left monitors the current flowing in each circuit. To its right is the box containing the LM334 constant current devices. On the top of this box are six ten-turn potentiometers (used here as variable resistors), which regulate the current and six switches that determine which circuit will be monitored by the DMM. **B:** Closer view of the chambers that shows in more detail the agar bridges and the coils of chloridized silver wire (arrow). **C:** Close-up of a chamber on the stage of a microscope. The bridges can be seen to terminate in blobs of agarose; these serve to make contact with the ends of the chamber. Within the Petri dish are seen wetted pieces of filter paper; they serve to increase the relative humidity, thus minimizing evaporation.

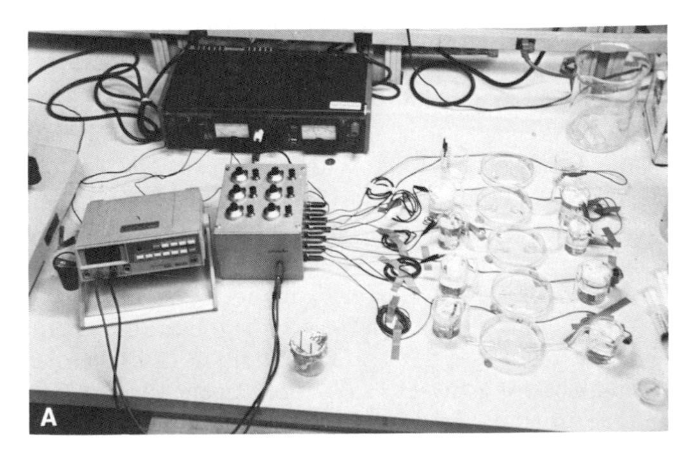
A

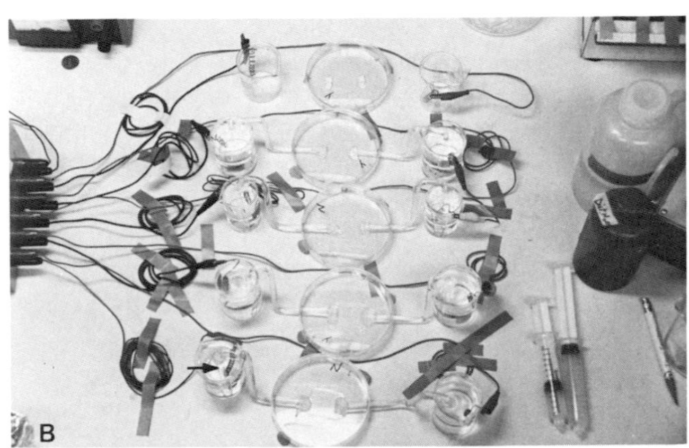
B

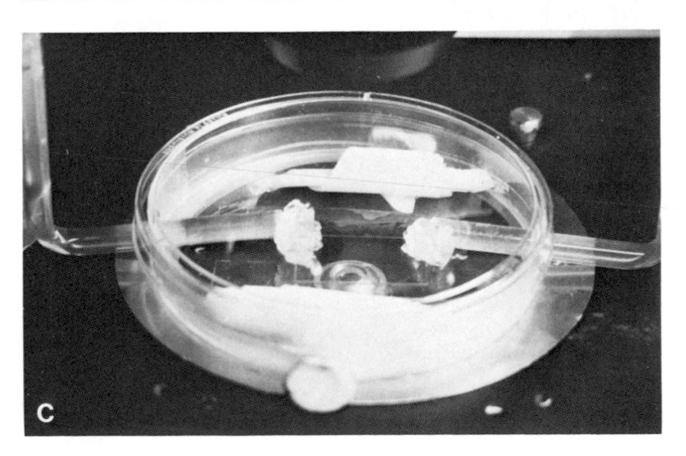
C

means of recording cells' responses to the applied fields can be employed, including 35-mm photography, cine-photography, and video recording. We are studying the responses of cultured spinal neurons to electrical fields and we take 35-mm photographs at intervals. The developed film is then projected by a photographic enlarger onto a digitizing pad. The location of a cell body and the position of each of the neurites on the cell body are entered into a computer as well as the position of the growth cone and the length of the neurite. This process is repeated for the next time point. Using software developed by John Cork and Ann Rajnicek in this lab, various parameters of growth with respect to the direction of the electrical field are calculated. This system is also suitable for analyzing cell migration or shape changes. Of course, the same kinds of analyses can be done manually from prints of the negatives, but the computer-based system greatly reduces the work and the cost. There are commercially available software packages for this type of analysis, so it is not necessary to write one's own software.

The apparatus described above was developed to meet the particular requirements of this laboratory. Undoubtedly, it will have to be modified and adapted for the specific needs of other experiments. The chambers may have to be put in an incubator, for example. If mammalian cells are to be observed on the stage of a microscope, some kind of incubator will have to be fitted on the stage. Arrangements may have to be made for perfusing medium though the chamber (Hinkle et al., 1981) or for longer-term exposures (Stump and Robinson, 1983), but the same criteria outlined earlier will have to be met.

ACKNOWLEDGMENTS

I am grateful to the members of my laboratory for sharing their unpublished results with me for this chapter and for commenting on the manuscript.

REFERENCES

Armbruster BL and Weisenseel MH (1983) Ionic currents traverse growing hyphae and sporangia of the mycelial water mold *Achlya debaryana*. Protoplasma 115:65–69.

Betz WJ, Caldwell JH, Ribchester RR, Robinson KR, and Stump RF (1980) Endogenous electric field around muscle fibres depends on the Na^+-K^+ pump. Nature 287:235–237.

Betz WJ, Caldwell JH, and Kinnamon SC (1984) Physiological basis of a steady endogenous current in rat lumbrical muscle. J. Gen. Physiol. 83:175–192.

Betz WJ, Caldwell JH, and Harris GL (1986) Effect of denervation on a steady electric current generated at the end-plate region of rat skeletal muscle. J. Physiol. 373:97–114.

Brawley SH and Robinson KR (1985) Cytochalasin treatment disrupts the endogenous currents associated with cell polarization in fucoid zygotes: Studies of the role of F-actin in embryogenesis. J. Cell Biol. 100:1173–1184.

Brownlee C and Wood JW (1986) A gradient of cytoplasmic free calcium in growing rhizoid cells of *Fucus serratus*. Nature 320:624–626.

Caldwell JH and Betz WJ, (1984) Properties of an endogenous steady current in rat muscle. J. Gen. Physiol. 83:157–173.

Chang DC (1986) Is the K permeability of the resting membrane controlled by the excitable K channel? Biophys. J. 50:1095–1100.

Dascal N, Gillo B, and Lass Y (1985) Role of calcium mobilization in mediation of acetylcholine-evoked chloride currents in *Xenopus laevis* oocytes. J. Physiol. 366:299–314.

Freeman JA, Manis PB, Samson PC, and Wikswo JP (1986) Microprocessor controlled two-and three-dimensional vibrating probes with video graphic: Biological and electrochemical applications. In Nuccitelli R (ed.): *Ionic Currents in Development*. Alan R. Liss, Inc., New York, pp 21–35.

Goodman NG (1945) *A Benjamin Franklin Reader*. Thomas Y. Cromwell Co., New York.

Gow NAR (1984) Transhyphal electrical currents in fungi. J. Gen. Microbiol. 130:3313–3318.

Gow NAR, Kropf DL, and Harold FM (1984) Growing hypae of *Achlya bisexualis* generate a longitudinal pH gradient in the surrounding medium. M. Gen. Microbiol. 130: 2967–2974.

Hille B (1984) *Ionic Channels of Excitable Membranes*. Sinauer Associates, Inc., Sunderland, MA.

Hinkle L, McCaig CD, and Robinson KR (1981) The direction of growth of differentiating neurons and myoblasts from frog embryos in an applied electrical field: J. Physiol. (Lond.) 314:121–135.

Jaffe LF (1966) Electrical currents through the developing *Fucus* egg. Proc. Natl. Acad. Sci. U.S.A. 56:1102–1109.

Jaffe LF and Nuccitelli R (1974) An ultrasensitive vibrating probe for measuring steady extracellular currents. J. Cell Biol. 63:614–628.

Jaffe LF, Robinson KR, and Nuccitelli R (1974) Local cation entry and self-electrophoresis as an intracellular localization mechanism: Ann. N.Y. Acad. Sci. 238: 372–389.

Kline D, Robinson KR, and Nuccitelli R (1983) Ion curents and membrane domains in the cleaving *Xenopus* egg. J. Cell Biol. 97:1753–1761.

Kropf DL (1986) Electrophysiological properties of *Achyla* hyphae: Ionic current studied by intracellular potential recording. J. Cell Biol. 102:1206–1216.

Kropf DL, Lupa MDA, Caldwell JH, and Harold FM (1983) Cell polarity: Endogenous ion currents precede and predict branching in the water mold *Achyla*. Science 220:1385–1387.

Kropf DL, Caldwell JH, Gow NAR, and Harold FM (1984) Transcellular ion currents in the water mold *Achyla*. Acid proton symport as a mechanism of current entry. J. Cell Biol. 99:486–496.

Lund EJ (1947) *Bioelectric Fields and Growth*. The University of Texas Press, Austin.

McGinnis ME and Vanable JW Jr. (1986) Electrical fields in *Notophtalmus viridescens* limb stumps. Dev. Biol. 116:184–193.

Noda M, Ikedo T, Suzuki H, Takeshima H, Takahasi T, Kuno M, and Numa S

(1986) Expression of functional sodium channels from cloned cDNA. Nature 322:826–828.

Nuccitelli R (1977) Ooplasmic segregation and secretion in the *Pelvetia* egg is accompanied by a membrane-generated electrical current. Dev. Biol. 62:3–33.

Nuccitelli R (1986) A two-dimensional vibrating probe with a computerized graphics display. In Nuccitelli R (ed.): *Ionic Currents in Development.* Alan R. Liss, Inc., New York, pp 13–20.

Nuccitelli R and Wiley LM (1985) Polarity of isolated blastomeres from mouse morulae: Detection of transcellular ion currents. Dev. Biol. 109:452–463.

Palade PT and Barchi RL (1977) On the inhibition of muscle membrane chloride conductance by aromatic carboxylic acids. J. Gen. Physiol. 69:879–896.

Poo M-M (1981) *In situ* electrophoresis of membrane components. Annu. Rev. Biophys. Bioeng. 10:245–276.

Robinson KR (1979) Electrical currents through full-grown and maturing *Xenopus* oocytes. Proc. Natl. Acad. Sci. U.S.A. 76:837–841.

Robinson KR (1985) The responses of cells to electrical fields: A review. J. Cell Biol. 101:2023–2027.

Robinson KR and Jaffe LF (1975) Polarizing fucoid eggs drive a calcium current through themselves. Science 187:70–72.

Robinson KR and Cone R (1980) Polarization of fucoid eggs by a calcium ionophore gradient. Science 207:77–78.

Robinson KR and Patterson JW (1983) Localization of steady currents in the lens. Curr. Eye Res. 2:843–847.

Ryan TA, Myers J, Holowka D, Baird B, and Webb WW (1988) Molecular crowding on the cell surface. Science 239:61–64.

Salkoff L, Butler A, Wei A, Scavarda N, Baker K, Pauron D, and Smith C (1987) Molecular biology of the sodium channel. Trends Neurosci. 10:522–527.

Scheffey C (1986a) Tutorial: Electric fields and the vibrating probe, for the uninitiated. In Nuccitelli R (ed.): *Ionic Currents in Development.* Alan R. Liss, Inc., New York, pp xxv–xxxvii.

Scheffey C (1986b) Pitfalls of the vibrating probe technique, and what to do about them. In Nuccitelli R (ed.): *Ionic Currents in Development,* Alan R. Liss, Inc., New York, pp 3–12.

Schreurs WJ and Harold FM (1988) Transcellular proton current in *Achlya bisexualis* hyphae: Relationship to polarized growth: Proc. Natl. Acad. Sci. U.S.A. 85:1543–1538.

Stump RF and Robinson KR (1983) *Xenopus* neural crest cell migration in an applied electrical field. J. Cell Biol. 97:1226–1233.

Thiel R, Schreurs WJ, and Harold FM (1988) Transcellular ion currents during sporangium development in the water mould *Achlya bisexualis.* J. Gen. Microbiol. 134:1089–1097.

Troxell CL, Schefey C, and Pickett-Heaps JD (1986) Ionic currents during morphogenesis in *Microsterias and Closterium.* In Nuccitelli R (ed.): *Ionic Currents in Development.* Alan R. Liss, Inc., New York, pp. 105–112.

Weisenseel MH, Nuccitelli R, and Jaffe LF (1975) Large electrical currents traverse growing pollen tubes. J. Cell Biol. 66:556–567.

Weisenseel MH, Dorn A, and Jaffe LF (1976) Natural H^+ currents traverse growing roots and root hair of barley *(Hordeum vulgare).* Plant Physiol. 64:512–518.

Woodruff RI and Telfer WH (1980) Electrophoresis of proteins in intracellular bridges. Nature

286:84–86.
Yoshioka K and Miyata Y (1983) Charges in the distribution of the extrajunctional acetylcholine sensitivity along muscle fibers during development and following cordotomy in the rat. Neurosci. 9:437–443.
Young GPH, Young JD, Deshpande AK, Goldstein M, Koide SS, and Cohen ZA (1984) A Ca^{2+}-activated channel from *Xenopus laevis* oocyte membranes reconstituted into planar bilayers. Proc. Natl. Acad. Sci. U.S.A. 81:5155–5159.

APPENDIX: SUPPLIERS OF VIBRATING PROBE COMPONENTS

1. The Vibrating Probe Company, 238 Faro Ave., Davis, CA 95616, (916)752-3152. Sells complete two-dimensional system, or components, including a specially designed two-phase lock-in amplifier that has built-in circuits for electroplating, capacitance measurements, calibrating current source, and two-phase power supply for probe.

2. Applicable Electrotechnics, 2 Sioux Rd., W. Yarmouth, MA 02673. Markets the differential preamplifier designed by Carl Scheffey.

3. Evans Electronics, P.O. Box 5055, Berkeley, CA 94705, (415)653-3083. Sells printed circuit modules for phase-sensitive detectors and phase control that can be used to construct a lock-in amplifier.

4. Microprobe, Inc., P.O. Box 87, Clarksburg, MD 20871, (301)972-7100. Source for parylene-coated stainless steel electrodes.

5. EG & G Princeton Applied Research, P.O. Box 2564, Princeton, NJ 08540, (609)452–2111. Source for a wide variety of lock-in amplifiers.

6. Stanford Research Systems, Inc., 460 California Ave., Palo Alto, CA 94306, (415)324-3790. Sells both single-phase and two-phase lock-in amplifiers.

7. Line Tool Co., 943 Kurtz St., Allentown, PA 18102, (215)434-5624. Supplier of a very useful X-Y-Z micropositioner that is handy for mounting and positioning vibrating probe driver units. Of course, other micropositioners can be used with good results.

8. Vernitron Piezoelectric Division, 232 Fobes Rd., Bedford, Ohio 44146, (216)232-8600. Supplies a variety of piezoelectric bimorphs for use as bender elements.

Electric Fields in Vertebrate Repair, pages 27–75

CHAPTER 2

Natural and Applied Currents in Limb Regeneration and Development

Richard B. Borgens

Center for Paralysis Research, Department of Anatomy, School of Veterinary Medicine, Purdue University, West Lafayette, Indiana 47907

INTRODUCTION

Anyone who has ever observed the regeneration of a salamander limb has probably marveled at the process. Perhaps the fascination this phenomenon holds for us lies in our own poor powers of tissue regrowth. We are awed by the fact (and perhaps are envious) that a complex vertebrate with limbs whose external and internal structure are so similar to ours can restore them after amputation, within a period of months. This process has traditionally interested developmental biologists, in part because it appears to mimic the process of limb formation during development and has been used as an experimental paradigm for such studies for many years.

We have been interested in the role that an endogenous steady current (and its associated electric field) plays in limb regeneration and limb development. Before discussing the role of electrophysiology in these matters, it would be wise to briefly discuss the processes of limb regeneration and limb development and what is known of their control factors.

REGENERATION OF AMPHIBIAN LIMBS

The process of limb regeneration is best observed in the tailed amphibia (adult and larval salamanders and newts) and in larval anura (toads and frogs). It should be noted that many of these animals possess extraordinary powers of tissue regeneration that are not limited to their extremities. For example, most salamanders can regenerate portions of their jaw, facial structure, and trunk, portions of their nervous system including their spinal cord, portions of their eyes, and portions of their internal organs (Goss, 1969; Singer et al., 1979; Stensaas, 1983; Rose, 1964; O'Steen, 1959).

After an extremity has been amputated experimentally (or lost or autotomized in nature), the wound surface is covered in a matter of hours by a "wound epithelium" that seals off the cut surface from the external world. This wound healing phase is similar to simple skin wound healing in that the epidermis moves in over the wounded surface by migration and then proliferates to form a thickened covering, sometimes called the apical cap (Singer and Salpeter, 1961; Thornton, 1968). [This multi-layered epidermal covering is anatomically similar to the thickened integument (the apical ectodermal ridge) that forms over the apex of developing limb buds in most vertebrates (but curiously, not in urodele amphibians).] Beneath the apical cap, parts of dying cells and other cytolytic debris are apparently transported away from the lesion by this wound epidermis. Thus the first response of the limb stump after damage is to seal off the raw surface and to debride the tissue beneath. Subsequently, healthy cells at the distal region of the stump's end begin to dedifferentiate and then to redifferentiate. In other words, they begin to lose the cytological characteristics that give them an identity (such as cartilage or muscle) and then to transform into a mesenchyme-like cell called a blastemal cell (Trampusch and Harrebomee, 1965). It is these pluripotent cells that will form the components of the developing limb. For years it was a controversial matter whether cells of the limb stump underwent this transformation, and if they did, what was their fate in the newly formed limb. Modern techniques are rapidly providing answers to these questions. For example, it is probable that a majority of blastemal cells are derived from cells local to the wound surface, as has been determined by the use of monoclonal antibodies that have been raised to both blastemal and presumptive progenitor cells in the stump (such as muscle) (Kintner and Brockes, 1984). It is now clear that "dedifferentiation" (Trampusch and Harrebomee, 1965) takes place and is the mechanism by which the bulk of the early blastemal cells are produced. It is probable that cells maintain their cytological identity during "redifferentiation." For example, muscle cells that dedifferentiate into blastemal cells *redifferentiate* into muscle. However, under experimental conditions, single cell types (such as cartilage or Schwann cells) may give rise to blastemal cells capable of reforming the entire missing appendage (Wallace et al., 1971, 1974). Transdetermination of cell type, although possible, is probably of minor importance to the process of limb regeneration in nature.

Role of the Apical Cap and Nervous Tissue

Traditionally, the two tissues suggested to provide control over the regeneration of the limb have been the wound epidermis (and its sequel, the apical cap epithelium) and the proximal segments of peripheral nerves left within the limb stump (Wallace, 1981a, b).

The epidermal covering of the stump not only seals off the wound and helps with the phagocytosis and removal of cellular debris, it is necessary for the process to be initiated. Limb regeneration cannot proceed in the absence of a wound epidermis, which can be demonstrated by placing a limb stump surface in an environment in which a wound epidermis cannot form (such as the peritoneal cavity) (Goss, 1956a). However, if an epithelium has already covered this surface prior to insertion into the body cavity, then regeneration will proceed (Goss, 1956b). Lastly, the wound epidermis imposes a directionality upon the outgrowing blastema. If the apical cap is surgically placed in an eccentric position on the stump, then the blastema and the resulting limb outgrowth will be so inclined (Stocum, 1985; Thornton, 1960).

The role of the nerve in limb regrowth has been extensively studied since the early 19th century (Todd, 1823). It is well known that a suitable supply of nervous tissue is required within the salamander's forelimb stump for normal regeneration. If the stump of a normal limb is maintained in a denervated state, no regeneration can occur (Wallace, 1981a). It is also interesting that this dependence on nerve is acquired during development. An ''aneurogenic'' amphibian larva can be produced by the surgical removal of its cranial ganglia and neural tube (Piatt, 1942). (They are kept alive by parabiosis with a normal larva.) Limbs of the aneurogenic animal will regenerate after amputation, in the absence of nervous tissue. However, these aneurogenic tissues can be allowed to experience the presence of nerve, after which they form a reversible dependence upon it (Thornton and Thornton, 1970).

The fundamental role of nervous tissue in limb regeneration has been the subject of intensive investigation for years. Many putative ''neurotrophic factors'' produced by nerves have been isolated and tested; however, no single growth factor (or combination of them) has yet been identified that will substitute for nerve in promoting limb regeneration. Glial growth factor (GGF) has been shown to stimulate Schwann cell proliferation—cells that have been identified as participating in the formation of the blastema. Furthermore, endogenous levels of GGF increase in the blastema and are lost after denervation (Brockes, 1984). This mitogenic factor may then satisfy at least some of the criteria for an important neural factor necessary for limb regeneration (Brockes, 1984). Nerves are not only necessary, they are necessary in a quantitative way. Singer once suggested that this relationship was a simple numerical one between nerve number and the limb stump's cross-sectional area (Singer, 1946). This notion was later altered to suggest that there was a critical *ratio* between the cross-sectional area of the limb stump and the cross-sectional area of the transected axons contained within it (Singer, 1965). It was also demonstrated that grassfrogs *(Rana)* were deficient in this nerve-stump relationship. Singer performed the famous experiment in which he surgically deviated the sciatic nerve of the hindlimb to the forelimb stump in postmetamorphic grassfrogs (Singer, 1954). These hyperinnervated stumps began to regenerate. The resulting structures were very hypomorphic; that is, they were deficient in internal structure and lacked limb pattern. In fact, in most ways these structures did not appear to be limbs at all. Even though these structures are described as limb "regenerates" by most authorities, Wallace is well within his rights in decrying the use of this term (Borgens, 1986). Here I will try to limit unjustified enthusiasm and refer to these and similar structures as "hypomorphic" regenerates.

Why Don't Frogs Regenerate Their Limbs?

Tadpoles lose their ability to regenerate their limbs as they progress through metamorphosis (Schotté and Harland, 1943). This loss occurs in a proximal-distal direction in their extremities and is nearly complete after metamorphosis. About 10% of barely postmetamorphic froglets retain a small degree of regenerative ability (hypomorphic regeneration only), while large, fully adult frogs (of the genus *Rana*) are usually nonregenerators (Singer, 1954). Certain other groups of anura possess various degrees of regenerative response to amputation (Dent, 1962; Goode, 1967; Scadding, 1981). This lack of regenerative ability in frogs and toads, when compared with salamanders and newts, has intrigued biologists for well over a century. A variety of hypotheses have been suggested for this loss of regenerative ability, not only during anuran ontogeny but also phylogeny. Here I will only mention the two most common notions, directing the interested reader to reviews of this subject (Polezhaev, 1946, 1972; Rose, 1945, 1964). The usual suggestion, supported in part by experimentation, is that neuronal factors (qualitative or quantitative) are deficient in these groups (discussed above). Another common suggestion is that a well-formed and precocious scar forms over the wound surface, along with a covering of full-thickness skin. This dense cicatrix, not found in salamanders and newts, is believed to prevent a critical communication between the underlying mesodermal tissues of the stump and the overlying epidermis. This theory would suggest that if the scar tissue is inhibited or reduced, then limb regeneration should occur in response to amputation in adult frogs. A variety of techniques have been used to do just this, the most common having been the applications of various caustic chemical agents; bathing the stump tip in salt solutions; and chronically traumatizing the stump end (by pin-pricking, for example) (Polezhaev, 1972; Rose, 1942, 1944, 1945). These techniques indeed seemed to induce a measure of regenerative response. However, it is not clear if the promotion of this limited regrowth was because of the action of these agents on the scar itself or some other mechanism (see below).

It has not been my intention to provide a complete review of this old and lengthy literature. I have omitted many new and interesting avenues of investigation in the field of limb regeneration, such as pattern formation studies, new research on the transdetermination of cells in the blastema, and possible immunological factors (Bryant et al., 1981; Kintner and Brockes, 1984; Sicard, 1985). I have instead tried to present the reader with a short discussion of a few basic tenets of limb regeneration as preparation for a more thorough appreciation of the role of electrical factors in the process.

"BIOELECTRIC POTENTIALS"—THE EARLY STUDIES

Studies of so-called wound potentials made with the galvanometer date back to the middle of the 19th century. These pioneering investigations of integumentary wounds were well performed on a variety of animals by the fathers of modern physiology and embryology—men such as DuBois-Reymond, Herlitzka, and Herlitzka's student, Gaetano Viale. Viale focused his interest on the role of surface-detected voltages in severed peripheral nerves and on wounds made to the tails of frog larvae (Viale, 1916). These measurements, made on amphibians, drew the interest of Alberto Monroy who, to my knowledge, was the first to correlate electrical changes following amputation of the urodele limb to subsequent regenerative events. Using a galvanometer, he measured changes in surface-detected potential differences during the regeneration of the limb (and tails as well) in adult Tritons (Monroy, 1941). He found that the stump tip was always positive with respect to the undamaged more proximal part of the forelimb. These potentials declined steadily after amputation over the 46 days of measurement (after an initial increase in positive potential). His studies were the first of their kind and have since been duplicated by other investigators using more modern high-impedance electrometers (Becker, 1960, 1961a, b; Lassalle, 1974a; Rose and Rose, 1974). Such distally positive surface voltages suggest that a current might leave the stump end and complete its circuit by returning to the undamaged portions of the extremity—perhaps to the body as well. These types of voltage measurements are indicative of a natural current flow, although their quantitative aspects are suspect (Borgens, 1982a, 1983).

Monroy also made an attempt to determine the locus of the biological battery generating the voltages that he measured. His methodology was to remove certain parts of the body, determining by this process of elimination that the intact skin of the limb stump was probably the source of the potential differences. This was not a surprising or novel deduction for Monroy, in view of the fact that this investigation evolved from earlier studies of "wound potentials" in animal skins, in which the integument was assumed to be the source of its own injury potential and current. This simple and elegant deduction was confused by a series of papers in the 1960s attempting to demonstrate that the nervous system is the source of surface-detected bioelectric potentials (Becker, 1960, 1961a, b; Becker and Marino, 1982; Becker and Seldon, 1985). Their series of reports is seriously flawed. In most cases, quantitative data were not even presented, or critical control operations were not performed. For example, denervated salamander limbs were claimed to lack surface potentials (Becker, 1961a). However, sham-operated control groups were not used for comparison. A suitable and necessary control would be the sham operation—exposing the brachial plexus surgically, but not severing it. Since the surgical approach to the plexus severs the skin of the back in an area adjacent to (or between) the surface measurement electrodes, this control operation would be essential to determine if the induced alterations in the bioelectric potential were due to *skin damage* or to the absence of nerves. I have discussed more flaws in these experimental studies elsewhere, and direct the reader to these reports (Borgens, 1982a, 1985; Borgens et al., 1979b). Unfortunately, the notion of this group, that externally measured "bioelectric potentials" are produced by the nervous system, has tenaciously hung on in certain less critical circles—partly because of its expansion into a more global approach to medicine, pathology, psychology, and even parapsychology and extrasensory perception (Becker and Seldon, 1985; Becker, 1977). Theories that have such all-encompassing scope seem to have appeal; however, in general (as in this particular case), they may have little scientific support.

Lassalle also investigated the source of translimb potentials in salamanders, and found support for Monroy's original supposition that surface potentials are driven by the integument, or "skin battery" (Lassalle, 1974a, b). Lassalle was aware of a large literature demonstrating that animal integuments possess large potential differences across themselves (inwardly positive by some 40–80 mV) (Benos and Mandel, 1978; Kirschner, 1970, 1973; Ussing, 1964). In amphibians, this transcutaneous potential difference is very sensitive to the Na^+ concentration in the pond water bathing the apical surface of the integument (Helman and Fischer, 1977). Altering the

Na^+ concentration of the measurement medium, ablation of the skin, and osmotic shock predictably altered Lassalle's measurements of surface potentials on intact and regenerating salamanders (Lassalle, 1974b). A correlation in the magnitude of transcutaneous short-circuited skin currents and the magnitude of the surface-detected voltages measured on the bodies of amphibians also added support to the notion that the skin is indeed the source of electrical changes induced by amputation (Fontas and Mambrine, 1977).

In summary, externally measured, distally positive translimb potential differences reported in the older literature suggested the presence of an integument—driven current flow traversing the stump—leaving the stump's end and returning to the proximal limb stump and body. These currents and voltages diminish with time after amputation.

DISSECTING THE LIMB STUMP CURRENT

Lionel Jaffe, Joseph Vanable, and I decided to apply modern techniques for the detection of steady biological currents to this problem. The noninvasive ultrasensitive vibrating probe system had just been developed in Jaffe's laboratory (Jaffe and Nuccitelli, 1974) (see Robinson, Chapter 1, this volume), and we applied it to the problem of the developmental electrophysiology of limb regeneration. Indeed, as the antiquated measurements of bioelectric potentials suggested, strong steady currents were driven through regenerating salamander limbs.

An instantaneous response to amputation of the limb was the production of an intense steady electrical current, driven out of the tip of the limb stump (Borgens et al., 1977a). This current returned to the undamaged expanse of stump proximal to the wounded surface (Fig. 1). The circuit path is provided by the conductive pond water the animal inhabits or the thin film of moisture on the bodies of semiterrestrial species. Densities of current ranged from 20 to over 100 $\mu A/cm^2$ soon after amputation in most salamanders. This current falls steadily with time and is variable in the rate of its decline between species (Borgens et al., 1984). A steady level of a few $\mu A/cm^2$ is achieved within a period of days, and this resting level is reasonably stable for a variable period thereafter. Recently, McGinnis and Vanable (1986a) have demonstrated that the decline in current is caused by the increasing resistance of the wound epithelium covering the stump's surface and is not a result of changes in the voltage across the skin (see below). The production of this stump current is ubiquitous among urodeles. We have sampled many urodeles of differing habitats as well as stages of development, including: the adult red-spotted newt, *Notophthalmus viridescens;* the Mexican axolotl,

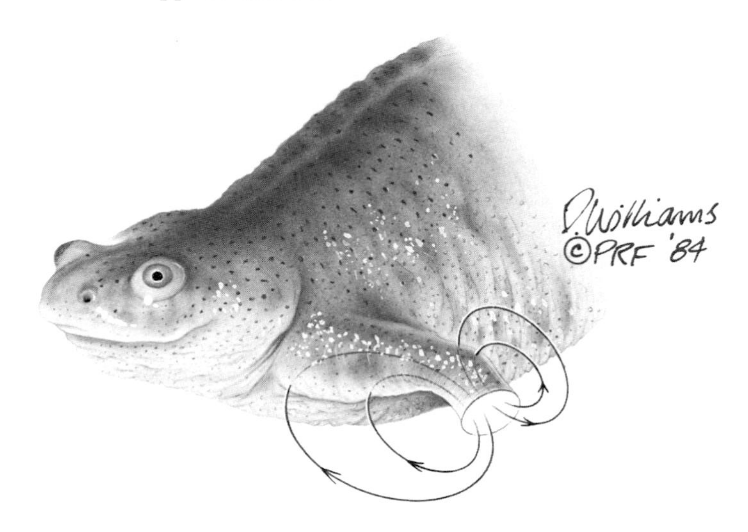

Fig. 1. Electric (ionic) currents are driven through the core tissues of a regenerating salamander stump. Currents return to the animal's tissues through pond water (aquatic species) or their conductive surface moisture (semiaquatic species). (Reproduced from Borgens, 1984b, with permission of the publisher.)

Ambystoma mexicanum; the adult tiger salamander, *Ambystoma tigrinum;* the adult West Coast newt, *Taricha torsosa;* the adult mudpuppy, *Necturus maculosus;* adult and larval Iberian newts, *Pleurodeles watl;* and a variety of adult plethodontid or forest salamanders, *Desomagnathus quadrimaculatus, Batrachoseps attenuatus, Aneides lugubris,* and *Ensatina eschscholtzi xanthoptica.* All generate skin-driven stump currents in response to amputation (Borgens et al., 1984).

A Skin Battery Is the Source

In contrast to the large currents leaving the stump face, very weak currents were found to enter the surface of most of the proximal limb stump (on the order of about 1 $\mu A/cm^2$). Weak incurrents are driven by the well-known transepidermal potential (TEP) known to exist not only in amphibians but most vertebrates, as well as man (Barker et al., 1982; Herlitzka, 1910; Kirschner, 1970). Skin batteries are internally positive by about 40–80 mV, and this voltage is usually associated with an inwardly directed active pumping of Na^+ across the integument from outside to inside. In adult amphibians, notably anura of the genus *Rana,* this transcutaneous potential is highly dependent on the concentration of Na^+ in the outside medium

(Borgens, 1985; Kirschner, 1970, 1973). If Na^+ is reduced, one sees a proportional decrease in the potential across the skin. If Na^+ is raised in the outside medium, the TEP is likewise increased. This dependence on external Na^+ provided us the opportunity to test the hypothesis that the skin is the electromotive source (EMF) driving charge out the stump end (as measured by our probe system). Using adult red-spotted newts, we measured the responses of stump currents to a modulation of the Na^+ concentration of the artificial pond water (APW) that served both as a measurement medium and maintenance medium for the animals. When Na^+ was replaced with choline, or was reduced to minute levels by simply removing NaCl from the APW, stump currents fell by more than 90% of their original values. If Na^+ was raised about fivefold in the measurement medium, we observed a fivefold increase in the stump current (Borgens et al., 1977a). These effects were usually rapid and reversible. Returning the animal to normal APW normalized current to about pretrial levels (Borgens et al., 1977a, 1984). The Na^+-dependent TEP is also sensitive to several pharmacological agents that specifically block Na^+ channels at the apical living layers of the epidermis (Kirschner, 1970, 1973). We have routinely used several of these compounds to reduce stump currents. Amiloride, the methyl ester of lysine (MEL), and benzamil are all effective in blocking the voltage across the skin and thus the stump current associated with it (Borgens et al., 1979c, 1984; Eltinge et al., 1986). The effects of these treatments are similar to the Na^+ modulation experiments in that after a return to normal APW and a short period of adjustment, stump currents usually return to near normal values (unless the skin is damaged by prolonged immersion in these agents).

We have also tested the ill-advised notion that nerves within the stump are associated with the outcurrents measured leaving it. We simply denervated limbs at the brachial plexus and then checked the density of current leaving the amputation surface. Curiously, these currents were slightly but consistently *increased* by denervation. This observation certainly leaves no doubt that nerves are not the source voltage for currents leaving the stump. However, these experiments drew our attention to the fact that various injured cells and tissues (nerve, muscle, and bone) possess their own injury currents that may decrease the net outcurrent leaving the stump's end. Since single cells are internally negative, a rupture in the membrane will permit a steady current to leak into the injured cell, driven by the resting potential across the uninjured membrane (Borgens, 1982a). In response to this local depolarization, the cell dies, or the rupture is sealed with new membrane. For example, current enters into the ends of severed axons as well as into bone fractures (fracture currents have a more complex origin, however; see

McGinnis, Chapter 6, this volume) (Borgens et al., 1980; Borgens, 1984a). These local injury *incurrents* would be expected to *reduce* the *net outcurrent* driven by the skin through the core tissues of the stump. The initial or immediate densities of current leaving the fresh amputation were found to be slightly less than the densities of current measured an hour or so later. At this time the inwardly directed cellular injury currents would have fallen off steeply. Therefore, by removing through surgical denervation one source of these injury currents—the peripheral nerve trunks—stump currents would be expected *to be slightly larger.* Altogether, we felt confident that these experiments demonstrated that the large stump currents had their source in the undamaged skin's transcutaneous voltage.

We were, of course, interested in the magnitude of the electrical field *within* the core tissues of the limb stump (where the blastema arises) associated with this current flow. We estimated that the field should have been on the order of a few tenths of a mV/mm, a very weak field when compared with the fields known to affect cells in culture (refer to Robinson, Chapter 1, Borgens, Chapter 4, and Vanable, Chapter 5, this volume) (Borgens et al., 1977a). Recent measurements using microelectrodes have demonstrated that the electrical fields within the stump's core tissue are much higher than expected—on the order of 6 mV/mm along the longitudinal axis of the proximal central core tissues and increasing to over 50 mV/mm at the stump's end (McGinnis and Vanable, 1986b). This is the primary area responsible for blastema formation, and furthermore suggests that tissues local to the stump tip may be *targets* of the field as well.

Stump Current in Nonregenerating Frogs

It is well known that adult frogs (especially *Rana*) do not regenerate their limbs. A reasonable question might be, "do adult frogs possess stump currents, and, if so, what would be their character?" Frogs (like most vertebrates) possess a transcutaneous potential. Therefore they should be expected to drive steady current out of the amputation surface. A curious difference in the anatomy of the limb between salamanders and adult frogs suggests that the character of the current leaving their stump might be strikingly different. As noticed by Schotté and Harland (1943), and as discussed by Rose (1944), the subepidermal lymph spaces begin to develop in frog larvae at about the same place on the limb as the ability to regenerate the limb *is lost during metamorphosis.* Rose felt that this large, fluid-filled space between the skin and underlying musculature would make it easier for the skin to migrate over the face of the amputation—closing the amputation face with a premature skin barrier. This process would perhaps inhibit the

Fig. 2. Ionic currents are predominantly driven through the subdermal lymph space in adult frogs. (A relatively small proportion does traverse the core tissues—not shown here for purposes of clarity.) In *Rana pipiens,* these circumferential densities of current are a characteristic response to limb amputation. Peak densities of current are observed leaving the *core tissues* in salamanders and newts, and are never observed leaving the peripheral regions of the stump. Compare with Figures 1 and 10. (Reproduced from Borgens, 1985, with permission of the publisher.)

process of regeneration as whole skin covering the amputation face is well known to inhibit regeneration in urodeles (Mescher, 1976). We viewed the lymph spaces associated with the loss of regenerative ability in a different light: such spaces would be effective low-resistance current shunts. Current driven by the overlying skin would be shunted through this pathway immediately below the skin and very little current would be driven through the relatively high-resistance musculature of the core tissues of the stump. It is the core tissues that provide the cellular foundation for the blastema of regeneration. Salamanders and newts do not possess this subdermal lymph sinus; thus current is driven through the core tissues of the stump. Actual measurements of stump currents in young postmetamorphic frogs with the vibrating probe supported this notion (Borgens et al., 1979d). Current densities were always highest leaving the lymph space in forelimb stumps (Fig. 2). Weak currents were found leaving the core tissues of the frog stump (averaging about 7 $\mu A/cm^2$). These currents were about four- to fivefold lower than the circumferential densities of current leaving the lymph space of the same limb stump. Additionally, we learned that the peak currents leaving frog stumps were generally weaker than those measured in newts (on the order of about 30 $\mu A/cm^2$). Thus currents were not only reduced in magnitude in postmetamorphic frogs (compared with salamanders and newts), but the topography of current flow between the two groups was different as well. These findings are in agreement with the notion that the skin-driven stump current is necessary for regeneration of the limb. A reduction in the current within the core tissues of the limb stump is associated with the lack of regenerative ability. Moreover, the development of the anuran lymph sinus, which is responsible for the shunting of current around but not through the stump's core tissues, is coincident with the loss of regenerative ability during metamorphosis. This finding suggests that if one could manipulate the electrophysiology of the limb stump in amphibians, then one might be able to manipulate its ability to regenerate.

ELECTRICAL MODULATION OF AMPHIBIAN REGENERATION

If current is indeed a control of the regenerative process, then modifying its character should modify regeneration in a predictable way. Stated another way, if current is *not* a control factor, then manipulation of the electrical environment of stump tissues should not affect regeneration at all.

We tested this idea in salamanders by using some of the same Na^+ channel-blocking agents discussed previously (amiloride and MEL) and by chronically maintaining salamanders in a culture medium low in Na^+ (Borgens et al., 1979c). In one experiment, large tiger salamanders were kept in moist but not wet culture conditions. Their forelimb stumps were swabbed daily with a dilute solution of 0.5 mM amiloride. This treatment resulted in a complete inhibition or marked distortion of limb regeneration in about one-half of the animals treated (Fig. 3). Of 33 total animals in this group, 12 were irreversibly inhibited from regenerating, showing the classical histological signs of arrested regeneration (scar tissues and callus capping the bone with whole skin covering the wound surface) (Borgens et al., 1979c). In other experiments, newts and tiger salamanders have been maintained in Na^+-deficient media. The responses of both species were similar: we observed a striking inhibition of regeneration for a variable period of time, and then the inhibited limbs began to regenerate. Not only did they regenerate, but they did so at a greatly accelerated rate. At this rate they eventually reached the same stage of development as the control population—sometimes even surpassing this group (Borgens et al., 1979c). What was also

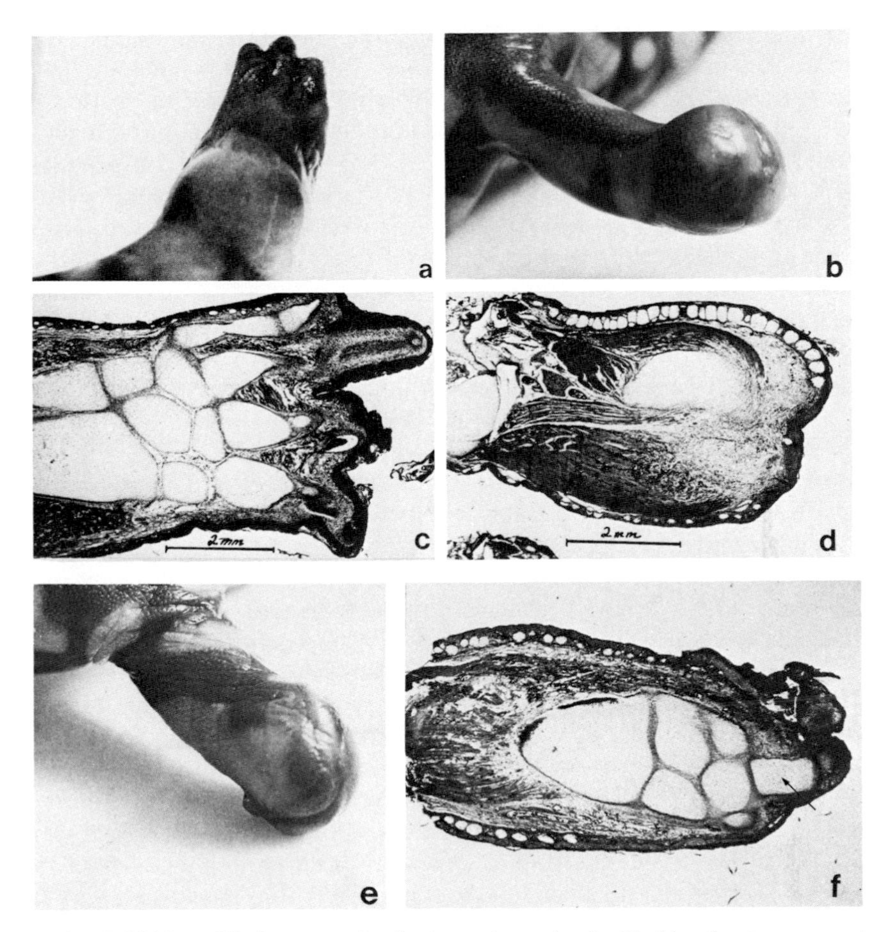

Fig. 3. Inhibition of limb regeneration in tiger salamanders by blocking the stump currents. **a:** External view of a tiger salamander forelimb regenerate, 11 weeks postamputation. **b:** External view of an experimentally treated limb stump showing inhibition of regeneration at 11 weeks, postamputation. Amiloride (a pharmacological agent effective in reducing the skin-driven stump current) was applied topically to this forelimb every other day following amputation. **c:** Photomicrograph of a histological section from the limb regenerate shown in a. **d:** Photomicrograph of a histological section from the forelimb stump shown in b. Note the callus formation in the distal stump tissues and lack of regenerated wrist and hand elements. **e:** External view of a malformed limb regenerate, treated with amiloride as in b and d. **f:** Photomicrograph of a histological section from this forelimb regenerate. Note there is only one metacarpal at the base of a single rudimentary digit. The carpals are essentially normal. (Reproduced from Borgens et al., 1979c, with permission of the publisher.)

interesting about this "escape" of inhibition is that, judged by our conventional understanding of limb regeneration, this escape should not have occurred. When limbs were analyzed histologically *during* the period of inhibition, their appearance was that of limb stumps that were irreversibly inhibited from regeneration. The stump tip was covered with full-thickness skin, a well-developed scar lay beneath, and a large callus capped the severed bone. This is typical of nonregenerating anuran limb stumps or the limb stumps of urodeles that have been inhibited from regenerating by a variety of methods. So atypical was the appearance of these stumps that we were unable to classify them using the normal staging methods, such as that of Iten and Bryant (1973) or Singer (1952). We remarked that the striking inhibition and accelerated escape suggested that something(s) necessary for regeneration was accumulating within the stump during this period of arrest (Borgens et al., 1979c).

Are there alternate hypotheses for this inhibition other than a current-mediated effect? One might suggest that these treatments caused the animals to be in negative salt balance (Na^+ depletion) and that such debilitated animals could not regenerate effectively. This is very unlikely, since amphibians are highly resistant to changes in the ionic strength of their media. Frogs can be cultured in deionized water for long periods of time without demonstrating Na^+ losses (McAfee, 1972). Indeed, we checked blood Na^+ levels in the larger Ambystomids used in these studies, and there were no differences between the experimental and control populations (Borgens et al., 1979c). Could the pharmacological agents be toxic to limb regeneration? This is doubtful, since three different chemical compounds have been used in our studies (amiloride, MEL, and benzamil), all associated with little mortality. Moreover, amiloride applications were made to less than 1% of the body area of the animal. Preliminary experiments suggested that a critical period for stump currents to be blocked was early in the regeneration process. Animals immersed in 0.5 mM amiloride regenerated as well as the controls if the treatment was postponed for about a week. This not only argued strongly for a "window of effect" but *against* a toxic response to amiloride as well (S. Datena, unpublished observations).

More troublesome is the question of why some of the experimental animals regenerated at all. In studies in which pharmacological agents were topically applied to limbs, it may have been that the animals molted, sloughing off the drug-treated skin. Under laboratory conditions, urodeles molt irregularly and frequently (Larsen, 1976). Moreover, there exists the possibility that the fresh exposed skin may have been insensitive to the application of the drugs. For example, when the electrical properties of frog skin removed soon after a molt are characterized, such skins are found to drive strong short-circuit currents that are peculiarly insensitive to amiloride (Nielsen and Tomilson, 1970). Also, animals may have washed off amiloride by rubbing along the droplets of condensation on the sides of their containers.

With regard to newts chronically immersed in a Na^+-depleted medium, we have evidence that they may have adapted to this condition and begun pumping other ions across their integument to support the transcutaneous voltage. We spot-checked these animals for the presence of stump currents with the vibrating probe and found that experimental animals indeed drove stump currents that were not Na^+-dependent (Borgens et al., 1979c). As it turns out, this plasticity is not uncommon in larvae of both urodeles and anura, and in adults of certain species in both groups (Borgens et al., 1983, 1984; Robinson and Stump, 1984). Under conditions of Na^+ deprivation, it is probable that amphibian skin does adapt in this way.

STIMULATION OF REGENERATION IN THE LIMBS OF ADULT FROGS

One type of experiment that has historically been used to test putative controls of limb regeneration has been to attempt to induce limb regeneration in large nonregenerating adult frogs (usually of the genus *Rana*). Such experiments testing the role of currents have been performed—a steady current being imposed within the forelimb stump tissues. The first to attempt this type of experiment was Smith, who placed a bimetallic rod into the terminal tissues of the stump (Smith, 1967). (Dissimilar metals in contact produce a weak electrochemical potential difference between them when placed in conductive medium.) This unit was assumed to impose a steady voltage (of about 0.7 V/mm) within the stump. It was found that these applications induced a modest degree of regeneration. At most, this regeneration consisted of an aggregation of "mesenchyme-like" cells and a bit of cartilage that formed in the distal stump. These kinds of responses are indeed scant and can be observed in untreated postmetamorphic frogs (Singer, 1954; Tomlinson et al., 1985). Moreover, since a sizeable number of control applications (small metallic rods fabricated from only one metal, or bimetallic rods that were insulated so as to be electrically quiescent) "regenerated" in a similar fashion, since the effects observed in response to the electrical treatments were independent of the *polarity* of the application, and since the effects of the electrically active rods could not be separated from the possible effects of the electrode products generated electrochemically at the site of the rod insertion, it was unclear if the responses observed in these frogs had anything to do with electricity at all (Smith, 1967).

Smith improved this experimental design and published another report claiming the induction of regeneration in frog limbs in response to an applied electric current. In the second experiment he used ingenious implantable stimulators, fashioned from tiny mercury cell batteries, small resistors, and Teflon-coated stimulating electrodes (multistranded stainless steel) (Smith, 1974). These stimulators were inserted under the skin of the back of the frog; the negative electrode was routed beneath the skin and through the musculature at the stump end to be anchored in this soft tissue. The stimulators drove about 100 nA total current. (This measurement would have fallen to only a fraction of this magnitude in a matter of minutes, once inserted into the animal, as the stimulating electrodes would have polarized.) In response to these applications, there was a more striking degree of hypomorphic growth than observed in the earlier experiment using bimetallic rods. Unfortunately, this creative attempt at demonstrating an *electrical* control of limb regeneration failed because of two major flaws. First, there was no control group and one wonders if a ''dead'' or sham stimulator with indwelling electrodes anchored in the distal soft tissues of the stump would have produced similar regeneration. It is well known that chronic physical irritation is sufficient to induce frogs to produce these hypomorphic growths (Polezhaev, 1972). Without the proper control group, one could not know what had occurred. Second, an extraordinary claim was made in this report, one that was unsubstantiated by the evidence presented. It was claimed that, in response to current, one frog regenerated a completely normal limb in all respects (Smith, 1974). The only evidence in support of this claim was a photomicrograph of apparently normal histology of the wrist area of an anuran forelimb. There were no pictures of the development of this limb, even though it was described to have taken 1 year to form. This one animal, widely believed to have been a case of ''mistaken identity'' (Wallace, 1981e) and presented in unrefereed symposium proceedings, mars this otherwise imaginative experiment. Lionel Jaffe, Joseph Vanable, and I decided to redo this experiment and to further improve on its design (Borgens et al., 1977b). We modified Smith's original stimulator design to incorporate wick-stimulating electrodes. These long silastic tubes were filled with amphibian Ringer's solution and a cotton thread. Current was carried to the stump's target tissues via a conductive solution of electrolytes (the Ringer's solution), and the thread helped keep the bore of the tube electrically conductive, even if gas bubbles should form within. This arrangement radically reduced the probability that electrode contaminants could reach the target tissues at the stump's surface. Therefore, whatever responses might be observed in the experimental group (with the cathode at the stump surface) would be current-mediated. We also incorporated two control groups into this experiment, one composed of animals implanted with sham stimulators and the other implanted with the positive electrode at the stump's tip. The responses to these applications were as follows: negative electrodes at the stump's tip induced a measure of hypomorphic forelimb regeneration in all animals. Positive electrodes produced varying degrees of stump *degeneration,* and, in all cases, sham stimulators had no effect whatsoever—stumps healed over with skin and scar as would normal amputations in these large grassfrogs (Figs. 4–7).

Since the negative electrode had been in contact with the stump tip and was effective in initiating outgrowth, we could be reasonably sure that this effect was current-mediated and not induced by electrode products, since the positive-charged electrode products liberated at the AgAgCl-wick electrode interface would be captured here and would not be driven down the wick toward the target tissues. However, electrode products produced at the *positive* pole would tend to be driven away, down the wick toward the target tissues, by electrophoresis. Thus we could not rule out the possibility that the degenerative effects observed after stimulation of the stump tissues by the distally positive electrode was a result of cytotoxic contaminants and not the applied current. One fact, however, argues against this hypothesis. If electrophoresis of electrode products caused this tissue degeneration, one would expect the damage to be greatest where these putative compounds would leave the bore of the wick electrode, i.e., where their concentration would surely be highest. This was not the observation, however. Tissue damage was relatively uniform throughout the forelimb stump; the most intense areas of damage seemed to be in the hard tissue and in muscles, sometimes at great distances from the stimulating electrode itself. We found no evidence of a focus of tissue degeneration near the tip of the electrode.

This experiment demonstrated that a current artificially imposed within adult frog limb stumps of the same *polarity* and imposed for a similar duration (weeks) to the endogenous current measured in salamander stumps is capable of initiating a regenerative response. These large adults would otherwise heal forelimb amputations with scar and skin. Moreover, a reverse polarity of current and field *may* be associated with *degeneration* of stump tissues.

A curious problem arose from these studies, however. Since the stimulating electrode was an AgAgCl-wick electrode, and the opposite pole was *platinum* in construction, it is certain that total currents (100–200 nA) declined to radically *lower* values within minutes after implantation, because of electrode polarization. Thus the regenerative effects in response to these

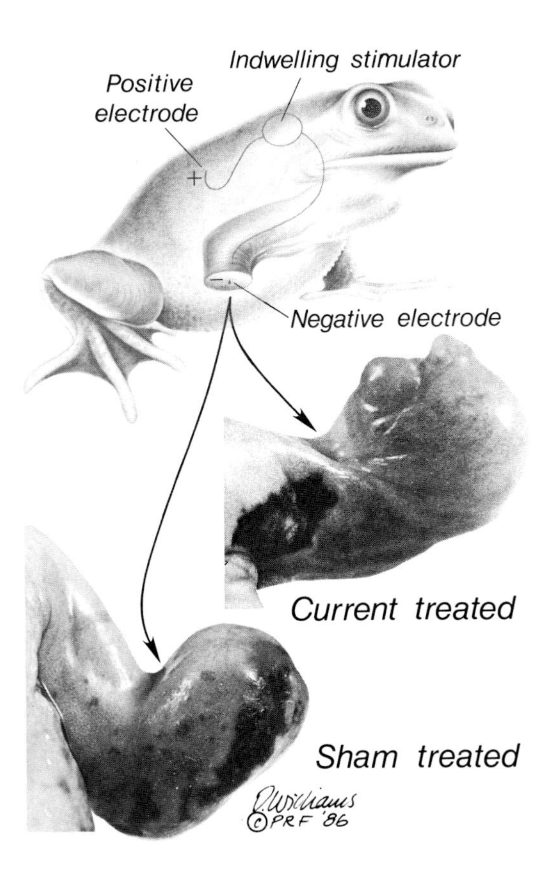

Fig. 4. Electrical stimulation of forelimb regeneration in adult frogs. This drawing demonstrates the general features of experiments in which indwelling stimulators initiate hypomorphic regeneration in large adult frogs. The stimulator is located in the dorsal lymph space, beneath the back skin. The positive electrode is situated beneath, while the negative electrode is routed through the core tissues of the distal stump tissues—the uninsulated portion of this electrode is retracted back in contact with the musculature. The stimulator units were implanted at the time of amputation and removed about 2 weeks later. The responses of forelimb stumps to a sham operation (implanted *inactive* stimulators) and *active*, distally negative applications are depicted by actual photographs. Compare with Figures 5–7 and 10. (Reproduced from Borgens et al., 1977b, with permission of the publisher.)

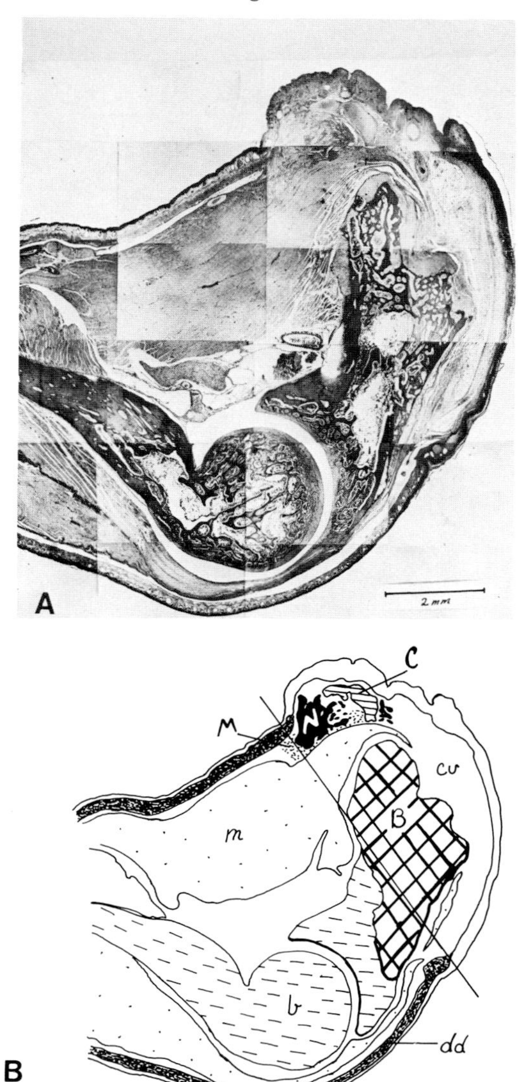

Fig. 5. **A:** Composite photomicrograph of the histology of a cathode-stimulated frog limb regenerate 9 months, 1 week postamputation. **B:** Tracing of composite. N, regenerated nerve; M, regenerated muscle; C, cartilage island; B, regenerated bone; m, original muscle; b, original bone; dd, dense dermis; cv, loose connective tissue and vascular tissue. The line represents the approximate plane of amputation. (Reproduced from Borgens et al., 1977b, with permission of the publisher.)

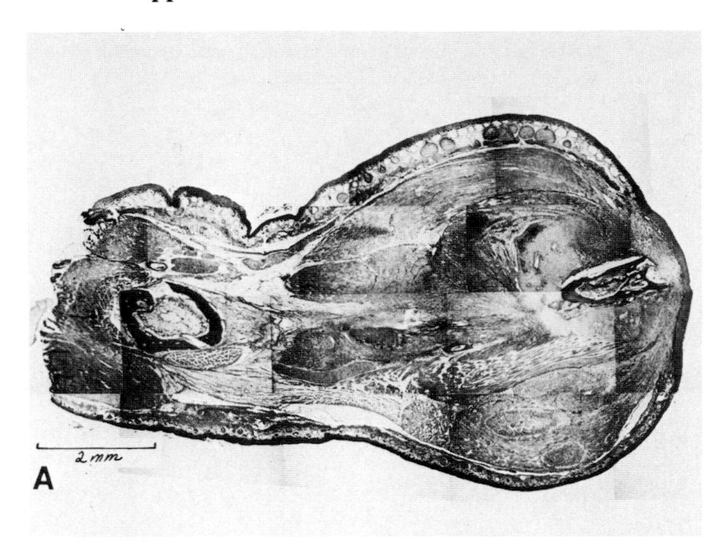

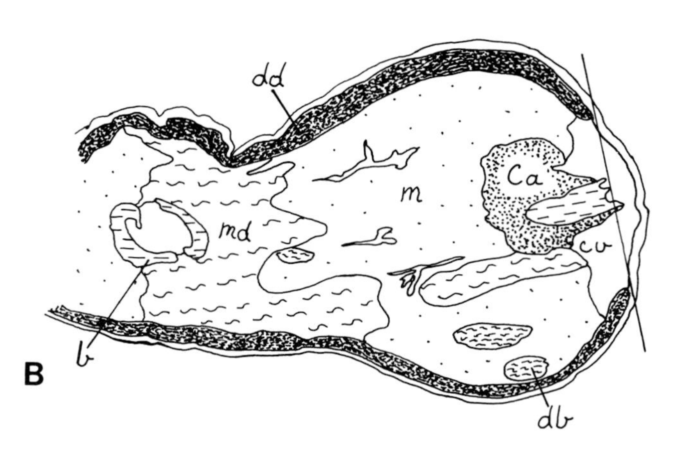

Fig. 6. **A:** Composite photomicrograph of an anodally stimulated frog limb stump, 4 months postamputation. **B:** Tracing of composite. Ca, callus; b, original bone; cv, loose connective tissue and vascular tissue; m, original muscle; md, area of muscular degeneration; db, pockets of cellular debris; dd, dense dermis. The distal remnant of the radioulna was not in continuity with the proximal section within the limb stump. Line indicates the approximate plane of amputation. (Reproduced from Borgens et al., 1977b, with permission of the publisher.)

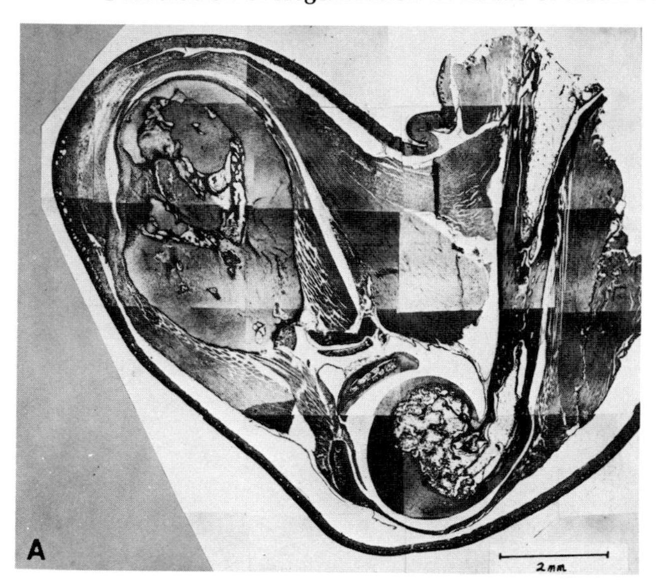

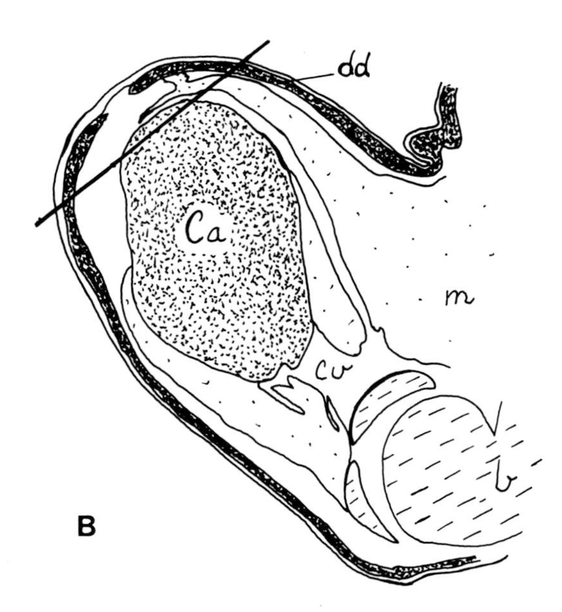

Fig. 7. **A:** Composite photomicrograph of the histology of a sham-treated frog limb stump healed over with skin and scar 6 months, 3 weeks postamputation. **B:** Tracing of composite. Ca, callus; b, original bone; m, original muscle; dd, dense dermis; cv, loose connective tissue and vascular tissue. The line represents the approximate plane of amputation. (Reproduced from Borgens et al., 1977b, with permission of the publisher.)

applied currents and voltages were within a range on the order of 10- to 100-fold weaker than either those normally measured in regenerating amphibians or the lower limits of voltages known to affect cells in culture (see Robinson, Chapter 1, this volume).

It should be remembered that our understanding of the lower limit of fields critical to development is still in its infancy. We have few guidelines here (see below). Since culture experiments are designed to permit observation of cell responses to fields within a matter of a few hours, they may well be overly conservative guides to absolute field parameters in vivo, where there are extended periods of time (days or weeks) for cells to perceive and respond to the steady voltage gradient about them. It is also possible that cells in vivo may be more sensitive to developmental cues (such as fields) than the same cells removed to culture.

We extended these studies to large adults of the African clawed frog, *Xenopus laevis,* which are known to regenerate their limbs hypomorphically as a natural response to amputation. Using similar stimulation regimens, we observed changes in the character of limb regeneration for a period up to 9 months postamputation (Fig. 8) (Borgens et al., 1979a). Within 2 months postamputation, the external appearance of the current-treated group was striking. Many of these regenerates appeared similar to the "paddle" or "palette" stage of development usually observed in salamanders and newts. Control (sham-treated) animals formed typical "spikes" of cartilage surrounded by skin (Dent, 1962; Goode, 1967; Scadding, 1981). The electrically treated growths went on to develop into limb-like structures and, in some cases, terminated in two to three digital-type structures (Fig. 8B). Histologically, all of the regenerates, both control and experimental, were hypomorphic (Fig. 9). In the experimentals, however, the cartilage core of each regenerate was perforated throughout its length by nervous tissue lying within lacuna. This was not observed in control animals.

It is interesting to compare the overall regenerative responses of these two different anuran species to the applied current. While electrically treated *Ranids* developed structures that were grossly malformed in external appearance, the histological differentiation in these stumps was rather well-developed: bone, muscle, nerve, and cartilage anlagen (possibly to carpals or metacarpals) formed in regenerates. In the clawed frog, the internal structure was always hypomorphic (typical of the "cartilage spikes" formed by this species). Curiously, these electrically induced structures were externally striking in appearance in some cases, and we refer to them as "pseudolimbs." These limb-like structures, formed in response to an applied electric field, challenge our conventional understanding of the relationship between

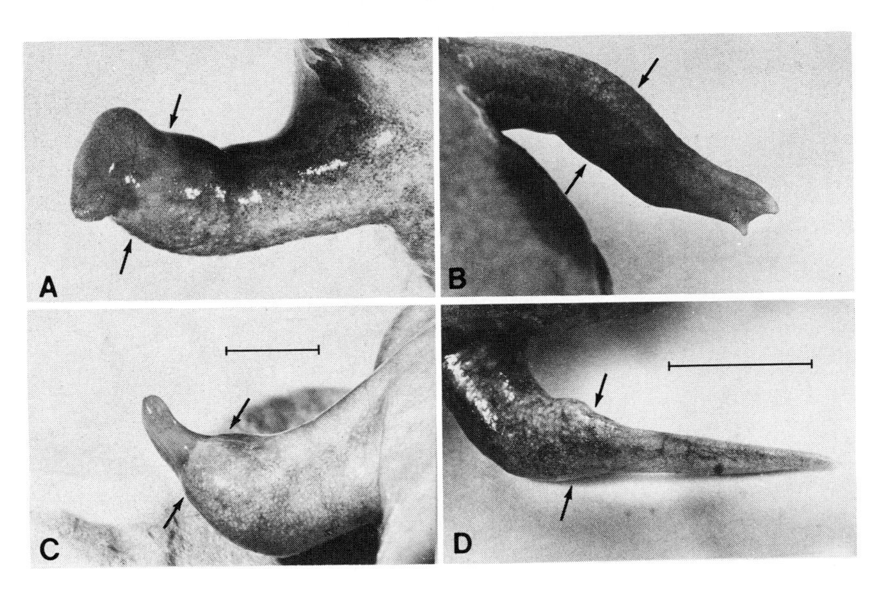

Fig. 8. External views of sham- and electrically treated *Xenopus* forelimb regenerates. **A:** Cathodally stimulated forelimb regenerate at 8 weeks postamputation. Note the paddle stage regenerate. **B:** Same animal as in A, at 26 weeks postamputation. Note the striking "pseudolimb" (refer to text) formed in response to the application of the electrical field. **C:** Sham-treated stump at 8 weeks postamputation. Note the forming "spike" (a slender tapering cartilagenous rod covered with skin). **D:** Same animal as in C, at 26 weeks postamputation. Note the fully formed "spike." These growths are a characteristic response to amputations in this species. Arrows depict the approximate plane of amputation. Scale bar for A and C = 5 mm, for B and D = 10 mm. (Reproduced from Borgens et al., 1979a, with permission of the publisher.)

developing structure and form (Borgens, 1983). The "pseudolimbs" were devoid of an internal organized multitissue structure, yet were molded into an external form similar to a normal extremity.

Although responses to applied fields were quite different between the two species, they shared one response in common: in all electrically-treated animals, gross amounts of nervous tissue formed in the regenerate. In the grassfrogs, up to 20% of all the regenerated tissues was nerve, observed in great, neuroma-like whorls within the regenerate (Borgens et al., 1977a). In the clawed frog, large amounts of nerve (continuous with the nerve trunks of the undamaged forelimb) were found to ramify throughout the cartilage core of the "pseudolimb" (Borgens et al., 1979a). Since nerve is not known to *penetrate* cartilage (once it has formed) and since chondrogenesis occurs first

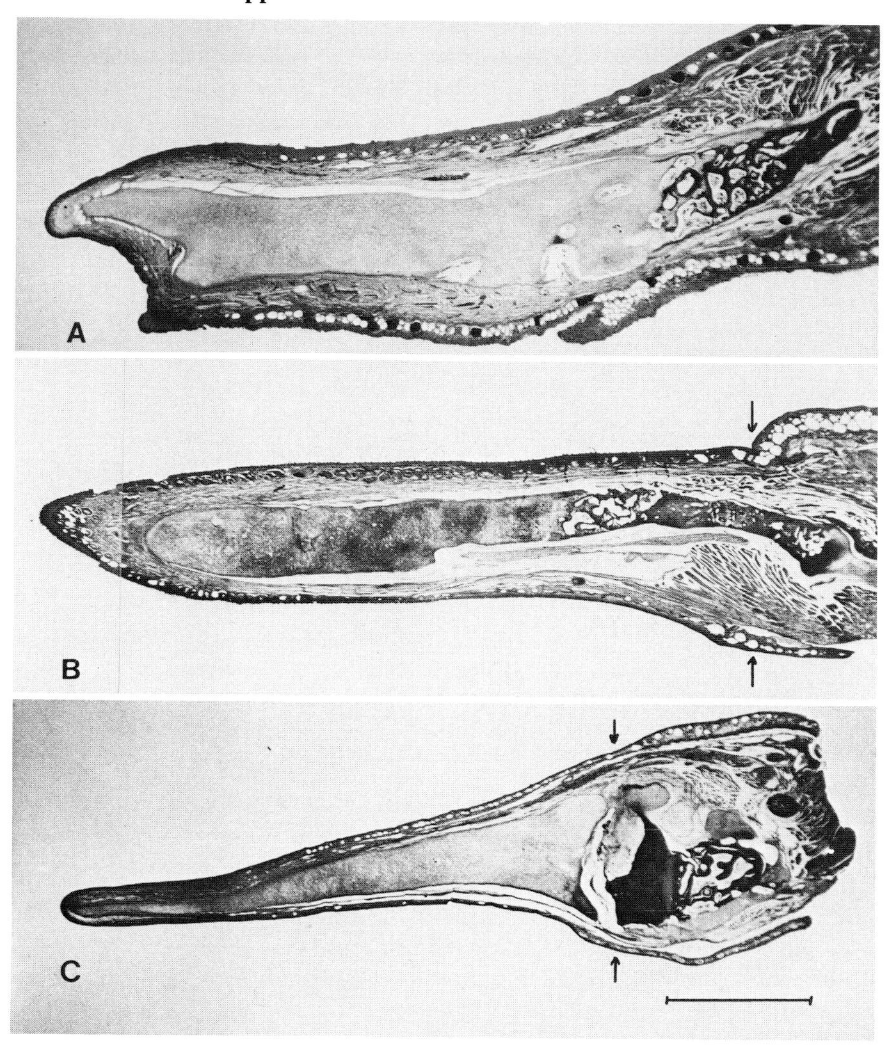

Fig. 9. Photomicrographs of histological sections from sham- and electrically treated *Xenopus* regenerates. **A:** Cathodally stimulated animal (the same as in Fig. 8A and C), 11 months, 2 weeks postamputation. The approximate plane of amputation is at the right-hand border of the photograph. Note the ossification of the proximal portion of the cartilage core of this regenerate and the lacunae filled with nerve trunks. **B:** Another cathodally stimulated animal 6 months, 3 weeks postamputation. The large nerve trunk ventral to the cartilage core of the regenerate was followed in other sections to course throughout the core to the end of the regenerate. Arrows mark the approximate plane of amputation. **C:** Sham-treated stump, 10 months, 2 weeks postamputation. Note the lack of nerve trunks in the regenerate and the absence of ossification of the cartilage core distal to the plane of amputation (arrows). Scale bar = 2mm. (Reproduced from Borgens et al., 1979a, with permission of the publisher.)

during differentiation of the regenerate, this suggests that the field's primary effect was to increase the rate as well as the amount of nerve regeneration in the electrically treated stumps (Borgens et al., 1979a). Both of these studies suggest: 1) that peripheral nervous tissue in vivo may respond in a striking way to the presence of an applied field (see Borgens, Chapter 4, this volume); and 2) that nervous tissue may be the direct *target* of the endogenous field within limb stumps in inducing changes in the development of the stump after amputation. If we recall the putative role of nerve in controlling limb regeneration, then this view suggests that natural fields (or applied fields) may promote a sufficient quantity of nerve growth into the stump to support limb regeneration (see below). Alternatively, or additionally, since the fields are actually steepest over the stump's terminal end, the apical cap epithelium may be a target of the field as well.

Overall, we have learned that imposing an electrical field of a polarity, magnitude, and perhaps duration similar to that of a regenerating salamander within the core stump tissues of an adult nonregenerating frog initiates regrowth of limb structures (Fig. 10).

IONIC CURRENTS

Development of Limbs

I have described several lines of evidence suggesting that the current leaving the urodele forelimb stump is relevant to its regeneration and is not just an epiphenomenon. Another observation providing circumstantial support for this notion is that endogenous local outcurrents *predict* the *emergence* of the limb in amphibians. Since the development and the regeneration of limbs share many similarities in the anatomy and physiology of development, it is satisfying that a steady ionic current predicts the exact focus of limb bud formation in both an anuran *(Xenopus)* and a urodele (the axolotl) (Borgens, 1983; Robinson, 1983). In the salamander, the hindlimb develops after the forelimb, in large, well-formed animals. Moreover, it takes several days for the hindlimb bud to form from a minute condensation of mesenchyme into a prominent protuberance. In the frog, the hindlimb forms first and proceeds through development in a matter of hours. In contrasting these two model systems, we can obtain a rather broad picture of the temporal and spatial distribution of current during limb ontogeny.

In the clawed frog, *Xenopus,* currents are detected leaving a focused area where the limbs will emerge in stage 43 (prebud) larvae (Robinson, 1983). The current accurately identifies the area of the emerging limb and declines steadily (from values on the order of 5–10 $\mu A/cm^2$) as the bud develops and

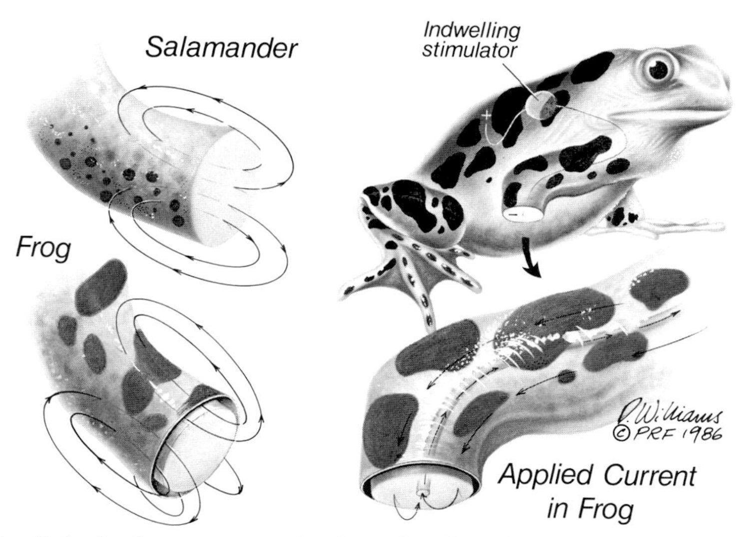

Fig. 10. Path of endogenous current in salamanders (through the core tissues of the stump), the endogenous current in frogs (predominantly through the subdermal lymph space found in adult frogs but not salamanders and newts), and the polarity and position of applied current pulled through the core tissues of an adult frog's limb stump via implanted batteries and electrodes (negative electrode at the stump tip). This *applied* current increases the magnitude of the endogenous currents naturally deficient in this area in adult frogs. Arrows indicate the direction of current flow within tissues, external environment, and the wick electrode. Compare with Figures 1 and 2.

becomes more prominent. In the axolotl, currents (on the order of 2–5 $\mu A/cm^2$) predict the locale of hindlimb formation 3–4 days *before* mesenchyme condensations become marked (Borgens et al., 1983). These currents decline steadily and become more focused about the developing bud in a way altogether similar to the pattern seen in *Xenopus* larvae (Fig. 11) (Robinson, 1983).

Both in Kenneth Robinson's laboratory and in our own, we have made an attempt to determine the ionic dependency of the current driven through this local limb-forming flank. These attempts have been unsuccessful. As I have discussed before, larval and neoteneous amphibians are apparently able to pump a variety of ions across their epidermis—the replacement of one ion (such as Na^+) is compensated for by an increased pumping of another ion (for example, Ca^{2+} or K^+). Thus any attempt to temporarily replace or block one component of the current is usually thwarted. In the clawed frog,

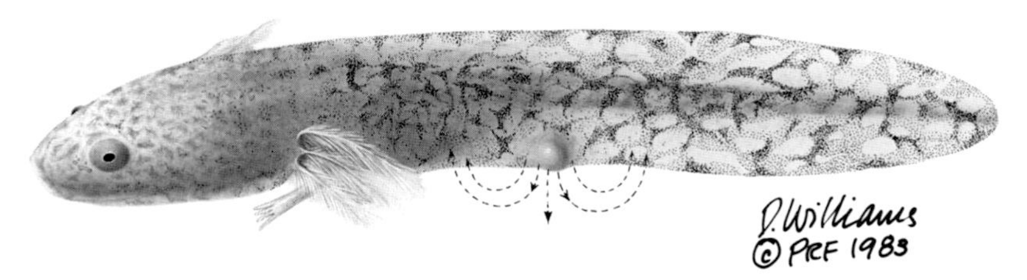

Fig. 11. An endogenous skin-driven current flows out of the "limb-forming" region of the larval amphibian flank. This current is *predictive* of subsequent limb bud formation and localizes about the developing bud. (Reproduced from Borgens, 1985, with permission of the publisher.)

there is an apparent limb bud-gill current loop, suggesting that an inward pumping of ions at the gill might contribute to the outcurrent measured leaving the limb bud. We detected no such inwardly directed current in axolotl gills, however. Weak inwardly directed currents were spread uniformly over nonlimb-forming integument. (This is the direction of ionic current flow one would expect, since amphibian integument actively takes up cations from their dilute aqueous environment.) The battery driving the bulk of current in the immature axolotl is the skin—as is the case with current generation in the limb stumps of the adults (Borgens et al., 1983). Such a "skin battery" also contributes to the net current leaving limb buds in *Xenopus* larvae as well (Murphy and Borgens, unpublished measurements).

Do Ionic Currents Induce Limb Bud Formation?

The measurements of limb bud currents provided us with a way to view not only the formation of limb buds, but developmental plaques in general—at least those condensations that form from an aggregation of cells beneath the embryonic integument. For example, limb buds and epibrachial plaques form from an aggregation of cells that migrate into a focused region of development from distant areas (Balinsky, 1970; Hinchliffe and Johnson, 1980). In limbs the early accumulations of mesenchyme arises from at least three identified sources: neural crest cells, somatic mesoderm, and the parietal layer of the lateral plate mesoderm. The currents (and their associated subepidermal voltage gradients) that predict the exact locus of bud formation provide a testable model that may explain the basis for these focused cellular accumulations, a rationale for the establishment of their spatial location on the embryo, and perhaps their subsequent innervation.

We have recently discovered that persistent outcurrents are associated with the development of the neural folds in amphibian embryos (whose development is characterized by a *dispersion* of neural crest cells). Predictive outcurrents *are not* measured in association with the development of the optic primodeum or auditory placode—which form after induction from a proliferation *of local cells* (see Borgens and McCaig, Chapter 3, this volume). The best studied of all of these model systems is the limb bud, a structure we shall examine in more detail to attempt an understanding of the role of endogenous fields during organogenesis.

The larval amphibian integument is composed of closely opposed cells, usually described as being two cell layers thick. (However, the numerous glandular Leydig cells and the overlapping of basal cells give the impression of three layers in most areas of the flank.) Apical cells are tightly opposed, with tight junctional complexes joining their most apical aspect. They reside over layers of basal cells and their processes and over large gland cells. All of these cells are joined by numerous junctional complexes: gap junctions, spot desmosomes, and hemidesmosomes (connecting each basal cell with the subjacent basal lamina). Extending into the cytoplasm from hemidesmosomes and spot desmosomes are vast arrays of tonofilaments. Beneath the basal lamina resides a well-ordered sublamellar matrix or basal lamella. This highly structured extracellular matrix is composed of a variety of macromolecules, such as glycosaminoglycans and palisades of collagen fibers whose architecture is strikingly arranged (Fig. 12a) (Borgens, 1984b; Borgens et al., 1987).

The physiology of embryonic amphibian integument is similar to adult skin in that it possesses a transcutaneous potential difference supported by active cation transport processes (McCaig and Robinson, 1980), which result in an inwardly positive voltage across the skin (in the tens of millivolts). This voltage is permitted by abundant tight junctional complexes between cells—an effective barrier to extracellular charge movement (current flow). This is a typical description of the anatomy and physiology of flank integument prior to local organogenesis. Our hypothesis is that the first detectable change in the local area of skin where a limb bud will form is a programmed drop in tight junction resistance. This makes this local area electrically leaky and results in short-circuit current flow (Fig. 13b). This electrical leak, focused in the area of limb formation, will dramatically lower the transepidermal potential in this local area, which we believe induces a disorganization of epidermal architecture. At the time of the current leak, there is a visible deterioration of the integrity of the skin in this focused area (Fig. 13c). Also evident is a complete deterioration of the basal lamella, whose strikingly organized character is nearly lost. All of the hemidesmosomes and many

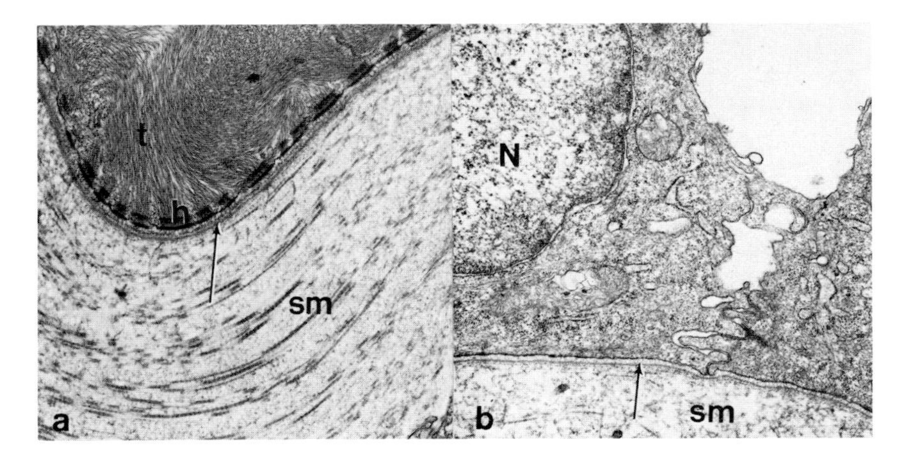

Fig. 12. Electron micrographs of the basal domain of the larval amphibian integument. **a:** Ventrolateral nonlimb-forming flank, adjacent to **b,** limb-forming flank (this area was about 800 μm caudal to a. In a, note the vast arrays of tonofilaments (t) within the basal domain of this basal cell and the numerous hemidesmosomes (h), the well-ordered sublamellar matrix (sm) is apparent beneath the basal lamina (arrow). In b, the basal lamina (arrow) is still present; however, the sublamellar matrix (sm) is characteristically disorganized. Note the absence of hemidesmosomes and tonofilaments and the large extracellular space between adjacent basal cells. N is the nucleus of one of these cells. Magnification ×18,000 for both panels. (Reproduced from Borgens, 1984b, with permission of the publisher; see also Borgens et al., 1987.)

arrays of tonofilaments likewise disappear—but only in the local area of the limb bud (Fig. 12b) (Borgens, 1984b; Borgens et al., 1987; Kelly and Bluemink, 1974). Our view is that the continuing dissolution of the skin's anatomy leads to further increases in current density leaking out of this locale, which in turn leads to a further deterioration in the anatomy of the local integument. This positive feedback loop subserves the production of an ever-increasing current leak and an ever-increasing voltage gradient beneath the epidermis in the area of the current leak (Fig. 13b, c) (Borgens, 1984b). Dying and sloughing apical cells would also contribute to the leakage current (and subepidermal voltage gradients) since, when such cells are lost, so are the tight junction complexes coupling them, producing a closed-circuit electrical leak (Fig. 14) (Borgens et al., 1987). Beneath this leaky epidermis, the voltage gradient would be negative with respect to areas more distant (Fig. 13c) (Borgens, 1984b). Since all developmentally active cells that have been tested migrate toward the negative pole of an applied field in tissue

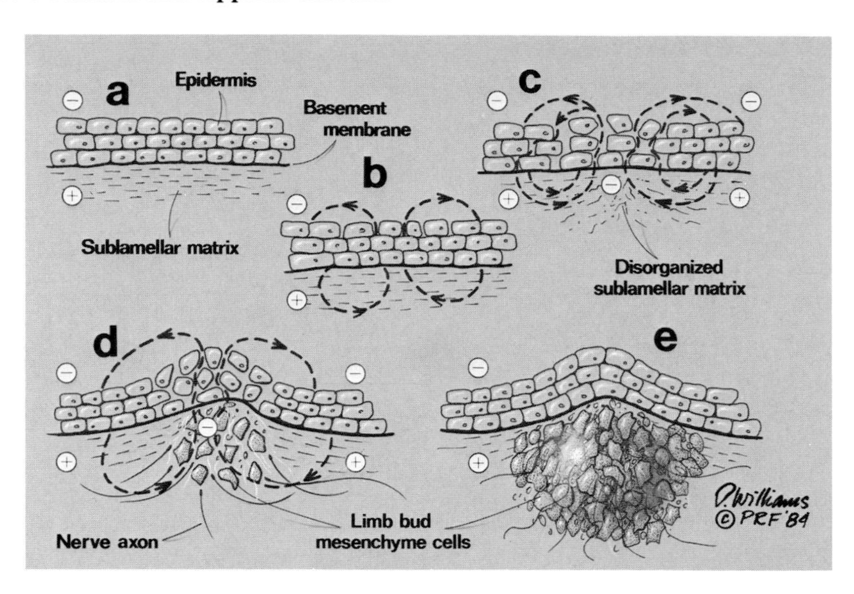

Fig. 13. Hypothetical scheme for the initiation of the limb bud. **a:** In limb-forming regions of the flank, epidermal cells are closely opposed and reside upon a well-formed basal lamina and a highly ordered sublamellar matrix. Integumentary cells are attached at their apical surfaces by tight junctions that form an effective barrier to extracellular charge movement between cells and help maintain an inwardly positive transepidermal potential difference (TEP) (in the tens of millivolts) produced by cellular ionic pumps. **b:** A preprogrammed drop in the resistance of the tight junctional complexes in a local area of presumptive bud formation allows a ''short-circuit'' current to flow outward through this electrically leaky area of integument. **c:** This short-circuit current would substantially reduce the TEP in this local area of flank integument, perhaps causing even more cellular disorganization and producing a subepidermal voltage gradient negative in this local area with respect to adjacent regions of flank. **d:** This subepidermal voltage gradient would produce coarse vectoral cues for migratory mesenchyme cells. (All such cells are known to migrate toward the negative pole of an applied field.) Since neurite growth is directed toward the negative pole of an applied field, this voltage gradient may also help direct the growth of neurites into the developing bud. **e:** As this mass of cells increases in density and as the anatomy of this region rapidly changes, the physiology of the bud may change as well—permitting a return to a more normal structural organization. For example, a recoupling of tight junctions at the bud's apical epidermis would shut down the leakage current and perhaps reverse the anatomical effects of the collapse of the TEP. Thus this putative positive feedback system (which produces changes in a local area of flank integument) may be self-limiting. (Reproduced from Borgens, 1984b, with permission of the publisher.)

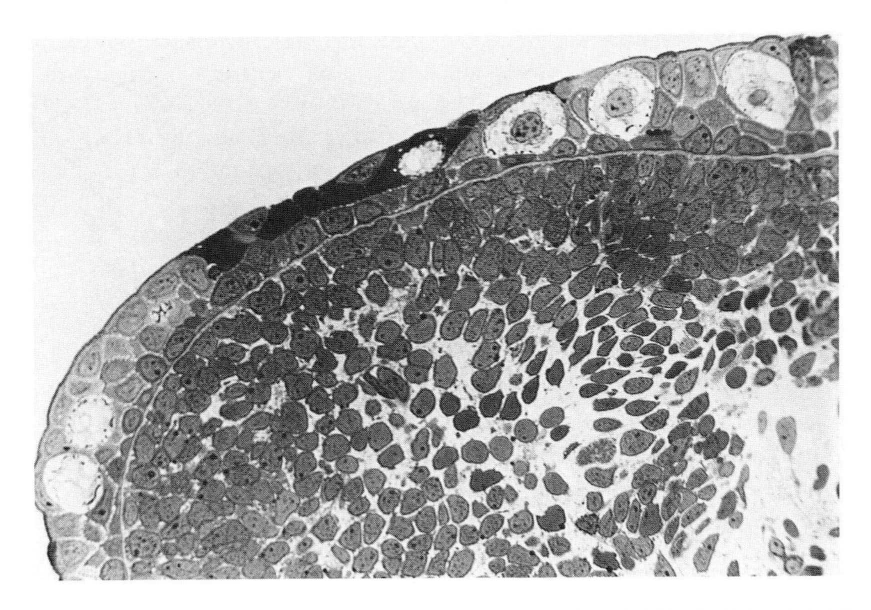

Fig. 14. Photomicrograph of a 1-μm-thick section of a larval axolotl hindlimb bud (toluidine blue). Note the intensely staining cells toward the apical aspect of the bud. These cells were subsequently determined to be desquamating by ultrastructural techniques and are characteristic of limb buds at all stages of their development. (Reproduced from Borgens et al., 1987, with permission of the publisher.)

culture, endogenous subepidermal voltage gradients may provide a coarse vector for migratory mesenchyme cells (Fig. 13d) (Robinson, 1985). Neurites also grow toward the negative pole of an applied field in vivo and in vitro (see Borgens, Chapter 4, this volume, and Borgens, 1986), suggesting that these natural voltage gradients beneath the larval integument may provide directional information for subsequent innervation of the limb bud as it becomes more pronounced (Fig. 13d). As the condensation of mesenchyme cells becomes more dense, the anatomy and physiology of the limb bud may change, returning to a more normal epidermal architecture. For example, a decrease in current density is measured leaving the limb bud as it develops (Borgens et al., 1983), possibly reflecting a "recoupling" or an increase in resistance of tight junctions between apical epidermal cells. This return to a more normal physiological state is furthermore suggested by a return of small measured *incurrents* (1 $\mu A/cm^2$) at the apex of prominent and well-developed buds (Borgens et al., 1983). We have also observed a return of some of the normal structural characteristics of integumentary cells [such as tonofilaments, hemidesmosomes, and the overall structural architecture of the basal lamella in these transitional areas (Borgens et al., 1987)]. Thus we envision a means by which a local preprogrammed change in larval integument leads to a self-limiting positive feedback system that can impart directional specificity to migratory mesenchyme cells and perhaps also to neurites. These local changes result in the apposition of mesenchymal and epidermal cells, setting the stage for subsequent differentiation.

DEVELOPMENTAL CURRENTS AND THE OLDER LITERATURE

If one accepts the premise that the endogenous currents and voltages are true controls of limb development and regeneration, then one should be able to interpret many older unreconciled observations (stemming from nearly 100 years of experimental manipulation of the limb system) on the basis of this theory. For example, many investigators have been intrigued by the maintenance or establishment of polarity in regenerating and developing limbs. To determine rules governing the expression of pattern and polarity in limbs, they have made surgical rotations of limbs, sometimes reversing portions of the limb stump 180°, notching limb stumps, ligating stumps, and in general surgically perturbing the alignment of the developing parts as much as possible (Wallace, 1981c, d). These procedures have essentially two things in common. First, when segments of the limb are realigned, there is always an expanse of mesodermal tissue exposed. For example, when one switches a segment of a limb 180°, the limb diameters are mismatched at the junction of the proximal and (now reversed) distal section. Second, wherever this mismatch occurs, an epithelium grows over this expanse of mesodermal tissue, and a limb usually arises from it. This limb may regenerate in an eccentric position, or several limbs may arise from multiple foci or incompletely joined portions of the extremity. The purpose of this experimentation was to show that whatever the orientation of a segment of salamander limb stump, only distal structures formed (the law of distal transformation). From our point of view, it is important to ask *why* structures formed at all. At any locus where only a wound epidermis covers exposed musculature, current will be driven out of this region by the battery of the adjacent intact integument. Thus current would be expected to leave these areas just as current leaves the amputation surface. Current would probably be driven out of as many locales as there are expanses of epithelium-covered surfaces.

Many investigators have tried to chronically irritate the stump tip of nonregenerating adult frogs in the hopes of initiating limb regeneration

(Polezhaev, 1972; Rose, 1944). The rationale has been that by disallowing the thickening of the wound epidermis, the production of a basal lamina, and a dense cicatrix (none of which are reported to occur at the tip of a urodele stump), certain necessary epidermal-mesenchymal interactions leading to limb regeneration would be permitted. Chronic trauma (pin-pricking) and caustic agents (nitric acid, iodine, and solutions of "hypertonic" salts) have been reported to have been applied to the stump end in frogs; they all induce a measure of hypomorphic outgrowth from an amputated frog forelimb stump. I would suggest that what these procedures have in common is that they maintain the stump end in a chronically traumatized state—which is to say, *an electrically leaky state*. Perhaps the normal architecture of the subdermal lymph space and the integrity of the subepidermal mesodermal tissues are disturbed by these procedures, resulting in higher densities of stump current being shunted through the *core tissues* near the end of the frog's forelimb stump. We envision this increased exodus of endogenous current (and a possible change in its path through the stump) as a common denominator precipitating regeneration in response to these different techniques.

The last technique mentioned above—the applications of salt solutions to the stump tip—calls for special mention. A close inspection of these famous experiments by Rose strongly supports the contention that the induction of regeneration is *current-mediated* and *not* mediated by *traumatization* or a disruption of normal wound healing mechanisms (Rose, 1942, 1944, 1945). Although it is usually reported that these sodium chloride applications were *hypertonic* or caustic in nature, a close inspection of these papers proves just the opposite. The best regeneration induced by application of salt solutions to frog stumps reported by Rose was in response to a concentration of 52 mM NaCl (Rose, 1942). This is hardly to be construed as "caustic," since it is much less than the salt concentration of Ringer's solution or body fluids. This fact is supported by Polezhaev's report of an induction of hypomorphic regeneration in frogs using a solution of 0.8% Na HCO_3 (about 95 mM Na^+) (Polezhaev, 1972). Once again, this concentration can hardly be construed to be hypertonic, since it is about the salt concentration found in Ringer's solution and in extracellular fluids. Although these concentrations seem too weak to be considered damaging, it is worth pointing out that they are about 20–50-fold higher than the concentration of salts found in the *pond water* environment in which amphibians live. As mentioned, the production of a short-circuit current by isolated amphibian skin in organ culture (and the production of skin-driven stump currents in the whole animal) is known to be dependent on the concentration of external Na^+ in the medium bathing the apical surface of the skin (Benos and Mandel, 1978; Borgens et al., 1977a; Ussing, 1964; Whitlembury, 1964). This relationship between external Na^+ concentration and the short-circuit current demonstrates typical saturation kinetics. That is, high concentrations of Na^+ in the bathing medium will boost the production of current to maximal values. The Km for the skin of *R. pipiens* is only 5 mM (the Km being the external concentration of Na^+ at which the Na^+ current is only half maximal). Thus, while it is probable that none of these applications caused damage to the stump tip, as they were intended to, the relatively high concentration of sodium (compared with pond water) would be expected to increase greatly the production of a current driven out of the stump's end by the skin battery and perhaps the relative proportion of current driven through the core tissues of the stump, beneath the lymph space.

While methods aimed at limiting the premature invasion of full-thickness skin has been viewed as a traditional way to induce a modest regenerative response in anura, closing off the amputation surface with full-thickness skin has been a time-honored means of *inhibiting* regeneration in salamanders (Mescher, 1976). We would suggest that the high-resistance skin flap would block the exodus of a density of current sufficient to support regeneration. It is interesting to note that wherever such a flap is incomplete (by accident or design), limbs will regenerate from these ''skinless'' areas. Their axis of growth is usually deviated according to the position of this expanse of epithelium (Mescher, 1976). Recall that I have discussed above the probability that an integumentary-driven current would exit such regions. We have discussed several ways in which investigators have attempted to induce regeneration of limbs in frogs as a means of understanding the controls of this process. The most famous of these techniques belongs to Marcus Singer, who hyperinnervated frog stumps, and to Oscar Schotté, who achieved the most extensive regenerative response in adult frog limb stumps by implanting several adrenal glands into the throat area of *Rana clamitans* (Singer, 1954; Schotté and Wilber, 1958). Schotté argued that the increased levels of adrenocorticoids would sufficiently suppress scar tissue formation in the forelimb stumps to allow a greater regenerative response. We feel that another contribution of the extra-adrenal glands would be to increase the concentrations of glucocorticoids and mineral corticoids, which are also known to boost the transepidermal potential in amphibians (Bishop et al., 1961; Myers et al., 1961). Such an increase would in turn boost the amount of endogenous current driven out the stump's end. We have already discussed that nerves respond exceedingly well to imposed fields in vivo or in vitro. In experiments in which frog forelimb stumps were stimulated with a weak DC field, we observed great amounts of nervous tissue in the hypomorphic regenerate. In some cases, 20% of all the regenerated tissues distal to the plane of amputation was nerve. This finding led us to suggest that an alternate (or additional) explanation for the regenerative response was that we had *electrically hyperinnervated* the stump end, achieving results similar to those of Singer who *surgically* hyperinnervated it (Singer, 1954).

The developmental currents that predict the exact position of limb formation in both frog and salamander larvae also suggest novel explanations for some older observations in experimental embryology. For example, if one makes a small skin incision in the flank area of a salamander and routes a nerve trunk beneath, then, depending on the location of the manipulation, an accessory limb will form (Bodemer, 1958, 1959). At distances far from limb areas, this procedure has limited (if any) success. Close to limb areas, accessory limbs can easily be induced. One can even mark out forelimb and hindlimb "fields" by this procedure. Indeed, even dorsal crest "fields" in the European crested newt can be so mapped (Wallace, 1981f). We have measured a peak current directly over the area of limb bud formation. This outcurrent falls off steadily with distance from the limb-forming region in any direction. Can these two profiles—a curve in peak limb-producing potency by experimental method, and a curve in the distribution of naturally occurring outcurrent predictive of limb development—be superimposable? It is certain that the surgical wounding of the flank will indeed produce an outcurrent at this place. Moreover, Bodemer found that simply supplying extra innervation beneath the skin in the salamander was not sufficient to induce supernumerary limbs. He had to wound this local area as well (Bodemer, 1958, 1959). We would suggest that the wounding supplies the necessary current leak for growth of the limb, and that the nerve supplies whatever nerves naturally supply *to permit* the developing structure.

All of the discussion above suggests support for the hypothesis that an endogenous current may be a necessary control (among several) for limb regeneration and generation. To my knowledge there has only been one experimental and theoretical challenge to this proposition. Lassalle has claimed that *surface*-detected voltages play no role in the regeneration process, using three basic arguments (Lassalle, 1979, 1980). First, I should mention that in principle I would agree—voltage gradients measured on the *outside* of a salamander's body (using antiquated surface measurement techniques previously discussed) are not controlling developmental processes that are taking place *within* the limb stump, in its core tissues. It is the voltage gradients *within the core tissues* of the structure that are critical to limb regeneration. However, Lassalle may have "thrown the baby out with the bath water," and his criticism should be discussed. Essentially, he reports that: 1) larval *Pleurodeles* and axolotls do not possess surface voltages; 2) such voltages in the adults are lost (or invert) during a molt; and 3) similar potentials are observed during skin wound healing—suggesting to Lassalle that they cannot be relevant to both processes (regeneration or skin wound healing).

1. Lassalle's inability to measure surface voltages on larval and adult *Pleurodeles* can be explained as follows: if the transcutaneous resistance was reasonably homogeneous on the *body surface,* then one would not be able to determine a potential at all with surface contact electrodes. Furthermore, the capricious nature of bioelectric potential measurements suggests that if artificial means were not employed to "optimize" these voltages, then the conventional techniques Lassalle employed would probably not be sensitive enough to sample shallow voltage gradients along the surface of a wet-bodied salamander. The assertion that larvae do not possess transepidermal voltages (or that they are greatly reduced in magnitude) stems from several older and technically deficient reports (Alvarado and Moody, 1970; Taylor and Barker, 1965). More modern experiments demonstrate conclusively that the skins of larval forms do indeed take up ions from the medium by an active process and do support a transepidermal potential (Cox and Alvarado, 1979; McCaig and Robinson, 1980). Lastly, we have measured stump currents in *larval* and adult *Pleurodeles* as well as in axolotls (Borgens et al., 1984). These species do not provide an exception in any way to the common electrical response of amphibian limbs to amputation.

2. Lassalle had also observed that surface potentials are lost (or invert) during the molting process, suggesting to him that they cannot be important to the regenerative process. There are known changes in the skin's permeability to water and perhaps in its ability to pump ions across itself during a molt (Larsen, 1976). This factor usually results in an increase in the conductance of the bathing media. In fact, several investigators use this increase in conductance as an indirect index of molting (Larsen, 1976). An increase in conductance of the moisture on the skin's surface would be associated with a decrease in its *resistivity,* since these two factors are inversely proportional. This fact would make it even *more difficult* to measure a surface potential difference under these conditions, since investigators who attempt such measurements have always had to *optimize* this resistance (for example, by drying the skin's surface) in order to obtain a measurable voltage.

3. Lassalle noted that skin wounds have the same general character of potential (externally positive at the lesion with respect to uninjured areas) as the surface voltages measured after amputation in salamanders (Lassalle, 1980). He feels that these two processes should not share the same putative electrical control and argues that naturally produced currents and voltages cannot then be important to either. We do not feel that there is any reason to believe that these phenomena are mutually exclusive. Wound healing processes are a part of the early stages of limb regeneration in salamanders and are observed at a time when current densities exiting the stump are the most intense. Experiments on the possible role of naturally produced fields in skin wound healing are in their infancy (Barker et al., 1982). At this writing it is probable, however, that the intense lateral voltages near a skin wound (~ 150 mV/mm) are involved in the process of cell migration and closure during mammalian wound healing (refer to Chapter 5, this volume).

MAMMALIAN REGENERATION

It is commonly believed that mammals do not regenerate their body parts; however, recently it has become apparent that this is not the case. In the last decade, there have appeared rather anecdotal reports of spontaneous regeneration of the tips of digits after accidental amputation in children (Douglas, 1972; Illingworth, 1974; Rosenthal et al., 1979). The "before and after" type of clinical photographs (Fig. 15) presented in the clinical literature initiated controversy among basic scientists interested in the regeneration of body parts. It has been difficult to believe that the apparent perfection of the regenerated fingertip, described as complete in most every detail (including the fingerprint), could have arisen from a complete amputation through the last distal phalanx when not accompanied by documentation of intervening stages of development.

I decided to test the notion that mammals could regenerate the tips of their digits using the mouse as a laboratory model. Here the sequence of events during regeneration could be followed, and the histogenesis of whatever structure reproduced could be determined (Borgens, 1982b). Indeed, mice regrow their digits (Fig. 16). There appear to be only three rules for the production of cosmetically perfect fingertips: first, the tip of the digit must not be covered with skin. This type of surgical procedure is a common treatment for digit amputations in humans, and it is apparently the reason why regrowth of fingers has not been a common clinical observation after treatment of accidental amputations. Interestingly, we should recall that a similar "skin flap" is sufficient to completely inhibit the regeneration

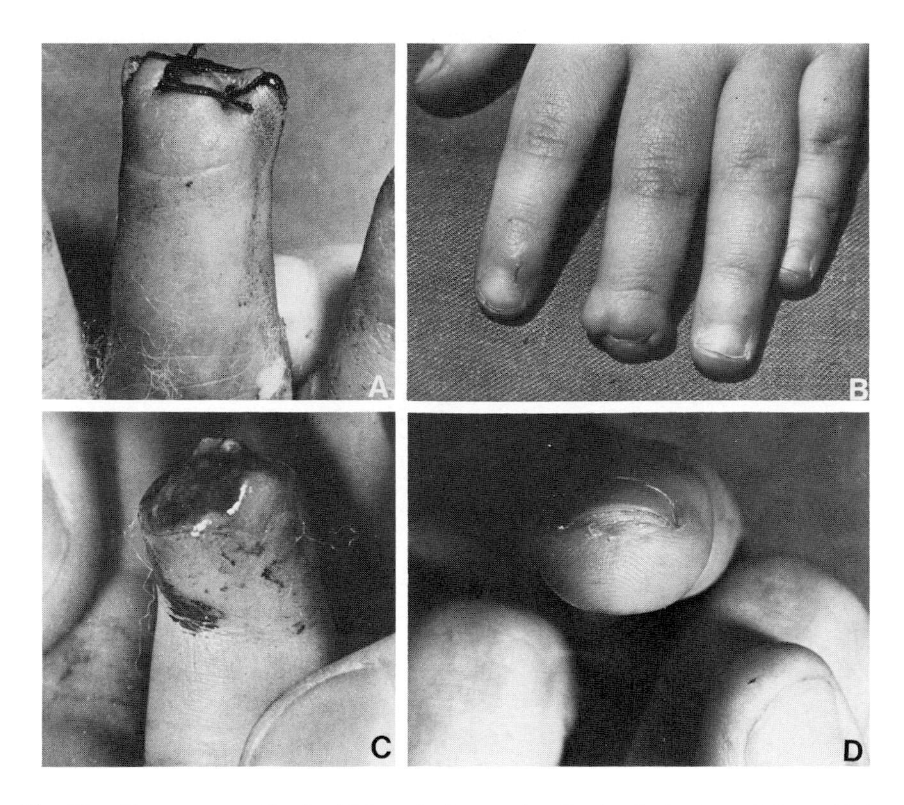

Fig. 15. Finger regeneration in children. **A:** Amputation distal to the last interphalangeal joint in a 3-year-old girl treated surgically. **B:** The resulting blunt finger stump 4 months after the accident. **C:** A similar amputation in a 5-year-old girl, treated conservatively, results in fingertip regrowth. **D:** This regenerated finger was complete and cosmetically perfect 12 weeks after the accident. (Photographs courtesy of C.M. Illingworth.)

of limbs in salamanders (Mescher, 1976). Second, the plane of amputation must fall distal to the last interphalangeal joint (LIJ). Amputations that pass through the LIJ always result in a healed and scarified blunt finger stump. Third, some of the nail bed epidermis must be present in the stump. Amputations that pass a few hundred micra distal to the joint sometimes result in incomplete regeneration (for example, a complete fingertip is formed, minus elements of the dorsal claw) (Borgens, 1982b). We have also measured a strong, steady skin-driven current leaving the amputation,

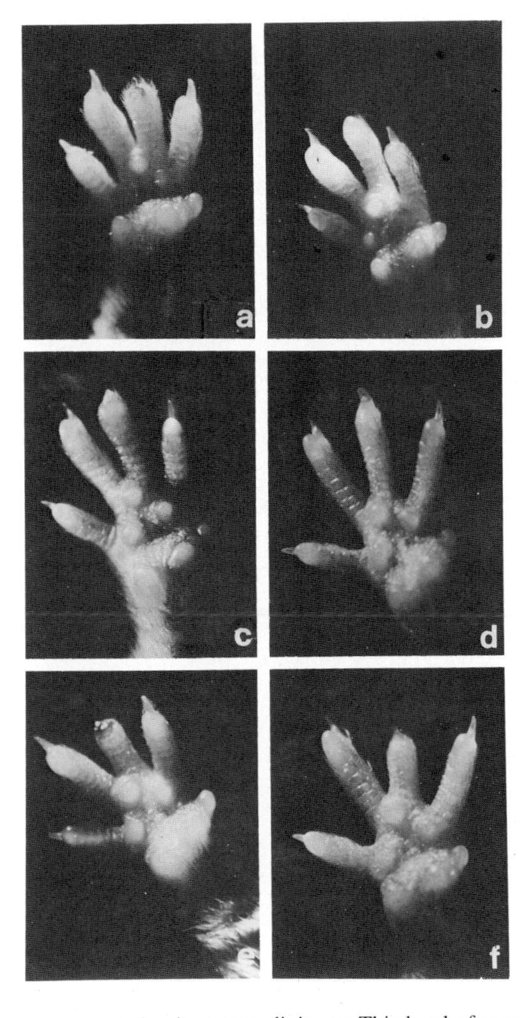

Fig. 16. Responses to amputation in mouse digits. **a:** This level of amputation passed about 100 μm distal to the last interphalangeal joint and resulted in incomplete regrowth of the distal toe. **b:** The toe at 5 weeks postamputation. This toe lacks a dorsal claw and the sole of the claw, but is otherwise complete. **c:** A similar amputation in another animal results in complete regrowth at 2 weeks, 5 days after amputation. Note the swollen appearance of the tip, and the emergence of the new claw. **d:** Same toe as c, at 4 weeks, 6 days after amputation. This regrowth was complete and was comprised of normal histological components. **e:** Amputation through the last interphalangeal joint or the distal one-third of the middle phalange always resulted in a blunt toe stump. **f:** Same toe as in e, at 4 weeks, 6 days postamputation. (c, d, e, and f reproduced from Borgens, 1982b, with permission of the publisher. Copyright 1982 by the AAAS.)

similar in character to the stump current previously discussed for salamanders (Borgens, unpublished measurements). Similar currents have been measured leaving children's fingertips after accidental amputation (Illingworth and Barker, 1980). We do not know if this current plays a role in fingertip regeneration, however, since we have not yet tested this notion critically.

Besides the regrowth of digits, there are other examples of regenerating structures in mammals: the cyclical regrowth of antlers in deer is a true epimorphic regenerative phenomenon (Goss, 1972), as is the regrowth of holes punched in the ears of certain mammals, such as the rabbit (Grimes and Goss, 1970) [but curiously not in cats or certain species of bats (Goss, 1981)]. It has been reported that the classical description of regeneration of the extremities in neonatal opossum (Mizell, 1968) may be flawed. Apparently it is extremely difficult to separate what is still *developing* in the neonate (while it is still in the marsupium) from what may actually regrow after amputation (Tassava and Fleming, 1981).

It is clear, of course, that major parts of a limb do not regenerate in the mammal. A variety of experiments have attempted to induce regeneration in the amputated mammalian limb, using methods that have produced limited success in the adult frog or lizard. For example, systemic glucocorticoids have been elevated by Schotté (Schotté and Smith, 1961) after his earlier success with amphibians (Schotté and Wilber, 1958); surgical hyperinnervation of the stump has been performed (Bar-Maor and Gitlin, 1961) after the technique of Singer (1954); salt treatments to the stumps have been made (Neufeld, 1980) after the fashion of Rose (1945); and electrical stimulation of rat stump tissues has been attempted (Becker, 1972; Becker and Spadaro, 1972; Libbin et al., 1979a,b) after the fashion of Smith (1974) and Borgens et al. (1977b). None of these applications has proved successful in inducing regeneration. The salt treatments of Neufeld (1980) did succeed in producing a recognizable blastema, and this was the extent of his well-documented claim. Becker (1972), Becker and Spadero (1972), and Libbin and coworkers (1979a,b), on the other hand, claimed to have produced "partial limb regeneration" by the imposition of electric current. However, an inspection of their data suggests that in fact no regeneration occurred in these instances. In Becker's paper (1972), as in the rest of subsequent reports, a growth response was defined, in part, as the appearance of an epiphyseal plate-like structure after amputation. The occurrence of this type of response (judging from the numbers reported in Table 1 of his paper) seemed to suggest differences between electrically treated and "control" applications. There are two difficulties with this notion. First, rats produce wonderfully orga-

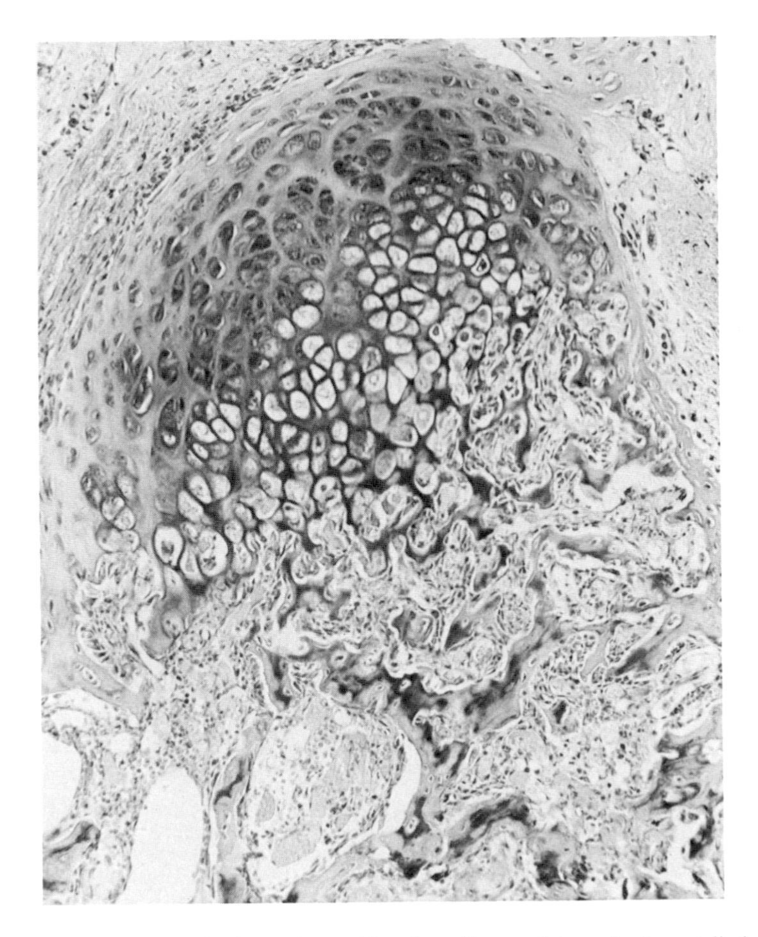

Fig. 17. Portion of a typical ectopic nodule of cartilage and bone in the rat limb. This calcification formed in the tendinous portion of a gastrocnemium muscle, which was minced 32 days previously. The upper part of the nodule is composed of hyaline cartilage; the lower, spongy bone. These nodules are strikingly similar in anatomy and sequence of ossification to epiphyseal plates. They are incorrectly used as evidence for claimed ''electrically induced limb regeneration'' in adult rats. (Photograph courtesy of B.M. Carlson.)

nized ''epiphyseal plate-like'' structures as a normal response to tissue trauma. Such organized cartilaginous and partly mineralized ossifications can even be produced in the absence of gross trauma by pinching or chronically irritating the Achilles' tendon (Fig. 17) (Carlson, 1970). Investigators such as Carlson (1970) have warned against the use of these structures as evidence supporting claims of epimorphic regeneration in the rodent. Second, in a subsequent report, Becker and Spadaro (1972) included more data on the numbers of "growth responses" than that described in the short note in *Nature* (Becker, 1972). In the second report, the bulk of the data (presented in their Table 1) appears identical to that reported in the first paper, except that there were more "growth response" cases listed for what was originally described as a "control" procedure. Taking these cases into account, there is no statistical difference between the electrically treated and control groups. Altogether there was no rigorous histological analysis in either of these reports. There was no description of how the exact plane of amputation was marked. Certain displayed structures described as "anlagen" (Becker, 1972) could have easily been fragments of bone in partial stages of cytolysis, and, lastly, the putative regeneration occurred in only 3–7 days—a time at which wound healing and débridement would just have been completed either in the rat or even in *salamander* limb stumps and far too short a time for restructuring of the limb to begin.

Libbin and coworkers attempted a more in-depth study, in which they first analyzed the rat limb's native response to amputation (Libbin et al., 1979a) and then compared these "innate" responses with the effects of imposed currents on rat limb amputation (Libbin et al., 1979b). What is curious about these two papers is that this group reported that there were no differences between the untreated rats (innate responses) and the electrically treated groups. After saying this, however, they seemed to believe still that "partial limbs" had regenerated in response to the application of the current. As I have remarked before (Borgens, 1985), one man's "partial limb" may be another's fracture callus. Altogether, none of these reports show (on the basis of any rigorous criteria) that any mammalian limb stump structure had formed in response to the application of an electric field other than would have formed anyway in the untreated animal. Thus, if we are to believe that current induces "partial limb regeneration" in the rat, we should likewise believe that rats partially regenerate their limbs as a normal consequence of amputation.

In closing, it is wise to recall that any damage to cells or integuments initiates a steady flow of electric (ionic) current. This current flow can be relevant to regeneration or incidental to it. In the case of the mammalian extremity, it is far from clear if endogenous or applied currents are related to the regrowth of the limb—or even of digits. In the amphibian, the evidence—both direct and circumstantial—suggests a causative role for stump currents in the initiation of limb regeneration.

SUMMARY

1. A steady, skin-driven, DC current leaves the end of regenerating amphibian limbs. A variety of techniques that reduce or inhibit this current in the salamander all disturb the normal process of regeneration.

2. Non-regenerating adult frogs also drive a current out the end of the limb stump. However, current densities are greatest just under the skin of the stump tissues, since most of the current is shunted through the subdermal lymph spaces found in adult frogs (but not in their larvae, or in salamanders). When the electrical fields within the core tissues of the adult frog stump are enhanced by implanted batteries and electrodes, a measure of limb regeneration is initiated.

3. A steady current also predicts the exact locus of limb bud formation in frog and salamander larvae. This current declines in magnitude as the bud becomes increasingly more prominent. At the apex of large buds (just prior to differentiation), the current reverses its direction. The position and character of this developmental current correlate with anatomical changes in limb-forming flank integument and predict the location of limb emergence.

4. These developmental currents and fields are thought to exert their influence via their action on nervous tissue and/or the wound epidermis (in regenerating limbs) and on migratory mesenchyme cells (forming the limb anlagen).

REFERENCES

Alvarado RH and Moody A (1970) Sodium and chloride transport in tadpoles of the bullfrog *Rana catesbeiana*. Am. J. Physiol. 218:1510–1516.

Balinsky BI (1970) Development of the ectodermal organs in vertebrates. In *Introduction to Embryology*, ed. 3. W.B. Saunders, Philadelphia, pp. 367–423.

Barker AT, Jaffe LF, and Vanable JW Jr. (1982) The glabrous epidermis of cavies contains a powerful battery. Am. J. Physiol. 242:R358–366.

Bar-Maor JA and Gitlin G (1961) Attempted induction of forelimb regeneration by augmentation of nerve supply in young rat. Plast. Reconstr. Surg. Trans. Bull. 27:460–462.

Becker RO (1960) The bioelectric field pattern in the salamander and its simulation by an electronic analog. IRE Trans. Med. Electron 7:202–207.

Becker RO (1961a) The bioelectric factors in amphibian limb regeneration. J. Bone Joint Surg. 43A:643–656.

Becker RO (1961b) Search for evidence of axial current flow in peripheral nerves of salamander. Science 134:101–102.

Becker RO (1972) Stimulation of partial limb regeneration in rats. Nature 235:109–111.

Becker RO (1977) An application of direct current neural systems to psychic phenomena. Psychoenergetic Systems 2:189–202.

Becker RO and Spadaro JA (1972) Electrical stimulation of partial limb regeneration in mammals. Bull. N.Y. Acad. Med. 48:627–641.

Becker RO and Marino AA (1982) *Electromagnetism and Life*. State University of New York Press, Albany.

Becker RO and Seldon G (1985) *The Body Electric: Electromagnetism and the Foundation of Life*. William Morrow and Company, New York.

Benos DJ and Mandel LJ (1978) Irreversible inhibition of sodium entry sites in frog skin by a photosensitive amiloride analog. Science 199:1205–1207.

Bishop WR, Mumback MW, and Scheer BT (1961) Interrenal control of active sodium transport across frog skin. Am. J. Physiol. 200:451–460.

Bodemer CW (1958) The development of nerve-induced supernumerary limbs in the adult newt, *Triturus viridescens*. J. Morphol. 102:555–581.

Bodemer CW (1959) Observations on the mechanism of induction of supernumerary limbs in adult *Triturus viridescens*. J. Exp. Zool. 140:79–99.

Borgens RB (1982a) What is the role of naturally produced electric current in vertebrate regeneration and healing? Int. Rev. Cytol. 76:245–298.

Borgens RB (1982b) Mice regrow the tips of their foretoes. Science 217:747–750.

Borgens RB (1983) The role of ionic current in the regeneration and development of the amphibian limb. In Fallon JF and Caplan AI (eds.): *Limb Development and Regeneration*, Alan R. Liss, Inc., New York, pp. 597–608.

Borgens RB (1984a) Endogenous ionic currents traverse intact and damaged bone. Science 225:478–482.

Borgens RB (1984b) Are limb development and limb regeneration both initiated by an integumentary wounding? A hypothesis. Differentiation 28:87–93.

Borgens RB (1985) Natural voltage gradients and the generation and regeneration of limbs. In Sicard RE (ed.): *Regulation of Vertebrate Limb Regeneration*. Oxford Press, New York, pp. 6–31.

Borgens RB (1986) The role of applied electric fields in neuronal regeneration and development. In Nuccitelli R (ed.): *Ionic Currents in Development*. Alan R. Liss, Inc., New York, pp. 239–250.

Borgens RB, Vanable JW Jr., and Jaffe LF (1977a) Bioelectricity and regeneration: Large currents leave the stumps of regenerating newt limbs. Proc. Natl. Acad. Sci. U.S.A. 74:4528–4532.

Borgens RB, Vanable JW Jr., and Jaffe LF (1977b) Bioelectricity and regeneration. I. Initiation of frog limb regeneration by minute currents. J. Exp. Zool. 200:403–416.

Borgens RB, Vanable JW Jr., and Jaffe LF (1979a) Small artificial currents enhance *Xenopus* limb regeneration. J. Exp. Zool. 207:217–255.

Borgens RB, Vanable JW Jr., and Jaffe LF (1979b) Bioelectricity and regeneration. Bioscience, 29:468–474.

Borgens RB, Vanable JW Jr., and Jaffe LF (1979c) Reduction of sodium dependent stump currents disturbs urodele limb regeneration: J. Exp. Zool. 209:377–386.

Borgens RB, Vanable JW Jr., and Jaffe LF (1979d) Role of subdermal current shunts in the failure of frogs to regenerate. J. Exp. Zool. 209:49–55.

Borgens RB, Jaffe LF, and Cohen MJ (1980) Large and persistent electrical currents enter the transected lamprey spinal cord. Proc. Natl. Acad. Sci. U.S.A. 77:1208–1213.

Borgens RB, Rouleau MF, and DeLanney LE (1983) A steady efflux of ionic current predicts hind limb development in the axolotl. J. Exp. Zool. 228:491–503.

Borgens RB, McGinnis ME, Vanable JW Jr., and Miles ES (1984) Stump currents in regenerating salamanders and newts. J. Exp. Zool. 231:249–256.

Borgens RB, Callahan L, and Rouleau MF (1987) The anatomy of axolotl flank integument during limb bud development with special reference to a transcutaneous current predicting limb formation. J. Exp. Zool. 244:203–214.

Brockes JP (1984) Mitogenic growth factors and nerve dependence of limb regeneration. Science, 225:1280–1287.

Bryant SV, French V, and Bryant PJ (1981) Distal regeneration and symmetry. Science 212:993–1002.

Carlson BM (1970) Relationship between the tissue and epimorphic regeneration of muscles. Am. Zool. 10:175–186.

Cox TC and Alvarado RH (1979) Electrical and transport characteristics of skin of larval *Rana catesbieana*. Am. J. Physiol. 237:R74–R79.

Dent JN (1962) Limb regeneration in larvae and metamorphosing individuals of the South African clawed toad. J. Morphol. 110:61–77.

Douglas BS (1972) Conservative management of guillotine amputation of the finger in children. Aust. Paediatr. J. 8:86–89.

Eltinge EM, Cragoe EJ Jr., and Vanable JW Jr. (1986) Effects of amiloride analogues on adult *Notopthalmus viredescens* limb stump currents. Comp. Biochem. Physiol. 84a:39.

Fontas B and Mambrine J (1977) Origine des differences de potential ciscutanées chez *Rana esculenta*. C.R. Acad. Sci. (Paris) Ser. D 285:229–232.

Goode RP (1967) The regeneration of limbs in adult anurans. J. Embryol. Exp. Morphol. 18: 259–267.

Goss RJ (1956a) Regenerative inhibition following limb amputation and immediate insertion into the body cavity. Anat. Rec. 126:15–27.

Goss RJ (1956b) The regenerative response of amputated limbs to delayed insertion into the body cavity. Anat. Rec. 126:283–297.

Goss RJ (1969) *Principles of Regeneration*. Academic Press, New York.

Goss RJ (1972) Wound healing and antler regeneration. In Maibach HI and Rovee DT (eds.): *Epidermal Wound Healing*. Year Book Medical, Chicago, pp. 219–228.

Goss RJ (1981) Tissue interactions in mammalian regeneration. In Becker RO (ed.): *Mechanisms of Growth Control*. C.C. Thomas, Springfield, IL, pp. 12–26.

Grimes LN and Goss RJ (1970) Regeneration of holes in rabbit ears. Am. Zool. 10:537–549.

Helman SI and Fischer RS (1977) Microelectrode studies of the active Na transport pathway of frog skin. J. Gen. Physiol. 69:571–604.

Herlitzka A (1910) Ein Beitrag zur Physiologie der Generation. Wilhelm Roux Arch. 10: 126–158.

Hinchliffe JR and Johnson DR (1980) *The Development of the Vertebrate Limb*. Clarendon, Oxford.

Illingworth CM (1974) Trapped fingers and amputated fingertips in children. J. Pediatr. Surg. 9:853–858.

Illingworth CM and Barker AT (1980) Measurement of electrical currents emerging during the regeneration of amputated fingertips in children. Clin. Phys. Physiol. Meas. 1:87–89.

Iten L and Bryant SV (1973) Forelimb regeneration from different levels of amputation in the newt, *N. viridescens:* Length, rate, and stages. Wilhelm Roux Arch. 173:263–282.

Jaffe LF and Nuccitelli R (1974) An ultrasensitive vibrating probe for measuring steady extracellular current. J. Cell Biol. 63:614–628.

Kelly RO and Bluemink JG (1974) An ultrastructural analysis of cell and matrix differentiation during early limb development in *Xenopus laevis*. Dev. Biol. 37:1–17.

Kintner CR and Brockes JP (1984) Monoclonal antibodies identify blastemal cells derived from dedifferentiating muscle in newt limb regeneration. Nature 308:67–69.

Kirschner LB (1970) The study of NaCl transport in aquatic animals. Am. Zool. 10: 365–376.

Kirschner LB (1973) Electrolyte transport across the body surface of freshwater fish and amphibia. In Ussing HH and Thorn NA (eds.): *Transport Mechanisms in Epithelia*. Munksgaard, Copenhagen, pp. 447–460.

Larsen IO (1976) Physiology of molting. In Lofts L (ed.): *Physiology of the Amphibia*, Vol. 3. Academic Press, New York, pp. 54–100.

Lassalle B (1974a) Characteristiques des potentials de surface des membres du triton *Pleurodeles watlii*, Michah. C.R. Acad. Sci. (Paris) Ser. D. 278:483–486.

Lassalle B (1974b) Origine epidermique des potentiels de surface du membre de triton *Pleurodeles watlii*, Michah. C.R. Acad. Sci. (Paris) Ser. D. 278:1055–1058.

Lassalle B (1979) Surface potentials and the control of amphibian regeneration. J. Embryol. Exp. Morphol. 53:213–223.

Lassalle B (1980) Are surface potentials necessary for amphibian regeneration? Dev. Biol. 75: 460–466.

Libbin RM, Person P, Papierman S, Shah D, Nevid D, and Grob H (1979a) Partial regeneration of the above elbow amputated rat forelimb. I. Innate responses. J. Morphol. 159:427–438.

Libbin RM, Person P, Papierman S, Shah D, Nevid D, and Grob H (1979b) Partial regeneration of the above elbow amputated rat forelimb. II. Electrical and mechanical facilitation. J. Morphol. 159:439–451.

McAfee RD (1972) Survival of *Rana pipiens* in deionized water. Science 178:183–184.

McCaig CD and Robinson KR (1980) The ontogeny of the transepiderm potential difference in frog embryos. J. Gen. Physiol. 76:14.

McGinnis ME and Vanable JW Jr. (1986a) Wound epithelium resistance controls stump currents. Dev. Biol. 116:174.

McGinnis ME and Vanable JW Jr. (1986b) Electrical fields in *Notophthalmus viridescens* limb stumps. Dev. Biol. 116:184–193.

Mescher AL (1976) Effects on adult newt limb regeneration of partial and complete skin flaps over the amputation surface. J. Exp. Zool. 195:117–127.

Mizell M (1968) Limb regeneration: Induction in the newborn opossum. Science 161: 283–286.

Monroy A (1941) Ricerche sulle correnti elettriche dalla superficie del corpo di tritoni adulti normali e durante la reigenerazione degli arti e della coda. Pubbl. Stn. Zool. Napoli [II] 18:265–281.

Myers RM, Bish WR, and Scheer BT (1961) Anterior pituitary control of active sodium transport across frog skin. Am. J. Physiol. 202:444–450.

Neufeld DA (1980) Partial blastema formation after amputation in adult mice. J. Exp. Zool. 213:31–36.

Nielsen R and Tomilson RWS (1970) The effect of amiloride on sodium transport in the normal and moulting frog skin. Acta Physiol. Scand. 79:238–243.

O'Steen WK (1959) Regeneration and repair of the intestine in *Rana clamitans* larvae. J. Exp. Zool. 141:449–460.

Piatt J (1942) Transplantation of aneurogenic forelimbs in *Amblystoma punctatum*. J. Exp. Zool. 91:79–101.

Polezhaev LV (1946) The loss and restoration of regenerative capacity in the limbs of tailless amphibia. Biol. Rev. 21:141–147.

Polezhaev LV (1972) *Loss and Restoration of Regenerative Capacity in Tissues and Organs of Animals*. Harvard University Press, Cambridge, MA.

Robinson KR (1983) Endogenous electrical current leaves the limb and prelimb region of the *Xenopus* embryo. Dev. Biol. 97:203–211.

Robinson KR (1985) The responses of cells to electrical fields: A review. J. Cell Biol. 101: 2023–2027.

Robinson KR and Stump RF (1984) Self-generated electrical currents through *Xenopus* neurulae. J. Physiol. 352:339.

Rose SM (1942) A method for inducing limb regeneration in adult anura. Soc. Exp. Biol. Med. 49:408–410.

Rose SM (1944) Methods of initiating limb regeneration in adult anura. J. Exp. Zool. 95: 149–170.

Rose SM (1945) The effect of NaCl in stimulating regeneration of limbs of frogs. J. Morphol. 77:119–139.

Rose SM (1948) Epidermal dedifferentiation during blastema formation in regenerating limbs of *Triturus viridescens*. J. Exp. Zool. 108:337–361.

Rose SM (1964) Regeneration. In Moore JA (ed.): *Physiology of the Amphibia*. Academic Press, New York, pp. 545–622.

Rose SM and Rose FC (1974) Electrical studies on normally regenerating, on X-rayed, and on denervated limb stumps of *Triturus*. Growth 38:363–380.

Rosenthal LJ, Reiner MA, and Bleicher MA (1979) Nonoperative management of distal fingertip amputation in children. Pediatrics 64:1–3.

Scadding SR (1981) Limb regeneration in adult amphibia. Can. J. Zool. 59:34–50.

Schotté OE and Harland H (1943) Amputation level and regeneration in limbs of late *Rana clamitans* tadpoles. J. Morphol. 73:329–363.

Schotté OE and Wilber JF (1958) Effects of adrenal transplants upon forelimb regeneration in normal and in hypophysectomized adult frogs. J. Embryol. Exp. Morphol. 6: 247–269.

Schotté OE and Smith LB (1961) Effects of ACTH and cortisone upon amputational wound healing processes in mice digits. J. Exp. Zool. 146:209–229.

Sicard RE (1985) Leukocytic and immunological influence on regeneration of amphibian forelimbs. In Sicard RE (ed.): *Regulation of Vertebrate Limb Regeneration*. Oxford Press, New York, pp. 128–145.

Singer M (1946) The nervous system and regeneration of the forelimb of adult *Triturus*. V. The influence of number of nerve fibers, including a quantitative study of limb innervation. J. Exp. Zool. 101:299–337.

Singer M (1952) The influence of the nerve in regeneration of the amphibian extremity. Q. Rev. Biol. 27:169–200.

Singer M (1954) Induction of regeneration of the forelimb of the postmetamorphic frog by augmentation of the nerve supply. J. Exp. Zool. 126:419–471.

Singer M (1965) A theory of the trophic nervous control of amphibian limb regeneration, including a re-evaluation of quantitative nerve requirements. In Kiortsis V and Trampusch HAL (eds.): *Regeneration in Animals and Related Problems*. North Holland, Amsterdam, pp. 20–32.

Singer M and Salpeter M (1961) Regeneration in vertebrates: The role of the wound epithelium. In Zarrow MX (ed.): *Growth in Living Systems*. Basic Books, New York, pp. 277–311.

Singer M, Norlander RH, and Egar M (1979) Axonal guidance during embryogenesis and regeneration in the spinal cord of the newt: The blueprint hypothesis of neuronal pathway patterning. J. Comp. Neurol. 185:1–22.

Smith SD (1967) Induction and partial limb regeneration in *Rana pipiens* by galvanic stimulation. Anat. Rec. 158:89–97.

Smith SD (1974) Effects of electrode placement on stimulation of adult frog limb regeneration. Ann. N.Y. Acad. Sci. 238:500–507.

Stensaas LJ (1983) Regeneration in the spinal cord of the newt, *Notopthalmus (triturus) pyrrhogaster*. In Kao CC, Bunge RP, and Reier PJ (eds.): *Spinal Cord Regeneration*. Raven Press, New York, pp. 121–149.

Stocum DL (1985) The role of the skin in urodele limb regeneration. In Sicard RE (ed.): *Regulation of Vertebrate Limb Regeneration*. Oxford University Press, Oxford, pp. 32–53.

Tassava RA and Fleming MW (1981) Preamputation and postamputation histology of the neonatal hindlimb: Implications for regeneration experiments. J. Exp. Zool. 215: 143–152.

Taylor RE and Barker SB (1965). Transepidermal potential difference: Development in anuran larvae. Science 148:1612–1614.

Thornton CS (1960) Influence of an eccentric epidermal cap on limb regeneration in ablystoma larvae. Dev. Biol. 2:551–569.

Thornton CS (1968) Amphibian limb regeneration. Adv. Morphogen. 7:205–250.

Thornton CS and Thornton MT (1970) Recuperation of regeneration in denervated limbs of amblystoma larvae. J. Exp. Zool. 173:293–301.

Todd TJ (1823) On the process of reproduction of the members of the aquatic salamander. Qu. J. Lit. Sci. Arts 16:84.

Tomlinson BL, Tomlinson DE, and Tassava RA (1985) Pattern-deficient forelimb regeneration in adult bullfrogs. J. Exp. Zool. 236:313–322.

Trampusch HAL and Harrebomee AE (1965). Dedifferentiation, a prerequisite of regeneration. In Kiortisis V and Trampusch HAL (eds.): *Regeneration in Animals and Related Problems*. North Holland, Amsterdam, pp. 341–374.

Ussing HH (1964) Transport of electrolytes and water across epithelia. Harvey Lect. 59: 1–30.

Viale G (1916) Le correnti di riposo nei nervi durante la degenerazione e la rigenerazione. Arch Fisiol. 14:113–131.

Wallace H (1981a) Nervous control. In: *Vertebrate Limb Regeneration*. J. Wiley and Sons, New York, pp. 22–52.

Wallace H (1981b) The mechanism of regeneration. In: *Vertebrate Limb Regeneration*. J. Wiley and Sons, New York, pp. 132–155.

Wallace H (1981c) Reversed limbs. In: *Vertebrate Limb Regeneration*. J. Wiley and Sons, New York, pp. 181–183.

Wallace H (1981d) Partial amputation surfaces. In: *Vertebrate Limb Regeneration*. J. Wiley and Sons, New York, pp. 184–188.

Wallace H (1981e) Metabolic changes. In: *Vertebrate Limb Regeneration*. J. Wiley and Sons, New York, pp. 99–117.

Wallace H (1981f) Regional and axial determination. In: *Vertebrate Limb Regeneration*. J. Wiley and Sons, New York, pp. 156–191.

Wallace H, Wessels S, and Conn H (1971) Radioresistance of nerves in amphibian limb regeneration. Wilhelm Roux Arch. 166:219–225.

Wallace H, Wallace BM, and Maden M (1974) Participation of cartilage grafts in amphibian limb regeneration. J. Embryol. Exp. Morphol. 32:391–404.

Whitlembury G (1964) Electrical potential profile of the toad skin epithelium. J. Gen. Physiol. 47:795–808.

Electric Fields in Vertebrate Repair, pages 77–116

CHAPTER 3

Endogenous Currents in Nerve Repair, Regeneration, and Development

Richard B. Borgens and Colin D. McCaig

Center for Paralysis Research, Department of Anatomy, School of Veterinary Medicine, Purdue University, West Lafayette, Indiana 47907 (R.B.B.) and Department of Physiology, University of Aberdeen, Marischal College, Aberdeen AB9 1AS, Scotland, United Kingdom (C.D.McC.)

INTRODUCTION

How is an elongating nerve fiber similar to an elongating pollen tube, sperm acrosome, or a germinating egg of the brown algae, *Fucus?* One can think of few anatomical characteristics shared by such plant and animal cells; however, nature being conservative, it would not be surprising if these cells shared certain similarities in their physiology of development (Fig. 1). In all of these apically elongating cells, a steady endogenous ionic current enters the growing tip (Robinson and Jaffe, 1975; Borgens et al., 1979; Jaffe, 1979, 1980b; Jaffe and Nuccitelli, 1977; Nuccitelli, 1978; Schackman et al., 1978). Moreover, this current sometimes *predicts* the locus of growth (see discussion by Harold, 1986). In *Fucus*, about 10 hours after fertilization and prior to germination, the polarity of the plant is established. One pole of the fertilized egg will become the thallus, the other the rhizoid. At the latter pole, the previously spherical cell begins to elongate, forming a bulge. At the two-cell stage, this polarization into rhizoid and thallus progenitor cells is set. Using the vibrating probe system, it was found that current enters the presumptive rhizoid. These currents were detected about 1 hour after fertilization and 6 hours prior to germination, accurately predicting the region of subsequent growth (Jaffe, 1979, 1981). A major ionic component of the early transcellular current is Ca^{2+}. Indeed, rhizoids will grow out toward the region of highest concentration when germinating in a gradient of calcium ionophore (Robinson and Cone, 1980). An applied electric field can also

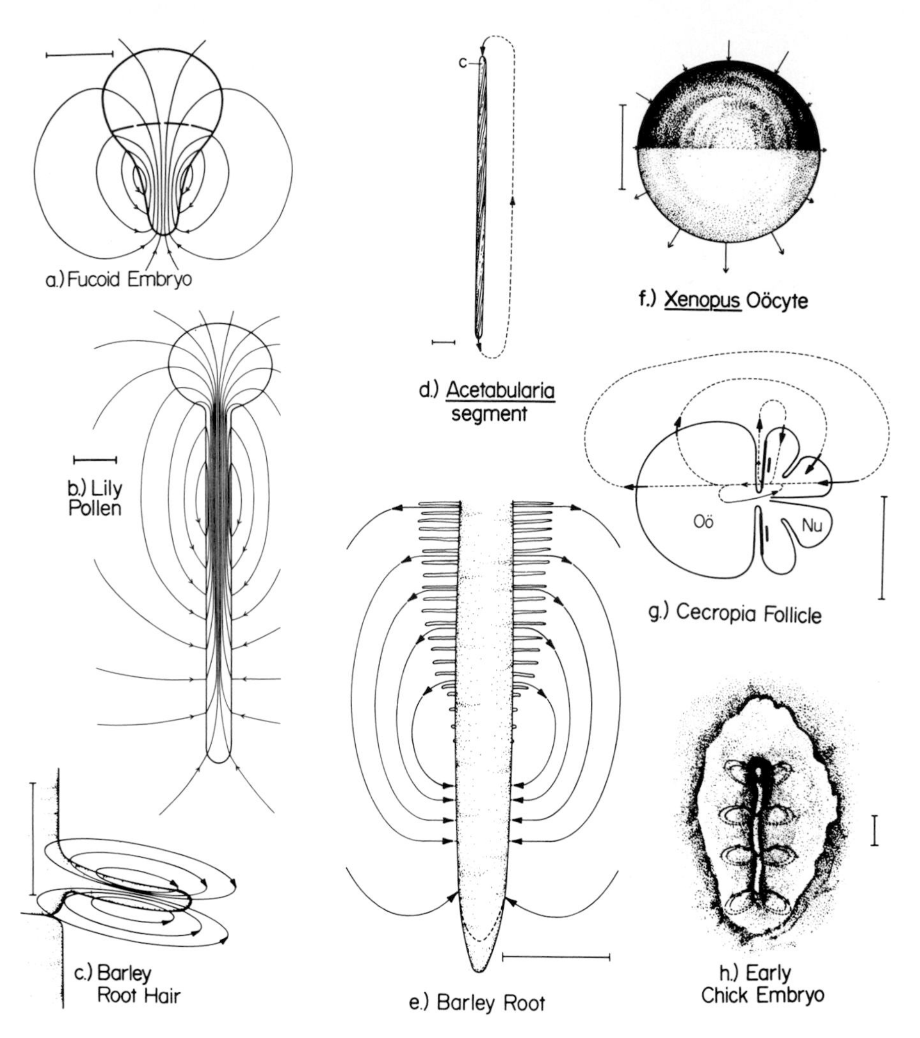

Fig. 1. Developmental currents in animals and plants. **a–e:** Currents associated with five apically growing plant cell systems. Current (peaking on the order of 1 $\mu A/cm^2$) enters the growing end in a, b, and c. The scale bar = 0.05 mm. For d and e, the scale bar = 1 mm. In d, the arrows indicate the direction of 100-second-long pulses of current entering and indicating the end of the anucleate stalk segment of *Acetabularia* that will eventually regenerate a new cap structure. These pulses of current have peak densities on the order of 10–1000 $\mu A/cm^2$. In e, current enters the elongating region of the barley root; densities of current are about 1 $\mu A/cm^2$. **f–h:** Three developing animal systems. Current enters the animal pole of the *Xenopus* oocyte (about 1 $\mu A/cm^2$ in density) and leaves the vegetal pole in about the same density. Current enters the nurse cell end of the (Nu) Cecropia oocyte-nurse cell complex. This transcellular current traverses the narrow cytoplasmic bridge coupling the two compartments and leaves the oocyte (Oö). Current is somewhat concentrated at its entry into the nurse cell; densities can reach 20 $\mu A/cm^2$. An endogenous current (about 30 $\mu A/cm^2$) leaves the chick embryo's primitive streak region. Scale bar in f, g, and h = 0.05 mm. (Reproduced from Borgens et al., 1979 with permission of the publisher.)

direct the establishment of polarity in these eggs (Chen and Jaffe, 1978). These observations suggest that the endogenous transcellular current in fucoid eggs is a cause, and not a consequence of growth. Can such endogenous currents (and their associated fields) actually be responsible for the positioning of intracellular components that lead to the fundamental asymmetries underlying the generation of pattern in organisms? Investigations of another simple developing system suggest that the answer is yes.

A steady endogenous current is driven through the nurse cell-oocyte complex of the Cecropia follicle. Current leaves the oocyte and enters the nurse cell; the circuit is complete with current traversing the cytoplasm of the nurse cell and oocyte via the small cytoplasmic bridge attaching them (Jaffe, 1979; Woodruff and Telfer, 1973). The oocyte is then electropositive and the nurse cell electronegative. This current is associated with a standing voltage gradient across these cells, within their cytoplasms and the cytoplasmic bridge (Woodruff and Telfer, 1973). Injected proteins (fluorescein-labeled serum globulins) that possessed an overall negative charge were driven across the intercellular bridge from the nurse cell to the oocyte but not in the opposite direction (Woodruff and Telfer, 1980). Injected electropositive proteins (fluorescein-labeled lysozyme) moved in the opposite direction—from the oocyte to the nurse cell. Additionally, if the net charge of the lysozyme moiety was changed to negative (by methyl carboxylation), then the polarity of movement switched in accordance with the polarity of the endogenous voltage gradient. Proteins that were essentially neutral (fluorescein-labeled hemoglobin and myoglobin) appeared to move with facility in either direction within the nurse cell-oocyte complex. These experiments (Woodruff and Telfer, 1980) provide convincing evidence that cytoplasmic self-electrophoresis can be a meaningful mechanism for the redistribution of cellular macromolecules. They also point to a complex interplay between endogenous electric fields and the better-characterized cellular transport systems that involve microtubules. We know nothing as yet of the ways in which these separable transport systems might interact, but their coexistence in the cytoplasm seems to ensure such interactions.

These growth currents and their associated intracellular and extercellular electric fields may thus be fundamental to the establishment of cell polarity, growth, and differentiation (see reviews by Jaffe 1979, 1980a,b, 1981; Borgens, 1984).

During organogenesis, dividing cells, elongating cells, and migrating cells—as part of the developing organism—are blended into precise patterns at precise locations by mechanisms of which we are still largely ignorant. This process can be visualized best in the developing nervous system, where unerring specificity of pattern is determined at many levels of integration: the cellular (i.e., synaptogenesis); the tissue (i.e., organ- or extremity-specific innervation); and the whole animal [i.e., differentiation of the central nervous system (CNS) and peripheral nervous system].

There are many factors that help provide cues for the initiation of neuronal growth and guidance. Some of the critical controls in neuronal development appear to be the expression and insertion of the mature Na^+ channel within embryonic membranes (Warner, 1985), widely distributed guidance cues (Harris, 1986), the presence of specific recognitional molecules inserted within cell membranes (Edelman, 1985; Kuwada, 1986), endogenous substrates of specific adhesivity (Letourneau, 1975, 1985; Eisen et al., 1986), and endogenous electrical fields (Hinkle et al., 1981; McCaig, 1986a, b, 1987). It is probable that the remarkable structuring of the nervous system during ontogeny, and perhaps the restructuring of some of its elements during regeneration, is controlled to differing extents, at different times, and in different places, by all of these (Purves and Lichtman, 1985). Here we shall focus on the role of endogenous electric (ionic) currents and their associated voltage gradients in the development and regeneration of nerves.

BIOELECTRIC POTENTIALS IN REGENERATING NERVES: EARLY INVESTIGATIONS

External voltage gradients have been measured along the outsides of severed peripheral nerve trunks for well over 100 years. These measurements were not originally carried out with the intention of uncovering mechanisms of growth control during axonal regeneration but were instead the only means possessed by early physiologists to investigate the nature of the action potential (AP) or "action current" (Cowan, 1934; Lorente de Nó, 1947a, b). An understanding of the nature of the membrane potential changes that subserve AP propagation was not to come until the integrated use of the oscilloscope, high-impedance differential amplification, and the ability to probe directly the cell interior with the microelectrode. Until this time the analytical instrument of choice was the low-input impedance galvanometer. With such instruments, early investigators were able to determine roughly the magnitude and polarity of voltage gradients sampled along the outsides of a variety of biological materials. These types of voltage measurements have been loosely referred to as "bioelectrical potentials"; however, it is wise to remember that these measurements were attempts by the fathers of modern physiology to understand conduction and musculature activity, as well as development. Efforts by such pioneers as E. DuBois-Reymond (1843) and Ramón y Cajal's student, Raphael Lorente de Nó (1947a, b), were physiological in scope and should be considered separate from the imaginative, but less rigorous investigations of later biologists such as Burr (Burr, 1932; Burr and Northrop, 1935) and, in modern times, Becker (Becker and Marino, 1982; Becker and Seldon, 1985) and Nordenstrom (1983).

Lorente de Nó speculated that such external voltage gradients may play a role in neuronal regeneration. He wrote that it was "thinkable that the permanent flow of demarcation current plays a role in the process of regeneration. . . . [Perhaps] the continued flow of the demarcation current into the last few millimeters of regenerating nerve is a mechanism by means of which energy is transferred to the regenerating end from points at some distance from it" (Lorente de Nó, 1947b, page 459). He and other investigators of this era (Cowan, 1934) reported that severed peripheral nerve trunks were negative at the site of the lesion with respect to undamaged regions. These voltages (reported to be in the tens of millivolts) declined steadily with time after the injury. Lorente de Nó reported that "demarcation potentials" (of a few millivolts) could still be observed 1 week posttransection, and correctly inferred that the polarity of these potentials demonstrated that a current should be flowing toward the cut end along the outside of the fiber and would enter its severed face. We will review these old measurements made on frog peripheral nerve in light of the modern literature; however, it would first be wise to consider the neuron's response to axotomy before discussing a role for the steady endogenous current induced by injury.

THE NEURON'S RESPONSE TO SEVERANCE

It is both convenient and instructive to describe the responses of the nerve cell to *transection* of its axon rather than to axonal compression or crush. This is somewhat arbitrary, since many (but not all) of the effects of crush and cut injuries are similar—indeed, severe compression of the axon eventually leads to a physical separation anyway. Transections facilitate understanding in that they yield two axon segments: the proximal segment, which is in continuity with the cell body and usually survives the insult to the neuron; and the distal segment, which, along with its terminal, undergoes Wallerian degeneration and is lost, although the myeline sheaths and basal laminae may persist. Even though many of the same observations described here occur in distal and proximal segments, our focus will be on the *proximal* segment unless especially noted. Therefore the use of the word *proximal* refers to positions near the cell body and *distal* refers to positions near the cut tip of the proximal segment of the severed axon.

Conventional wisdom might suggest that when a neuron is injured the damage would be rather restricted to the focus of the lesion. Quite to the contrary. Not only does the entire neuron respond to an injury of its fiber, but the entire neuronal network, of which it is a part, may experience physiological and anatomical reverberations resulting from trauma to only this one cell (Barron, 1983; Kelly, 1981; Tweedle, 1971). Neurons that project terminals onto the damaged cell's dendritic compartment, as well as neurons that have been isolated downstream, will probably experience various changes in their anatomy and physiology (Watson, 1976). The biochemical changes occurring throughout the neuron, the classical chromotalytic response, and other *cell body* responses to axotomy have been well described (Cragg, 1970; Grafstein, 1975; Veraa et al., 1979) and will not be reviewed here in their entirety. Rather, we will focus on the morphological and electrophysiological consequences of axotomy, concentrating on the local insult to the nerve fiber.

The immediate effect of transection is, of course, an intermingling of the axoplasm and the extracellular environment. There is a coincident circumferential collapse of the membrane at the face of the axon's open bore, and extrusion of axoplasm, a great influx of the majority ions of the extracellular environment (principally Na^+ and Ca^{2+}), and a loss of K^+ from the cell. This exchange of ions causes a dramatic alteration of the ionic composition of the axoplasm in the local region of the damage. As we will disucss below, this influx is not just a passive event in which ions flow down their electrochemical gradient. It is an active, metabolically driven movement of charge that will persist for perhaps weeks after the injury, well after the tip of the axon has sealed with new membrane. This new membrane may form on the order of 1 hour or less in some nerves and many hours in others (Kao et al., 1983; Lucas et al., 1985). Some small myelinated axons have been reported to continually break down at their tip for a period of days (Kao et al., 1983). As Lucas et al (1985) summarize, " . . . even after 100 years the question of the completeness of that resealing is still under debate. Undoubtedly a major factor in the reported differences in the degree and duration of these (injury) potential changes after transection is the order of magnitude difference in the diameters of vertebrate and invertebrate neurites [studied]" (page 248). After the tip has sealed, it also begins to swell. These club-shaped endings have been associated with continued destruction of the axon, culminating in a (sometimes extended) period of retrograde degeneration. Since the initial observations of Ramón y Cajal (1928), these swollen tips have been referred to as "retraction bulbs" or "terminal clubs." Swollen endings are many times connected to the more proximal healthy axon of normal caliber by a thin, thread-like, connecting piece of membrane, which eventually degrades, leaving only the club-shaped ending or "free ball" (after Ramón y Cajal), which eventually is degraded by catabolic processes (refer to Fig. 6B and C, Borgens, Chapter 4, this volume). This entire process—the formation of terminal swellings, the disassociation of the damaged tip, and retrograde degeneration of the axon—is many times referred to as "dieback," not to be confused with "Wallerian degeneration," the complete degradation of the distal axonal segment that has been separated from the soma by transection. There are, of course, physiological consequences of the injury: conduction is lost for some distance proximal to the cut end, as a result of the alteration of the ionic composition of the axoplasm, the physical collapse of the membrane and it's retrograde depolarization, and a retrograde demyelination. New membrane is formed at the severed end as damage to the axolemma is eventually repaired. Although conduction is reacquired, its characteristics are different. There is apparently a process of membrane maturation in which membrane ionic selectivity gradually returns. A change in the ionic basis of conduction apparently reflects this process. Initially AP conduction is Ca^{2+}-dependent near the damaged tip, while Na^{+}-dependent conduction is gradually reacquired (Meiri et al., 1981). Thus the regeneration of new axonal membrane appears to recapitulate development, at least in terms of the voltage-sensitive conductance channels that the membranes possess (Spitzer and Lamborghini, 1976). Axons that survive injury intact are also physiologically impaired. Myelinated axons (as studied in cat spinal cord) are limited in their conduction velocity and frequency of propagation, and display a tendency for conduction block at physiological temperature (Blight, 1983, 1985). The anatomical responses to transection are also profound and share many similarities in all classes of axons.

One striking morphological response to transection is the production of curious "zones" at the tip of the cut axon (Fig. 2). These zones form as a consequence of both a reorganization of cytoplasmic constituents as well as ongoing degenerative events within the axon terminal. Axoplasm exposed to the extracellular environment is degraded, becoming first granular in appearance and then clear and watery. This zone of clear cytoplasm extends proximally from the cut tip for a few hundred micrometers. More proximally is a second zone characterized by marked accumulations of organelles and fragments of organelles, such as mitochondria, lysosomes, and various vesicles (Watson, 1976; Zelena, 1969; Zelena et al., 1968). Mitochondria in this region are swollen. The organization of the organelles residing in this second zone appears random. This zone grades into a more proximal third

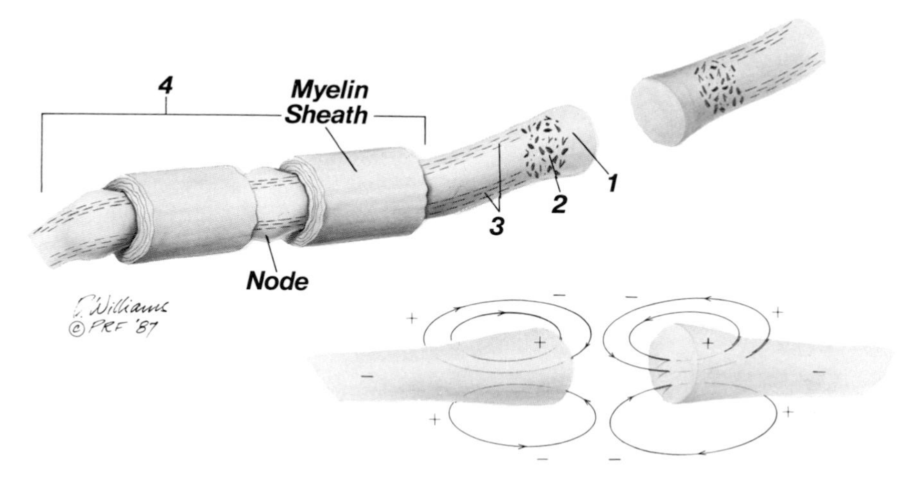

Fig. 2. In a myelinated axon, major regions of the axoplasm near the cut tip may form "zones" in response to axotomy. Region 1 is an area of clear and watery cytoplasm; region 2 is an area of organelle accumulation and fragments of organelles—inclusions in this zone are usually abnormal in character and randomly arranged; region 3 is an area of organelle accumulation where cellular inclusions (principally mitochondria) demonstrate sometimes striking orientations. Mitochondria, for example usually lie with their long axis parallel with the long axis of the nerve axon. Region 4 is an area of organelle depletion, except at the nodes, where accumulations are usually observed. The inset shows the direction of current flow in the proximal and distal segments of a severed axon. Note that current enters the cut tip in both segments and that the polarity of the *intracellular* voltage gradient is positive near the cut ends as well. (Reproduced from Borgens RB [1988]: Voltage gradients and ionic currents in injured and regenerating axons, Adv in Neurology 47:51–66, with permission of Raven Press, New York.)

zone in which the organelles are ordered. Mitochondria are more typical in their appearance, and they are arranged with their long axis parallel with the long axis of the nerve fiber. Furthermore, they are aggregated circumferentially beneath the membrane, while the core region of the axon is depleted of them. Watson (1976), impressed with this ordering, described the arrangement of mitochondria in this region as if they were "trains." In general, there are less organelles in the third zone than in the second, more immediately distal zone. Furthest from the place of transection is the fourth zone, one which may extend proximally many millimeters. This is an area of organelle depletion. The fact that organelles are depleted in this most proximal zone suggests that their accumulation in more distal regions is because of an actual *movement* and not a *resynthesis* (for example, of mitochondria). Moreover, these migrations of cellular constitutents are truly "bidirectional." They occur in the degenerating distal segment as well as the proximal axonal segment. In the proximal segment of the axon, mitochondria move in a proximal-distal direction (toward the cut end). In the distal segment, mitochondrial movement is distal-proximal in direction [toward the cut end as well (Zelena, 1969)]. In terms of the internal electrical polarity of the axon segments, mitochondrial movement in both cases is from regions of negativity toward more positive (less negative, depolarized) areas.

ELECTRICAL RESPONSES TO AXONAL INJURY

In a preceding chapter we described a flow of current that leaves a lesion made to an animal's integument. This is because animal integuments are internally positive with respect to the outside. Since, by convention, current flow is defined to be in the direction that positive charge moves, then current would leave skin wounds (refer to Borgens, Chapter 2, and Vanable, Chapter 5, this volume). Actual measurement has demonstrated this to be true. With respect to an individual cell, most animal cells are internally negative by some 50–90 mV; thus by the same convention one would expect current to enter a hole made in a cell membrane (Borgens, 1982). The voltage across the membrane is supported by ionic pumps, part of the living membrane. If the cell remains alive, one would expect to observe current driven into the cell lesion acutely and to persist for a variable time. This hole in our hypothetical membrane is an open leak—a low resistance pathway for current to flow. When this lesion is sealed, the persistence of injury current would depend on the character of the new membrane—its ionic selectivity and its leakage resistance. This general description of current flow into a cellular lesion models the situation found in severed or otherwise damaged nerve fibers in particular. Current would be expected to enter the open bore of the transected axon (or the focus of a crush or compression injury as well), associated with a distally negative voltage gradient *outside* the proximal segment of the axon, and a distally positive voltage gradient *inside* it. As discussed previously, demarcation potentials measured by the classical physiologists in fact demonstrated that the ends of severed peripheral nerve trunks are negative with respect to more proximal undamaged regions. Demarcation potentials measured in frog peripheral nerve were on the order of 30–40 mV and declined by 90% by 1 day postinjury (Lorente de Nó, 1947a, b).

It is probable that only a portion of the total injury current traversed the galvanometer (used in these early measurements). Thus the extracellular voltages associated with current flow reported in older investigations were probably quite conservative. Since the nerve fiber may have an unusually large surface area of healthy membrane, and the axonal transection would expose a leakage pathway of a comparably small cross-sectional area, it would follow that extremely large densities of current might converge on the severed end of an interrupted fiber. The extracellular voltage gradients associated with this endogenous current might be steep or shallow, depending on the degree of cellular packing about the axon, since the magnitude of an extracellular voltage gradient is inversely proportional to the cross-sectional area of the extracellular space. In summary, the older measurements of demarcation potentials are interesting, and in most cases correct with regard to the polarity of currents and fields surrounding transected nerve fibers—or nerve trunks. They offer little quantitative information concerning the current and fields produce by axonal injury. Actual measurements of current entering injured axons were performed with a vibrating probe for the measurement of extracellular current (Jaffe and Nuccitelli, 1974) in 1980 (Borgens et al., 1980). These measurements were made on identifiable axons of the ammocoete larvae of the lamprey. The larval lamprey CNS is extremely well suited for this type of investigation for the following reasons: the lamprey CNS does not possess an intrinsic vascularization; thus it can be completely removed from the animal and placed in simple organ culture [where it survives for up to 1 week (Borgens et al., 1980)]. This isolated brain-cord preparation is ideal for measurements using the probe system. The brain possesses giant identifiable reticulospinal neurons that project large axons (50–60 μm in diameter) down the spinal cord. Thus in some cases measurements of current could be compared between the same types of cells in different animals (Rovainen, 1967; Wood and Cohen, 1979, 1981).

INJURY CURRENTS

Currents in Giant Axons

Extremely large densities of current enter the transected lamprey spinal cord (Fig. 3) (Borgens et al., 1980). These currents (0.3–0.7 mA/cm^2) are relatively homogeneous across the face of the transection. Small bands of incurrents and outcurrents were measured along the lateral surfaces of the spinal cord back to the brain. We have since determined that the small focused incurrents observed at regular intervals along the lateral margin of the cord correlated with the severed spinal roots, snipped while removing the brain-spinal cord preparation from the animal to the measurement chamber (Borgens, unpublished measurements). A peak of current density could be observed when moving the vibrating electrode across the open bore of any one giant axon when crossing the face of the cord transection. As the probe

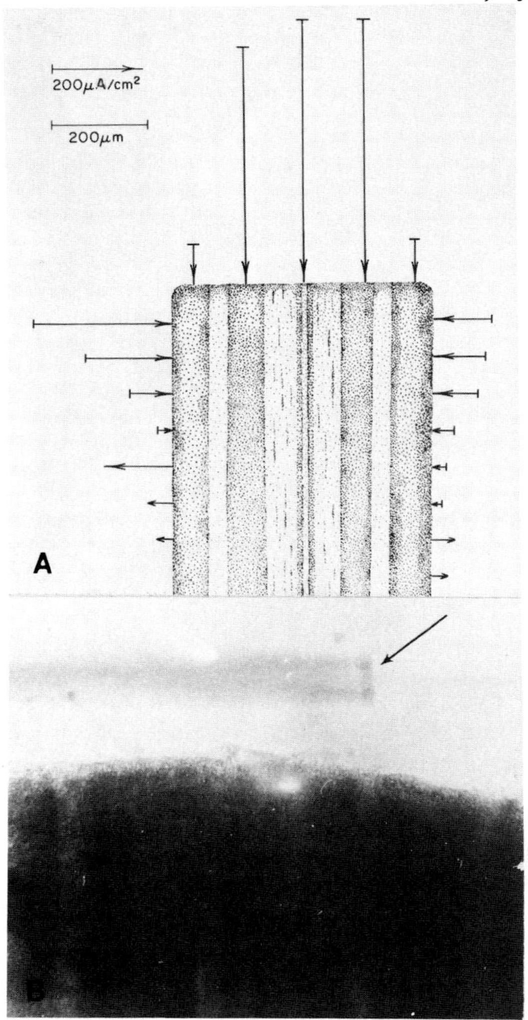

Fig. 3. **A:** Vibrating probe scan of current entering or leaving the cut end of a lamprey spinal cord (4 hours posttransection). Giant axons within the cord are shown as light bands parallel with the long axis of the spinal cord. The lengths of the arrows depicting current direction are proportional to the density of current. Note that current enters the severed end (and the severed axons contained within). **B:** Photograph of the proximal end of a living spinal cord transected 15 minutes previously. The laterally located Mauthner giant axons and the more medial Müller axons are visualized as light bands running along the long axis of the spinal cord. The arrow indicates the 20μm tip of a vibrating electrode—at the extremes of it is a 30-μm excursion. The probe is vibrating at about 400 Hz. (Reproduced from Borgens et al., 1980, with permission of the publisher.)

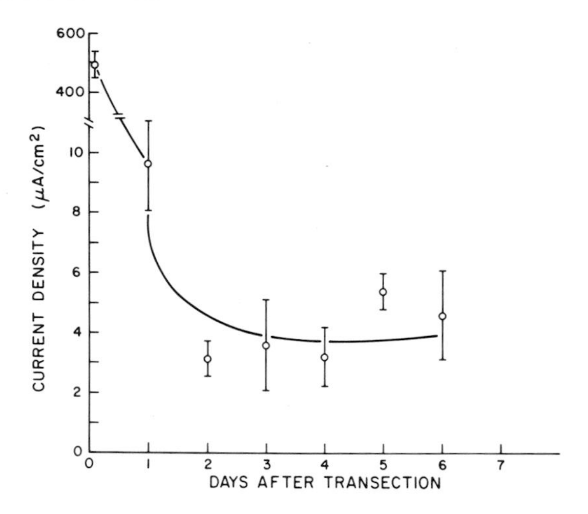

Fig. 4. Decline of current density entering the transected lamprey spinal cord. Note the relatively stable plateau of current entering the cord after the first day posttransection. These data were obtained from serial measurements on six animals. (Reproduced from Borgens et al., 1980, with permission of the publisher.)

passed from the center to the edge of such an axon, the current fell off in magnitude, demonstrating that current does indeed enter the cut end of the individual transected nerve fiber. Overall, the early and enormous currents fall off steeply within 1 hour after transection. Twenty-four hours after severance, currents were on the order of 20 $\mu A/cm^2$ and further declined to smaller yet steady levels of about 5 $\mu A/cm^2$ (Fig. 4). This density of current persisted to the end of the first week posttransection, at which time we discontinued measurements (Borgens et al., 1980). It is interesting that the initial decline in current density closely parallels the decline in the magnitude of "demarcation potentials" reported by early physiologists.

Current flow in biological systems represents a flow of charge. The injury current in nerves would probably be carried by the majority of ions in the extracellular fluids, for example, Na^+, Ca^{2+}, and Cl^-. This fact was demonstrated by substitution experiments in which a measurement of current density was made in the standard measurement medium (lamprey Ringer's solution), followed by a measurement made in a test medium in which one particular ion was replaced with a nontransportable equivalent (for example choline replacing Na^+; Ca^{2+} replaced with Mg^{2+}; Cl^- replaced with isethionate or methanesulfonate). Specific ionic channels were also blocked or poisoned (for example, the Ca^{2+} channel poisoned with La^{3+}). It is important to note that such modulations act specifically on the pump site (presumably in the axon membrane) and not at the leak site (which is the injury) where extracellular current measurements are actually made. Although the leakage current is ionically complex, changes in its magnitude and/or polarity reflect the effects of the experimental modulation of the ionic pumps, located in adjacent healthy membrane. These tests confirmed that the bulk of the current was carried by Na^+, and that much of the balance was carried by Ca^{2+}, while the end of the cut axon was not sealed with a new membrane. For a short time *after* the formation of such a new membrane, we inferred that Na^+ and Ca^{2+} conductances were still high and roughly proportionate to the ratio of Na^+ and Ca^{2+} in the external medium (Borgens et al., 1980). This notion suggested that the new membrane was ionically nonselective. We had further suggested that the dramatic fall in current within the first hour posttransection may reflect a ''resealing'' of the axon's cut end (Borgens et al., 1980). This explanation can be only partially correct, for the reason stated above: the new membrane appears to reflect the same character as an open ''leak'' for hours after its formation (Meiri et al., 1981). An additional explanation for the steep decline in current density is that the total current produced by the healthy membrane is concentrated where driven into the axon's newly severed tip (of modest cross-sectional area). With time after axotomy, there is a continuing, substantial, and retrograde collapse of membrane more proximal to the tip, producing a leakage pathway of an ever-increasing surface area; thus the density of current (as measured entering the open bore) would fall proportionately. It is certain that for the first few hours after the insult, this active flow of charge would produce a dramatic change in the ionic composition of the axoplasm. The concentration of Na^+ and Ca^{2+} would be expected to increase by orders of magnitude. Elevated Na^+ within the cytosol would be expected to induce mitochondria to release sequestered Ca^{2+} (Carafoli and Crompton, 1976). Additionally, since about 30% of the injury current entering the severed fiber is carried by Ca^{2+} (Borgens et al., 1980), this should result in a Ca^{2+} gradient within the local area of the severed tip. Since Ca^{2+} is not very mobile inside of cells [being easily bound within cytoplasm (Baker, 1972)], the peak concentration of Ca^{2+} should be observed at its entry point and decline in concentration proximally.

These first studies of the injury current in nerve permitted informed speculation concerning the nature of ionic conductances. Simultaneously, others performed direct measurements (using conventional microelectrode techniques) that supported and complemented the vibrating probe study.

Meiri and her colleagues studied the membrane properties of the tips of severed giant axons in the cockroach CNS (Meiri et al., 1981). They found an immediate decrease in membrane potential and input reistance at the axonal tip coincident with transection. Action potential amplitudes and conduction were impaired and eventually lost near the axon terminal within the first 2 hours posttransection. Thereafter, there was a slow recovery of these membrane parameters, including progressive (and characteristic) changes in ionic conductances. Early in the recovery process (in axons that were demonstrated to have sealed tips), ionic conduction was nonselective. Even though there was a new membrane at the severed end, the tip still behaved as an open leak, as we had suggested (Borgens et al., 1980). Selectivity was observed to return gradually, and the greatest increases were measured to occur in Na^+ and Ca^{2+}. By 8 days posttransection, all of the ions tested (Na^+, Ca^{2+}, K^+, and Cl^-) had regained their normal conduction characteristics (Meiri et al., 1981). This ''maturation'' of the membrane at the injured tip was coincident with a return of AP propagation, which (as mentioned) was first dependent on Ca^{2+} followed by a return to the normal Na^+-dependent spiking [a maturation process observed in *developing* neurons as well (Bixby and Spitzer, 1984; Spitzer, 1981)]. These two studies complement each other in two ways: they both demonstrate a dramatic alteration in the ionic physiology of the axon's terminus following injury, and they also demonstrate that the initial conductances that may affect changes within the axoplasm are in Na^+ and Ca^{2+}.

Anatomical Responses to Injury

Can we relate the dramatic changes in the morphology of the damaged *local axon compartment* to the intense endogenous current entering it in response to injury? It is certain that there would be both ionically mediated effects of the injury current as well as voltage-mediated effects. These two associated effects of net current flow entering the nerve fiber may help explain the deterioration of the local axoplasm near the lesion, the phenomenon of axonal dieback, perhaps the interesting formation of zones at the ends of interrupted fibers, and the migration of organelles. We can also speculate on the role such currents may play during axonal outgrowth.

It is certain that the immediate consequences of the injury current in nerves would be axonal destruction. As discussed, the local cytoplasm adjacent to the lesion experiences a dramatic increase in the concentration of cytoplasmic Ca^{2+} and Na^+. We have presented an instructive model (Borgens et al., 1980) suggesting that by 2 days posttransection in lamprey giant fibers, about 3.4 coulombs of charge enter each cm^2 of the cut tip. This hypothetical ''front'' of previously extracellular charge should also penetrate the severed cord and the axons contained within by about 3 mm within this time. Thus a significant portion of the axonal terminus should experience striking alterations in the ionic composition of the axoplasm. The immediate effect of this influx of cations is predictable. Since the integrity of the cytoskeleton is dependent on intracellular Ca^{2+} (Lasek and Hoffman, 1976), high $[Ca^{2+}]$ would be directly responsible for the dissolution of the cytoplasm at the plane of transection. In high $[Ca^{2+}]$, microtubules are depolymerized, tubulin is precipitated, and neurofilaments degenerate (Hyman and Pfenninger, 1985; Lasek and Hoffman, 1976; Schlaepfer, 1983). We have discussed that the degeneration of axoplasm and its organelles is most severe at the tip of the axon's proximal segment and less severe at positions closer to the cell body. It is likewise certain that the concentration of Ca^{2+} in the axoplasm would be greatest at the plane of transection and less at positions more proximal. It is no surprise, then, that a $[Ca^{2+}]$ influx should mediate axonal degeneration and dieback. This view has been championed by others, notably Schlaepfer and his colleagues (Schlaepfer, 1974, 1977, 1983; Schlaepfer and Bunge, 1973). They have prevented the Wallerian degeneration of distal segments of transected neurites in culture by maintaining the cultured neurons in a medium much reduced in Ca^{2+}, or by the addition of Ca^{2+} chelating agents to the culture medium (Schlaepfer, 1977). Exposing such ''preserved'' neurites to a medium normal in $[Ca^{2+}]$ precipitated fiber degeneration. Conversely, degeneration of isolated segments of rat peripheral nerve can be accelerated by maintenance in a culture high in $[Ca^{2+}]$. Another demonstration of the destructive effects of Ca^{2+} within nerve fibers was demonstrated by placing desheathed segments of peripheral nerve into a media containing a Ca^{2+} ionophore (Dupont A23187) and Ca^{2+} ions. This ionophore allows a local influx of Ca^{2+} at exposed axonal membranes, which in turn caused a degeneration of the axoplasm in this local area (Schlaepfer, 1977). Not only does high $[Ca^{2+}]$ affect the architecture of the cytoplasm but it would also have broad repercussions on other cell processes, such as axoplasmic transport. Transport itself is intimately tied to the architecture of the cytoskeleton (Lasek and Hoffman, 1976); thus its collapse spells the cessation of transport function as well (Ochs et al., 1977). In a later section we will review the role of intraaxonal Ca^{2+} in *growth processes* in axons. It is probable that this ion plays a fundamental role in defining the stability of the local axon tip after damage, regulating (probably within narrow limits of concentration) *regenerative* as well as *degenerative* responses to damage. Since axonal degeneration at the cellular level is associated with the electrophysiological consequences of injury, one might suggest that the degree of axonal dieback may be modulated artificially by electrical means. We have suggested that one early effect of an exogenously applied field might be to limit the degree of retrograde degeneration (Borgens et al., 1980). An applied *distally negative* electrical field should produce a "counter-current" or "bucking voltage" within the terminal of a severed axon residing in this field. Since the cut end is highly permeable to current flow (with a lowered input resistance), an exogenously applied current should be driven through the end of a proximal nerve segment in a proximal-distal direction (within the axoplasm). The *endogenous injury current* flows in a distal-proximal direction within the axon. Therefore the imposed current and its associated voltage gradient would be opposite in polarity to the endogenous current and its associated voltage gradient *inside the axon* (Fig. 5).

We have hypothesized that the overall destructive effects of the early injury current may be weakened or reduced as a consequence of the applied field. Our notion has been put to an experimental test. Distally negative and distally positive fields were applied across transected lamprey spinal cords, and the degenerative responses of identified axons were compared with a sham-treated group (Roederer et al., 1983). Within the first week posttransection, axonal dieback was reduced by *distally negative* fields and *enhanced* by *distally positive* fields. A problem arises, however, in interpreting these results quantitatively. The enormous densities of current entering the axon within the first 2 hours may peak on the order of 1 mA/cm^2 in density. It is not clear to what degree an exogenously applied total current of only 10 μA could effect such a large ionic influx.

This numerical mismatch may be critically important in a temporal sense. If, as we are suggesting, endogenous currents play a role in organelle transport, it may be of fundamental importance to have very large, but relatively brief current inflow into the newly severed axons. Such currents may be causal in both the buildup of mitochondria close to the tip, and in rapid transport and incorporation of materials for new membrane. (This may be especially true, given that the microtubule-based axoplasmic transport system is compromised by the damaging levels of Ca^{2+} influx). Mitochondrial accumulation and the ability of mitochondria to sequester Ca^{2+} may act to limit the damage caused by lesion-induced Ca^{2+} influx. The rapid formation of new membrane may reduce the mismatch between applied and endogenous currents. By 1–2 days after lesions to the lamprey cord, applied currents of 10 μA may be large when compared with the current still entering the site of damage. The massive injury current that does obvious damage to the axon can be viewed in the above sense as simultaneously instigating

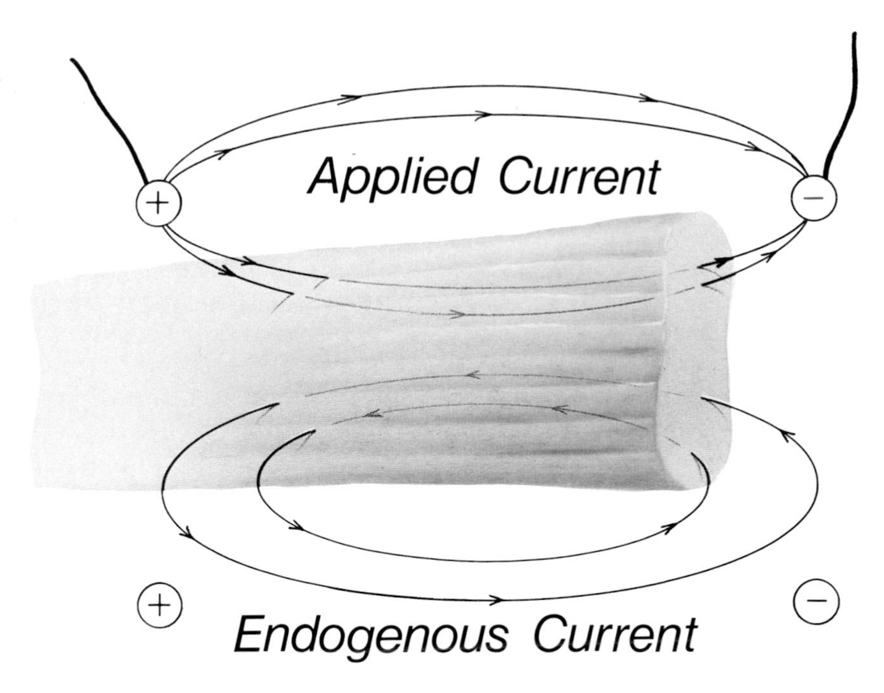

Fig. 5. Polarity of endogenous and applied voltage gradients (and current flow) at the tip of a severed axon. The endogenous injury current (lower portion of the axon) shows that current enters the cut tip and flows in a distal-proximal direction *inside* the axon. In the upper part of the drawing, an applied extracellular voltage (from a distally negative electrode pair) is of the same polarity as the extracellular component of the endogenous current; however, the *applied* current would penetrate the axon (because of its lowered input impedance near the cut tip) producing current flow in the reverse direction (proximal-distal) inside the axoplasm.

immediate protective responses, fundamental to subsequent regeneration. Whether dieback would be prevented in lamprey cord if current were applied only 1–2 days after section, i.e., when the large inward currents have subsided, is not known. Once the net endogenous current is dominated by the applied current (distally negative), regeneration of axons may be promoted in much the same way as development is stimulated toward an external cathode. Greater numbers of filopodia may be produced, and the depolarized tip may undergo a controlled influx of Ca^{2+} sufficient to activate gel-sol conversion of cytoplasm and actin-based contractile events. Localized incorporation of new membrane in vesicular form may also occur. By this stage, intracellular Ca^{2+} levels would be insufficient to cause continued degeneration.

Degenerating axons are many times identified by their swollen terminals. These "retraction bulbs" are found on most classes of axons following acute injury and persist for a variable time afterwards. It has been commonly assumed that swollen tips were due primarily to an interruption in axoplasmic transport and to an influx of extracellular water at the cut tip. Friede (1964) suggested that these swellings resulted instead from electrophoretic mechanisms. He tested this notion by locally depolarizing areas of rat peripheral nerve (with KCl) and observed that focal areas of electronegativity (current entering the nerve) were associated with axonal swelling. He suggested also that such focal swellings were independent of axonal transport (although acute depolarization may not easily be disentangled from a primary or secondary effect on transport mechanisms). Friede also claimed swellings occurred at known sites of membrane depolarization. He attempted to rule out osmotic effects by other experiments: using lactic dehydrogenase, malic dehydrogenase, and DPN diaphorase enzyme histochemistry as well as the acid Schiff protein determination histochemistry. With these techniques he calculated the changes in concentration of these enzymes at the ends of severed fibers. These enzymes increased substantially in swollen areas of the axons but were not seen to decrease in concenteration per unit volume. If the swellings had occurred because of an influx of extracellular water, one would expect the opposite finding. Overall, Friede's experiments suggested to him that the axonal swelling in response to damage was involved with electrophoretic or electrophysiological mechanisms. His experiments, while inconclusive and only suggestive, were one of the first challenges to the dogma that interruption of axoplasmic transport always dictates the character of axonal swellings.

COULD INTRAAXONAL VOLTAGE GRADIENTS BE RESPONSIBLE FOR ORGANELLE REARRANGEMENT?

There is little information concerning the movement of cellular organelles in response to natural or applied cytoplasmic fields. One can reasonably suggest that since organelles possess a net charge they should be able to be electrophoresed by an appropriate intracellular electric field (Voss et al., 1968). As will be discussed, charged receptors, protruding from the plasma membranes of cells, can be so moved (Jaffe, 1977; Jaffe et al., 1974; Orida and Poo, 1978; Poo and Robinson, 1977; Robinson, 1985); thus it requires no great leap in logic to suggest that cellular inclusions may do the same. The conventional explanation for the ordered movement of organelles within injured nerves is that it is (once again) mediated by axoplasmic transport mechanisms. This notion is, at the least, an incomplete explanation for ordered movement. For example, the movement of organelles, principally mitochondria, is truly bidirectional. That is, organelles move in a proximal-distal direction in the proximal axon segment, and in a distal-proximal direction in the distal axon segment (Zelena, 1969). Moreover, these movements occur at about the same rate and commence minutes after injury. Thus, in the proximal segment, anterograde transport processes should mediate the movement, while in the distal segment, a retrograde transport system would. Since these two independent transport systems are known to differ greatly in their *rates* of transport, it seems unlikely they could subserve transport of the same organelles at the same rate but in different directions. Alternately, or additionally, organelles within nerves may be moved by a standing voltage gradient imposed across them within the axoplasm. This field would be associated with the intense injury current flowing into the end of the severed fiber and through the axoplasm itself for some distance proximal to the injury. The polarity of the intracellular field would be *distally positive* in the proximal axonal segment and proximally *positive* in the *distal segment;* therefore negatively charged components, e.g., mitochondria, should move *toward the lesion* in both segments (Figs. 2 and 5) (Borgens, 1982).

In axons, we have noted that the negatively charged dye Lucifer yellow tends to accumulate near points of injury in damaged axons (Borgens, unpublished observations). Meiri and coworkers (1981) found that the intracellular marker cobalt (positively charged) would not stain the terminals of severed axons within the first 4 hours after transection. They explained this lack of staining as a possible leakage of cobalt from the cut end. However, during this same time, the injury current entering the axon would be most intense. Perhaps the distally positive intraaxonal voltage gradient may have helped exclude the cobalt? Cytoplasmic rearrangements of *organelles* has been observed in other cellular systems, perhaps in association with changes in the ionic conductances of their membranes. The apical-basal architecture of thyroid cells can be radically altered to a biapical form by exposing cultured cells to a medium in which the serum content has been raised from 0.5% to 5% (Nitsch and Wollman, 1980a, b). This gross change in the ionic character of the bathing media should induce changes in the conductance of ions across the membrane hypothesized to occur in association with an induced current flow through and around these cells (Jaffe, 1981). There is recent evidence that F-actin microfilaments, mitochondria, and ribosomes may all accumulate subjacent to the anodal-facing membrane of embryonic muscle cells, which have been exposed to a small applied electric field (McCaig and Dover, 1989) (Fig. 6). These events are obvious within 5 hours

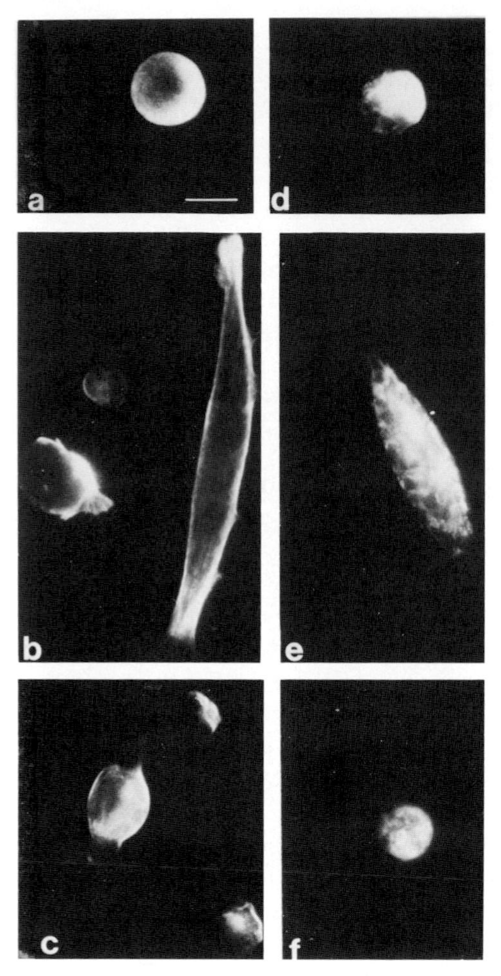

Fig. 6. **a–c:** Rhodamine phalloidin specific staining of polymerized F-actin. **a:** Spherical myoball (from disaggregated somites of *Xenopus laevis*) after a 6-hour exposure to 105 mV/mm. **b:** Perpendicularly oriented myoblast after 4½ hours at 150 mV/mm. **c:** Elongating myoblast after 30 minutes at 200 mV/mm. In each case F-actin-associated fluorescence is predominant on the anodal-facing side of cells (right). **d–f:** Rhodamine 123 is a specific stain for mitochondria. Here it was used to label live cells exposed to an electric field. **d:** Spherical myobal (from *Xenopus* somites) after 5 hours at 110 mV/mm. **e:** Perpendicularly oriented myoblast after 5 hours at 110 mV/mm. **f:** Spherical myoball after 50 minutes at 200 mV/mm. In each case, the fluorescence associated with mitochondria is predominant on the anodal-facing side of the muscle cells (at right). Scale bar = 25 μm throughout. (Reproduced from McCaig and Dover, 1989, with permission of the publisher.)

of exposure to a field of approximately 100 mV/mm but are present in less striking form after only 1 hour. Spherical myoblasts begin to elongate perpendicular to an applied field after about 1 hour in culture (Hinkle et al., 1981); therefore this process occurs within a time scale consistent with a role in determining cell orientation.

The reasons for these organelle asymmetries are unclear. An applied electric field will exist both as a voltage drop along the outer surface of a cell and, because of the resistance of the cell membrane, as a much smaller intracytoplasmic potential gradient. Cell surface receptors for the plant lectin concanavalin A and also acetylcholine receptors accumulate on the cathodal-facing surface of these cells (Poo and Robinson, 1977; Orida and Poo, 1978), probably by electroosmosis (McGlaughlin and Poo, 1981). Some cellular constituents, for example, microfilaments, appear tethered to the inner aspect of these mobile integral membrane proteins (Rees and Reese, 1981). The relative contribution of receptor movements on the cell surface, as opposed to the more direct effects of small intracytoplasmic voltage drops on the distribution of organelles, remains to be determined.

INJURY CURRENTS AND THE RESPONSE OF THE SOMA TO AXOTOMY

When axonal damage occurs near the soma, the neuron's response to injury is more intense. This fact can be demonstrated by noting the frequency of cell death as a function of the distance between the soma and the plane of fiber transection. The neuron is more apt to die if the axon is severed very close to the cell body (Barron, 1983; Cragg, 1970; Grafstein, 1975). Other responses of the neuron to injury (such as chromatolysis) are more severe if transections fall close to the cell body. It is probable that this process is mediated by a partial depolarization of the cell body membrane. Cragg (1970) has calculated that when a large-diameter axon (10 μm) is transected 1 mm from the cell body, a depolarization results, reducing the resting potential of the somatic membrane to about 30 mV. A small-diameter fiber (1 μm) severed 1 mm from the cell body, or the aforementioned 10-μm-diameter fiber severed at 10 cm from the cell body, will both depolarize the cell body membrane by only 2 mV. Thus the caliber of the axon and the distance it is severed from the soma determine the magnitude of depolarization of the soma, which, if sufficient, might result in cell death. The above figures also underscore the restricted nature of the intraaxonal fields associated with the endogenous current. Applying conventional cable analysis, one can demonstrate that cytoplasmic fields associated with axonal injury should be reasonably localized within the last few millimeters of the tip (Fig. 7).

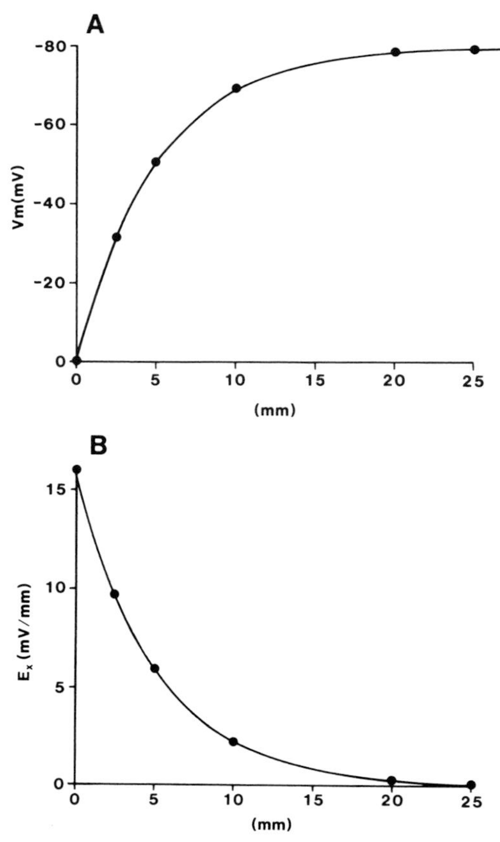

Fig. 7. Calculation of the membrane voltage (Vm) and cytoplasmic electrical field (E_x) with distance from the cut end of an idealized giant fiber (50 μm diameter). **A:** Depolarization of the membrane. **B:** Fall-off of the internal electrical field. Cable analysis suggests that 10 mm from the tip of the severed axon the axoplasmic electrical field is negligible and the membrane potential only slightly depolarized. These calculations assume a length constant of 5 mm, uniform membrane conductances, and an infinitely conductive external medium. These idealized conditions provide support for the notion that electrical disturbances to the axon caused by damage would be restricted to the local area of the lesion. (Reproduced from Borgens RB [1988]: Voltage gradients and ionic currents in injured and regenerating axons, Adv in Neurology 47:51–66, with permission of Raven Press, New York.)

Transections of the axon made very close to the soma might endanger the viability of the neuron in another manner as well. We have already discussed the fact that the Ca^{2+} component of the injury current may be involved in the destruction of the axon's cytoskeleton. If the proximal axon segment is short enough, the Ca^{2+}-mediated degradation of the cytosol and its architecture might reach proximally to compromise the cell body cytoplasm. Not surprisingly, similar explanations can be forwarded for the neuron's response to transection of dendrites. Using laser microbeam cell surgery, Lucas and colleagues (1985) found that the severance of dendrites (average diameter ~ 3 μm) 50 μm, 100 μm, and 150 μm from the cell body (cultured mouse spinal neurons) were associated with 30%, 53%, and 70% probabilities of neuronal survival, respectively. At distances over 30 μm from the cell body, there was little probability of cell death.

Even though severance of the axon close to the soma may induce more striking degenerative responses, indeed even cell death, under certain circumstances, such transections may also (and curiously!) produce *growth* responses. These growth-promoting effects can be seen not only in axons (Hayton and Kellerth, 1987) but in dendrites as well (Hall and Cohen, 1983; Linda et al., 1985). Hall and Cohen (1983) have reported that in lamprey giant reticulospinal neurons, transections of axons close to the cell body can induce extensive sprouting and growth from the dendritic compartment. What is even more interesting is that the total amount of linear growth appears to be regulated within narrow limits by the neuron. That is, the amount of growth was the same whether it came from the axon, the dendrite, or a *combination of these* (Hall and Cohen, 1983). Moreover, the character of the dendritic sprouting and growth was very "axon-like." The fibers projected away from the dendritic tree, staying well outside the dendritic compartment, and grew linearly in the rostrocaudal direction for long distances. A similar phenomenon has been observed in the mammalian CNS: "dendraxons" form in response to a transection of the intramedullary portion of motor axons in the spinal cord of the adult cat (Linda et al., 1985). I have suggested that endogenous currents entering the ends of growing nerves may be involved (or may even precipitate) their development. Perhaps currents that should normally leak into the tip of the proximal axon segment might be driven into the dendritic compartment as well, if the axon stub is close enough to the enveloping dendritic tree. The dendrites are known to undergo marked physiological changes after axotomy (as does the rest of the neuron). They are replete with synaptic sites and usually possess a lower-input impedance than axonal membrane (Llinas and Sugimori, 1979). It may be that the injury current could converge on these dendritic processes as well. It would be interesting to see if one could measure steady current entering the intact dendrites as it usually does at the cut surface of the short axon segment. Such a circuit would constitute a true transcellular current.

Having just introduced the notion of a transneuronal current (traversing the entire cell) after injury, it may be instructive to consider under what conditions this may occur in injured axons. As discussed above, such a transcellular current is only plausible if the axon is severed very close to the cell body. A transneuronal current is less likely to exist if the axon is severed at some distance from the soma. In a previous paragraph we have estimated that currents and voltages associated with injury would be quite localized to the axon tip. This suggestion holds *if* the pump sites driving the current are localized to the axonal membrane (our present understanding). If, however, pump sites are localized in the somatic compartment or the dendritic compartment and the leak is the axonal injury, then a transneuronal current becomes a real possibility. All that is needed to support an endogenous current through the entire cell is to have the ionic pumps and leaks (or persistently low-resistance channels in the anatomically intact axolemma) separated to opposite poles of the cell. Such "bifacial cells" (DeLoof, 1985) are not uncommon—many epidermal cells have been demonstrated to segregate "pump and leak" to separate poles. This segregation may not only subserve certain physiological duties (such as undirectional movement of Na^+ across epithelia) but may subserve an architectural polarization as well (Almers and Stirling, 1984). We have no knowledge of specific pumps highly restricted to various areas of the neuron. We do have information concerning the segregation of specific leak sites, however. Voltage-gated Ca^{2+} channels are numerous in some growth cone membranes and greatly restricted in number in the axonal membrane (Anglister et al., 1982), yet they are also plentiful in the dendritic compartment (Llinas and Sugimori, 1979). Perhaps the neuron may have the machinery for such a separation of both pump and leak. Nerves would then theoretically have the capacity to drive a transneuronal current supporting a transneuronal voltage gradient. Such an endogenous current, initiated at the instant of injury to the axon, could provide the signal (Cragg, 1970) for injury that spreads throughout the entire neuron.

DEVELOPMENTAL CURRENTS IN SINGLE NERVES

We began this chapter with the notion that the growing ends of neurons may be similar in several ways to other apically growing cellular systems, both in plants and animals. What many unicellular systems (that produce a local process or fiber) have in common is that growth is usually initiated—even predicted—by an endogenous current entering the presumptive growth point (or the tip of the growing process) (Borgens et al., 1979; Jaffe, 1977, 1980a,b, 1982). There is no information on this point regarding the initiation of axonal outgrowth. The so-called ''injury'' current that enters the tip of a recently severed axon, however, might be eligible for this list. We have mainly restricted this discussion to the ways this endogenous electricity is involved with the immediate consequence of injury—retrograde axonal degeneration. Why not focus on subsequent axonal growth? The major impediment is that we lack a rigorous investigation of endogenous currents in nerve growth cones. We note one claimed measurement of minute Ca^{2+} currents entering the ends of elongating neurites in culture (Freeman et al., 1985). No presentation of the quantitative data or physiological records of any individual measurement accompanied this report; thus it should be considered as incomplete until full documentation is made available. On the basis of the rationale already presented, it is highly probable that such growth cone currents exist, however. Some growth cones are rich in voltage-regulated Ca^{2+} channels (Anglister et al., 1982); there is evidence of elevated concentrations of intracellular Ca^{2+}—perhaps Ca^{2+} gradients (Connor, 1986); Na^{+} and Ca^{2+} currents have already been reported in axotomized neurons (Borgens et al., 1980), and such tip currents (induced by axotomy) continued to enter the end of the axons for many days *after* a new limiting membrane has formed (Borgens et al., 1980). All of this should be tempered by recent observations, using optical techniques, in which Ca^{2+} transients were seen in the soma and initial segment, but *not* in the neurite or growth cone of electrically active leech Retzius neurons, grown on concanavalin A. More interestingly still, when the same cell type was grown on laminin, Ca^{2+} transients were measurable at both neurite and growth cone levels (Ross et al., 1988). This work raises the important concept that the nature of the substrate on which a nerve is growing may determine the activity, or even the very presence of some voltage-gated channels in certain localized regions of the nerve. We have not wished to overemphasize the Ca^{2+} ion as the sole regulator of growth activity within the neuron, although it is certainly the most interesting from the standpoint of endogenous charge movement. Ca^{2+} is suggested to play a crucial role as a regulator of cytoplasmic fluidity (Abercrombie and Hart, 1986) and membrane addition in growing neurons (Llinas, 1979) and in cell motility in a number of cellular systems including nerves (Nuccitelli et al., 1977; Porter, 1976); it also controls the stability of the cytoarchitecture (Lasek and Hoffman, 1976). Recently, direct measurements of elevated $[Ca^{2+}]$ have been made in cultured rat diencephalon cells using the fluorescent probe Fura-2 (Grynkiewicz et al., 1985) and digital imaging techniques (Connor, 1986). Intracellular free calcium was found to be high (about 500 nM) in growth cones in elongating neurites. Fibers that had ceased growth were observed to have levels of about 30–70 nM (perhaps below resting levels?) in their terminal ends. Thus a focused high cytosolic [Ca^{2+}] is correlated with growth cone advancement (Connor, 1986). Indirect evidence for high levels of Ca^{2+} in the cytoplasm of the growth cone is based on the observation that there are no neurofilaments in growth cones. It is suggested that this is the result of the presence of a Ca^{2+}-activated protease that can degrade these filaments. Since the filaments are found throughout the axon up to the base of the growth cone (but not within it), axoplasmic [Ca^{2+}] is inferred to be greatly reduced when compared with the cytoplasm of the growth cone proper (Lasek and Hoffman, 1976). If leupeptine (an agent that inhibits the Ca^{2+}-activated protease) is injected into neurons, then accumulations of neurofilaments are observed within the axon terminal (Roots, 1983). Ca^{2+} has been implicated in the orientation response that nerves show to a gradient of nerve growth factor (NGF) (Gunderson and Barrett, 1980). The turning behavior of developing nerves toward the cathode of an applied field may also depend on local Ca^{2+} levels on either side of the growth cone (McCaig, 1989). Nerve galvonotropism is a Ca^{2+}-dependent event, being inhibited by the Ca^{2+} channel blocker cobalt, but not by the organic blockers diltiazem and nifedipine (Fig. 8). The involvement of Ca^{2+} in nerve orientation is also suggested by experiments in which reversed galvanotaxis occurs under conditions expected to alter the balance of Ca^{2+} entry on either side of the growth cone. Nerves simultaneously exposed to cobalt and to the Ca^{2+} ionophore A23187 turned toward the anode (Fig. 9), (McCaig, 1989). Under certain circumstances, A23187 is negatively charged and may be incorporated into the membrane at the growth cone, predominantly anodally. This process could lead to a local influx of Ca^{2+} on the anodal side and thus to the local activation of gel-to-sol conversion of cytoplasm, actin-based contractile processes, and anodal incorporation of new membrane vesicles, culminating in turning toward the anode. Such an explanation implies that in the normal situation, nerves turn toward the cathode because their endogenous Ca^{2+} channels either accumulate cathodally by electrophoresis/electroosmosis, or are locally activated cathodally because of electric field-induced membrane depolarization (Cooper and Schliwa, 1986).

There is growing evidence that Ca^{2+} channels may be distributed highly asymmetrically even within the confines of the neuronal soma and that in some regions, channels may be aggregated into "hot spots," with variable properties of inactivation (Thompson and Coombs, 1988), all of which means that *local* levels of Ca^{2+} in the one neuron may be regulated very carefully for specifically localized physiological reasons. It should be noted that the involvement of extracellular Ca^{2+} in nerve galvanotaxis is an

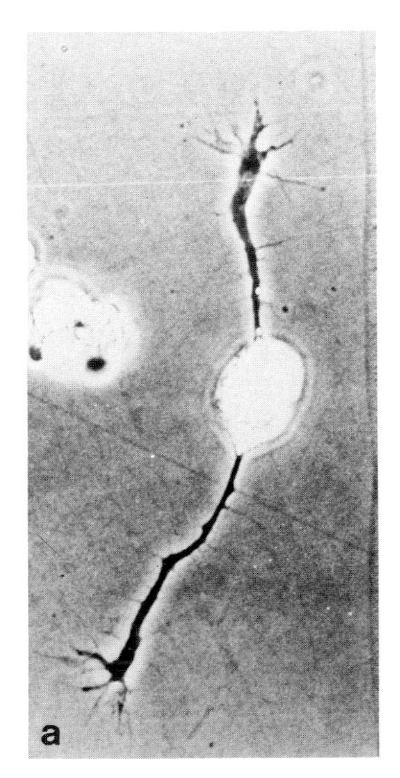

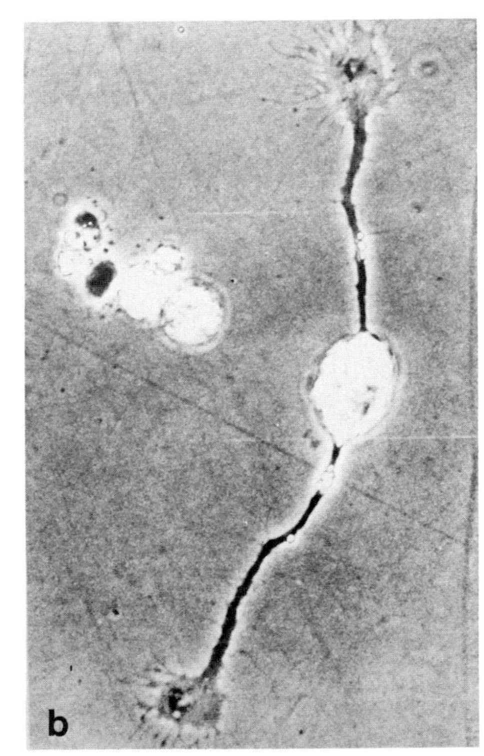

Fig. 8. Lack of response of *Xenopus* embryonic neurites to the electric field in the presence of Co^{2+}. (Compare with Figure 2, Borgens, Chapter 4, this volume.) **a:** Control nerve. **b:** Same nerve after 14 hours in Co^{2+} (5 mM/L) and exposed to 110 mV/mm. Neurites have elongated very slowly without orienting. The growth cones appear expanded and firmly adherent to the dish. Cell diameter is roughly 20 μm. Cathode is at left.

unsettled issue, since it is known that nerve growth toward a cathode continues in essentially Ca^{2+}-free media (Robinson and Muncy, 1986).

In addressing these issues, one must be careful not to confuse intracellular and extracellular requirements for Ca^{2+}. Neurons can develop in a media very low in Ca^{2+}, suggesting that Ca^{2+} in the external media may not be the sole or necessary source supporting growth and/or elongation (Bixby and Spitzer, 1984; Robinson and Muncy, 1986). Intracellular Ca^{2+} stores may be more than sufficient to either support growth or the stability of the intracellular architecture. For example, mitochondria serve as a calcium reservoir [im-

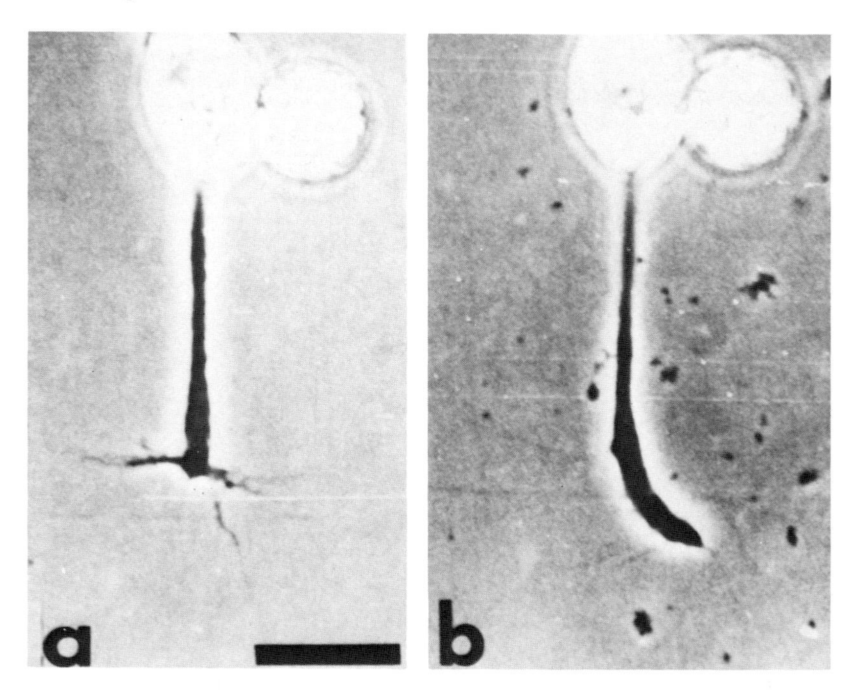

Fig. 9. Nerve response to an electric field in the presence of calcium ionophore and Co^{2+}. **a:** Control nerve. **b:** Same nerve after 84 minutes in A23187 (15 μM/L), plus Co^{2+} (5 mM), plus 110 mV/mm. Nerve has elongated and oriented toward the *anode* at right. Scale bar = 20 μm. (Reproduced from McCaig, 1989, with permission of the publishers.)

bibing free Ca^{2+} or releasing it (Carafoli and Crompton, 1976)]. One might reasonably suggest that the elevated numbers of these organelles in the terminals of interrupted fibers may subserve this purpose as well as increase the capacity for energy-dependent anabolic processes associated with membrane addition. Presently there are two emerging schools of thought interpreting the actual role of intracellular Ca^{2+} in growing neurites. One view holds that intracellular [Ca^{2+}] is elevated over resting levels at the nerve tip when growth is *terminated* or *temporarily inhibited* (Lasek and Hoffman, 1976). This notion is based on the abundant evidence of the Ca^{2+}-mediated disassembly of cytoskeletal proteins and the uncoupling of transport processes. It is also implied by the inhibition of a cycle of growth after AP propagation—known to increase Ca^{2+} in the presynaptic terminal as well as the growth cone (Cohan and Kater, 1986; Kater et al., 1988). The other view is that elevated [Ca^{2+}] *precipitates growth* (Anglister et al., 1982; Hyman and Pfenninger, 1985;

Llinas, 1979) and amoeboid filopodial extension (Nuccitelli et al., 1977) and is necessary for neuronal membrane insertion (Yawo and Kuno, 1985).

Different aspects of neuronal activity may even be controlled by different levels of intracellular Ca^{2+} (Mattson and Kater, 1987; Kater et al., 1988). Both the activity of growth cone filopodia and nerve elongation per se were suppressed by induced increases in Ca^{2+} entry (e.g., A23187). However, conditions that partially reduced Ca^{2+} entry (low levels of channel blockers or low Ca^{2+} media) suppressed growth cone motility but increased rates of neurite elongation. These authors explain their observations by suggesting that internal Ca^{2+} is optimal for growth cone motility in tissue culture, but above optimal levels for nerve elongation.

This suggestion is consistent with the view that the suppression of growth, resorption of nerve processes, and elongation *and* degeneration of nerve fibers is finely tuned to the *balance* of Ca^{2+} within the axon terminal, which is probably exquisitely tuned to within a narrow range of concentration. A more global view of neurogenesis suggests that it is difficult to separate growth and degenerative mechanisms in producing the net patterning of the nervous system (Cowan et al., 1984). It is certain that the calcium ion may play a pivotal role in regulating these processes. Although the relationship of the Ca^{2+} component of endogenous currents in nerve has been stressed, this is not meant to exclude the possible role of other ions (minor components of the current) in growth processes. We simply have little information with which to make informed speculations about their putative roles. Moreover, we do not wish to unduly stress the *ionically* mediated events associated with the natural current entering the proximal stump of a growing or regenerating neurite. There would certainly be *voltage*-mediated effects associated with the current flow as well.

One such voltage-mediated effect would probably be the ''self-electrophoresis'' (Jaffe, 1977; Poo and Robinson, 1977) and/or ''electroosmosis'' (McLaughlin and Poo, 1981) of macromolecules that are inserted into the plasma membrane and that possess charged components protruding from the axolemma into the *extracellular* space (perhaps into the *cytosol* as well) (Jaffe, 1977). The extracellular and intracellular fields associated with an endogenous current flow (or a voltage gradient imposed across the cell by the experimenter or within the embryo by another adjacent cellular current source) may well influence movement of these molecules toward or away from the axon terminal. Macromolecules (which may be driven within the plane of the membrane) important to nerve development might be various *receptors* to a variety of extracellular matrix components. These components are now recognized to be important to nerve motility, perhaps even inducers of nerve development. Neuronal pathfinding occurs in part through recognition of substrates of various adhesiveness or chemical affinity such as laminin or fibronectin (Letourneau, 1975; Sanes, 1985) or to specific molecular components of other axonal membranes that may provide a template for directed growth (Goodman et al., 1985; Kuwada, 1986). This process of "recognition" is mediated by receptors residing within the membrane (and amenable to redistribution). Receptors important to development, such as NGF or similar morphogens active in the CNS (Varon et al., 1988), may likewise be aggregated at nerve terminals by endogenous fields. Is there any evidence to support these notions? Various receptor-macromolecule complexes have already been demonstrated to move in the presence of an applied electric field in nerve and muscle cells in culture (Orida and Poo, 1978; Poo and Robinson, 1977). Altogether, it is certain that endogenous currents and voltages can affect the organization of macromolecules (and probably organelles), the ionic composition of the cytosol, and the overall architecture of the cytoskeleton. To what extent these proposed mechanisms are involved in nerve repair and elongation is still in need of experimental clarification and remains a fertile area for investigation.

ENDOGENOUS FIELDS IN THE EMBRYO

At the beginning of this chapter, we suggested that steady electrical fields may be one control (of several) affecting the patterning of the early embryo. Here we will consider the patterning of the nervous system proper.

It is clear from amphibian studies that steady currents are driven through the entire embryo at various times and places (Borgens et al., 1983, 1985; Borgens, 1989; Robinson, 1983). One of the earliest detectable fields is one associated with a large current leak at the blastopore, with smaller densities entering the rest of the *Xenopus* embryo (Robinson and Stump, 1984; Borgens, 1989). Outcurrents are also associated with the development and closure of the neural folds (Figs. 10 and 11). However, outcurrents are not predictive of otic placode development or the formation of the optic primordium (Borgens, unpublished measurements). These developing structures are characterized by a local proliferation of cells after an inductive event—their early development is not associated with a large immigration of cells into these regions. We suppose, then, that endogenous fields may be pertinent to developing structures associated with *migration* or *dispersion* of cells during embryogenesis. The closure of the neural folds (forming the neural tube) is coincident with a dispersion of cells (neural crest) from this general area (LeDouarin, 1985; Thiery et al., 1985) and the production of poineer neurites (projecting laterally to rudimentary targets). It is also known that a dissolu-

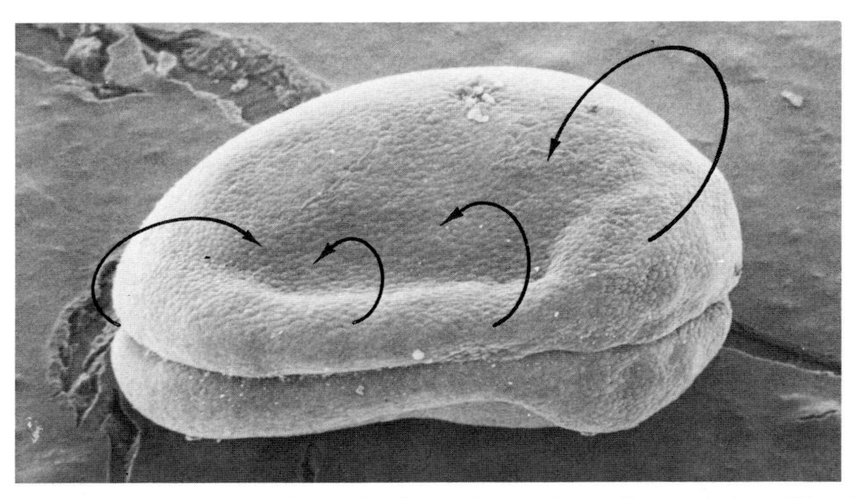

Fig. 10. Scanning electron micrograph of an early neural stage *Xenopus* embryo. (Neural folds have yet to completely close forming the neural tube.) The directions of currents leaving the neural folds are depicted by arrows at the cranial enlargement and entering most of the rest of the body of the embryo. A large current leak is also present at the blastopore (caudally located and not in this view). The magnitude of endogenous currents measured leaving the area of neural development peaked on the order of 2–4 $\mu A/cm^2$. (Borgens, unpublished measurements.)

tion of tight junctions of the neuroepithelium of the neural folds is coincident with neurulation (Decker, 1981), which may be expected, since such cells are loosening these connections prior to invagination. This fact may explain the leakage of current *out* of the lateral walls of the neural folds in stage 15–18 *Xenopus*. Current is also known to enter the floor of the neural plate prior to closure at stage 19 (Robinson and Stump, 1984). Thus a current loop exists *within the embryo* where the lateral domains are negative with respect to the midline in addition to more caudal regions (since current leaves the blastophore as well). However, these externally detected, outwardly directed neural fold currents disappear by stage 19, coincident with closure (Fig. 11). It may be possible that the neuroepithelium (now invaginated) is still generating current. Such a notion is supported by *Xenopus* ''half-embryo'' measurements, in which an endogenous wound current leaving the transected embryo is observed to be markedly reduced when measurements are made over the bifurcated and exposed base of the neural tube (Rajnicek, 1988). A current entering the base of the neural tube (driven by its own inwardly

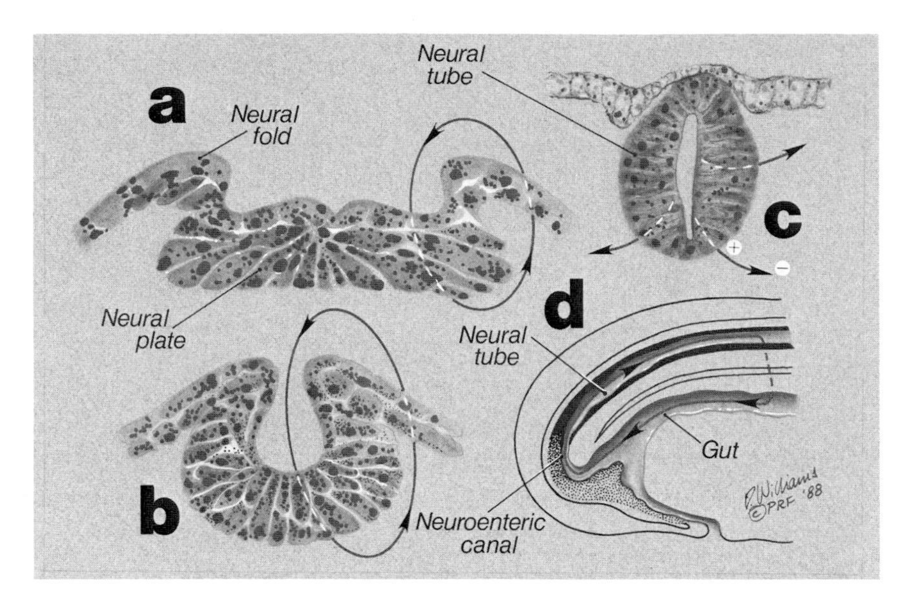

Fig. 11. Endogenous current through *Xenopus* neurulae. **a,b:** An inward ionic current, driven by the integument, leaks out of a low-resistance pathway provided by uncoupled epidermal cells of the lateral walls of the neural folds. **c:** The external component of this flow of charge (as measured with the vibrating electrode) disappears with closure of the neural folds (forming the neural tube). **d:** We infer that a significant portion of the current completing this circuit returns to the lumen of the neural tube via the archenteron and neuroenteric canal after closure of the neural folds. In all of the above, the endogenous extracellular electrical fields associated with this flow of charge would be laterally negative with respect to regions near the neural tube.

negative potential difference) would be expected to reduce the net outward current (driven by the embryonic epidermis) leaving the open wound. If there still exists an extracellular current driven through the embryo by the neural tube, it may have a role in the establishment of gross pattern. This hypothesized voltage gradient within the embryo may both operate as an inductive agent and provide a coarse vector for the dispersion of neural crest cells [which begin their migration after the closure of the neural fold at stage 19 (LeDouarin, 1985; Nieuwkoop and Faber, 1967)]. Interestingly, neural crest cells are known to migrate toward the cathode in an applied electric field (Stump and Robinson, 1983; Cooper and Keller, 1984). Lateral voltage gradients, negative at points distant from the neural tube, may likewise provide a coarse guide for the projection of neurites. It is interesting in this regard to consider the differing behavior of embryonic amphibian neuroblasts and myoblasts. As will be discussed in more detail in Chapter 4 (Borgens, this volume), neurites align themselves parallel with the long axis of an applied electric field (Hinkle et al., 1981; Patel and Poo, 1982). Myoblasts (originally an accidental contaminant of these cultures) develop a bipolar axis of symmetry *perpendicular* to the long axis of the field, cellular contacts even being formed between this primitive nerve and muscle (Hinkle et al., 1981) (see Fig. 2, Borgens, Chapter 4, this volume). This symmetry mirrors that which is seen in the developing embryo. Thus the "*morphogenetic* field," as described by the classical embryologists, may in fact have a physiological basis as an *electric* field.

If we are to assume that endogenous electrical fields are involved in the patterning of the nervous system, it is prudent to consider certain limitations. Only the grossest vectorial information can be provided by a field in a volume conductor (such as an embryo), i.e., rostrocaudal or mediolateral. Fields lack the directional specificity to control highly specific target cell-neuron contacts at distances over a few tens of micrometers. However, the final honing of a growth cone to a specific cell—or portion of a cell—when the gap between them is very small may be influenced by electrical fields. For example, steady currents are driven through the developing neuromuscular junction of muscle cells (Betz, 1986). Local electrical fields, confined to this extracellular environs (and this is the only case to have been examined experimentally), could possibly be involved in the establishment of synaptic connection. We have briefly outlined other controls of the formation of specific nerve connections in the introduction to this chapter. It is possible that they are all components of a multifactorial system governing neurogenesis, neuronal path finding, and synaptogenesis.

SUMMARY

1. Axotomy results in an immediate flow of ionic current into the locus of damage. In lamprey reticulospinal axons, current densities are immediately very large ($\sim$ 100–1000 $\mu A/cm^2$) but rapidly fall to stable levels of about 5 $\mu a/cm^2$. This plateau is persistent. Current continues to enter the end of a severed axon for days after the injury.

2. The ionic component of this injury current is mainly Na^+ and Ca^{2+}. The net effect of the current is to raise intracellular $[Ca^{2+}]$, thus implying that the early phase of endogenous charge movement after injury is involved in the process of retrograde axonal degeneration.

3. An endogenous current probably enters the growing or regenerating tip of a neurite. Indirect evidence suggests that this current may have a large

Ca^{2+} component, may lead to a Ca^{2+} gradient within the axoplasm of the growing tip, and may be implicated as a control of growth and elongation.

4. The developing nervous system is derived from an epithelial syncytium. This epithelium drives a steady current through itself and the whole embryo. A voltage gradient within the embryo may perhaps both serve as an initiator of neural crest cell dispersion and provide gross information for the establishment of neurite projection and a rough template for the geometric relationships to other forming tissues.

REFERENCES

Abercrombie RF and Hart CE (1986) Calcium and proton buffering and diffusion in isolated cytoplasm from myxicola axons. Am. J. Physiol. 250:C391–C405.

Almers W and Stirling C (1984) Distribution of transport proteins over animal cell membranes. J. Membr. Biol. 77, 169:185–186.

Anglister L, Farber IC, Shahar A, and Grinvald A (1982) Localization of voltage-sensitive calcium channels along developing neurites: Their possible role in regulating neurite elongation. Dev. Biol. 94:351–365.Baker PF (1972) Transport and metabolism of calcium ions in nerve. Prog. Biophys. Mol. Biol. 24:177–223.

Barron KD (1983) Comparative observations on the cytologic reactions of central and peripheral nerve cells to axotomy. In Kao CC, Bunge RP, and Reier PJ (eds): *Spinal Cord Reconstruction.* Raven Press, New York, pp. 7–40.

Becker RO and Marino AA (1982) *Electromagnetism and Life.* State University of New York Press, Albany.

Becker RO and Seldon G (1985) *The Body Electric: Electromagnetism and the Foundation of Life.* William Morrow and Company, New York.

Betz WJ, Caldwell JH, Harris GL, and Kinnamon SC (1986) A steady electric current at the rat *neuromuscular synapse.* In Nuccitelli R (ed.): *Ionic Currents in Development.* Alan R. Liss, Inc., New York, pp. 205–212.

Bixby JL and NC Spitzer (1984) Early differentiation of vertebrate spinal neurons in the absence of voltage-dependant Ca^{2+} and Na^{+} influx. Dev. Biol. 106:89–96.

Blight AR (1983) Axonal physiology of chronic spinal cord injury in the cat: Intracellular recording *in vitro.* Neuroscience 10:521–543.

Blight AR (1985) Computer simulation of action potentials and after potentials in mammalian myelinated axons: The case for a lower resistance myelin sheath. Neuroscience 15:13–31.

Borgens RB (1982) What is the role of naturally produced electric current in vertebrate regeneration and healing? Int. Rev. Ctyol. 76:245–298.

Borgens RB (1984) Are limb regeneration and limb development both initiated by an integumentary wound? Differentiation 28:87–93.

Borgens RB (1985) Natural voltage gradients and the generation and regeneration of limbs. In Sicard RE (ed.): *Regulation of Vertebrate Limb Regeneration.* Oxford University Press, New York, pp. 6–31.

Borgens RB (1988) Voltage gradients and ionic currents in injured and regenerating axons. Adv in Neurology 47:51–66.

Borgens RB, Vanable JW Jr., and Jaffe LF (1979) Bioelectricity and regeneration. Bioscience 29:468–474.

Borgens RB, Cohen MJ, and Jaffe LF (1980). Large and persistent electrical currents enter the transected lamprey spinal cord. Proc. Natl. Acad. Sci. U.S.A. 77:1209–1213.

Borgens RB, Rouleau MF, and DeLanney LE (1983) A steady efflux of ionic current predicts hind limb development in the axolotl. J. Exp. Zool. 228:491–503.

Burr HS (1932) An electro-dynamic theory of development. J. Comp. Neurol. 56:347–371.

Burr HS and Northrop FSC (1935) The electro-dynamic theory of life. Qu. Rev. Biol. 10:322–333.

Carafoli E and Crompton M (1976) Calcium ions and mitochondria. In Duncan CJ (ed.): *Symposium of the Society for Experimental Biology: Calcium and Biological Systems*, Vol. 30. Cambridge University Press, New York and London, pp. 89–115.

Chen TH and Jaffe LF (1978) Effects of membrane potential on calcium fluxes of *Pelvetia* eggs. Planta 140:63–67.

Cohan CS and Kater SB (1986) Suppression of neurite elongation and growth cone motility by electrical activity. Science 232:1638–1640.

Connor JA (1986) Digital imaging of free calcium changes and of spatial gradients in growing processes in single, mammalian central nervous system cells. Proc. Natl. Acad. Sci. U.S.A. 83:6179–7183.

Cooper MS and Keller RE (1984) Perpendicular orientation and directional migration of amphibian neural crest cells in DC electrical fields. Proc. Natl. Acad. Sci. U.S.A. 81:160–164.

Cooper MS and Schliwa M (1986) Motility of cultured fish epidermal cells in the presence and absence of direct current electric fields. J. Cell. Biol. 102:1384–1399.

Cowan SL (1934) The action of potassium and other ions and the injury potential and action current in nerve. Proc. R. Soc. Lond. [Biol] B115:216–260.

Cowan WM, Fawcett JW, O'Leary DDM, and Stanfield BB (1984) Regressive events in neurogenesis. Science 225:1250–1255.

Cragg BB (1970) What is the signal for chromatolysis? Brain Res. 23:1–21.

Decker RS (1981) Disassembly of the zonula occludens during amphibian neurulation. Dev. Biol. 81:12–22.

De Loof A (1985) Minireview: The cell as a miniature electrophoresis chamber. Comp. Biochem. Physiol. 80A:453–459.

DuBois-Reymond E (1843) Vorlaufiger abrifs einer untersuchung uber den sogenannten froschstrom und uber die electromotorischen fische. Ann. Phys. U. Chem. 58:1–48.

Edelman GM (1985) Molecular regulation of neural morphogenesis. In Edelman GM, Gall WE, and Cowan WM (eds.): *Molecular Regulation of Neural Morphogenesis*. John Wiley & Sons, New York, pp. 35–59.

Eisen JS, Myers PZ, and Westerfield M (1986) Pathway selection by growth cones of identified motoneurones in live zebra fish embryos. Nature 320:269–272.

Freeman JA, Manis PB, Snipes GJ, Mayes GN, Samson PC, Wikswo JP Jr., and Freeman DP (1985) Steady growth cone currents revealed by a novel circularly vibrating probe: A possible mechanism underlying neurite growth. J. Neurosci. Res. 13:257–283.

Friede RL (1964) Axon swelling produced in vivo in isolated segments of nerves. Acta Neuropathol. 3:229–237.

Goodman CS, Bastiani MJ, Raper JA, and Thomas JB (1985) Cell recognition during neuronal development in grasshopper and *Drosophila*. In Edelman GM, Gall WE, and Cowan WM (eds.): *Molecular Bases of Neural Development*. John Wiley & Sons, New York, pp. 295–316.

Grafstein B (1975) The nerve cell body response to axotomy. Exp. Neurol. 48:32–51.

Grynkiewicz G, Poenie M, and Tsien RY (1985) A new generation of Ca^{2+} indicators with greatly improved fluorescence properties. J. Biol. Chem. 260:3440–3450.

Gundersen RW and Barrett JN (1980) Characterization of the turning response of dorsal root neurites toward nerve growth factor. J. Cell. Biol. 87:546–554.

Hall GF and Cohen MJ (1983) Extensive dendritic sprouting induced by close axotomy of central neurons in the lamprey. Science 222:518–520.

Harold FM (1986) Transcellular ion currents in tip-growing organisms: Where are they taking us? In Nuccitelli R (ed.): *Ionic Currents in Development*. Alan R. Liss, Inc., New York, pp. 359–366.

Harris WA (1986) Homing behaviour of axons in the embryonic vertebrate brain. Nature 320:266–268.

Hayton L and Kellerth J-O (1987) Regeneration by supernumerary axons with synaptic terminals in spinal motoneurons of cats. Nature 325:711–714.

Hinkle L, McCaig CD, and Robinson KR (1981) The direction of growth of differentiating neurons and myoblasts from frog embryos in an applied electric field. J. Physiol. 314:121–131.

Hyman C and Pfenninger KH (1985) Intracellular regulators of neuronal sprouting: Calmodulin-binding proteins of nerve growth cones. J. Cell Biol. 101:1153–1160.

Jaffe LF (1977) Electrophoresis along cell membranes. Nature 265:600–602.

Jaffe LF (1979) Control of development by ionic currents. In Cone RA and Dowling JE (eds.): *Membrane Transduction Mechanisms*. Raven Press, New York, pp. 199–231.

Jaffe LF (1980a) Calcium explosions as triggers of development. Ann. N.Y. Acad. Sci. 339:86–101.

Jaffe LF (1980b) Control of plant development by steady ionic currents. In Spanswick RM, Lucas WJ, and Dainty J (eds.): *Plant Membrane Transport: Current Conceptual Issues*. Elsevier/North Holland/Biomedical Press, Amsterdam, pp. 381–388.

Jaffe LF (1981) The role of ionic currents in establishing developmental pattern. Philos. Trans. R. Soc. Lond. [Biol.] B295:553–566.

Jaffe LF (1982) Developmental currents, voltages, and gradients. In Subtelny S and Green PB (eds.): *Developmental Order: Its Origin and Regulation*. Alan R. Liss, Inc., New York, pp. 183–215.

Jaffe LF and Nuccitelli R (1974) An ultrasensitive vibrating probe for measuring steady extracellular currents. J. Cell. Biol. 63:614–628.

Jaffe LF and Nuccitelli R (1977) Electrical controls of development. Annu. Rev. Biophys. Bioeng. 6:445–476.

Jaffe LF, Robinson KR, and Nuccitelli R (1974) Local cation entry and self-electrophoresis as an intracellular localization mechanism. Ann. N.Y. Acad. Sci. 238:372–389.

Kao CC, Wrathall JR, and Kyoshimov K (1983) Axonal reaction to transection. In Kao CC, Bunge RP, and Reier PJ (eds.): *Spinal Cord Reconstruction*. Raven Press, New York, pp. 41–57.

Kater SB, Mattson MP, Cohan C, and Connor J (1988) Calcium regulation of the neuronal growth cone. Trends Neurosci. 11:315–321.

Kelly James P (1981) Reaction of neurons to injury. In Kandel ER and Schwartz JH (eds.): *Principles of Neural Science*. Elsevier/North Holland, Amsterdam.

Kuwada JY (1986) Cell recognition by neuronal growth cones in a simple vertebrate embryo. Science 233:740–747.

Lasek RJ and Hoffman PN (1976) The neuronal cytoskeleton, axonal transport, and axonal growth. In Goldman R, Pollard T, and Rosenbaum J (eds.): *Cell Motility*. Cold Spring Harbor Laboratory, Cold Spring Harbor, NY, pp. 1021–1049.

Le Douarin NM (1985) *In vivo* and *in vitro* analysis of the differentiation of the peripheral nervous system in the avain embryo. In Edelman GM, Gall WE, and Cowan WM (eds.): *Molecular Bases of Neural Development*. John Wiley & Sons, New York, p. 163.

Letourneau PC (1975) Possible roles for cell-to-substratum adhesion in neural morphogenesis. Dev. Biol. 44:77–91.

Letourneau PC (1985) Axonal growth and guidance. In Edelman GM, Gall WE, and Cowan WM (eds.): *Molecular Bases of Neural Development*. John Wiley & Sons, New York, pp. 269–294.

Linda H, Risling M, and Cullheim S (1985) "Dendraxons" in regenerating motoneurons in the cat: Do dendrites generate new axons after central axotomy? Brain Res. 358: 329–340.

Llinas R (1979) The role of calcium in neuronal function. In Schmitt FO and Worden FG, (eds.): *The Neurosciences: Fourth Study Programs*. MIT Press, Cambridge, MA, pp. 555–571.

Llinas R and Sugimori M (1979) Calcium conductances in Purkinje cell dendrites: Their role in development and integration. Prog. Brain Res. 51:323–334.

Lorente de Nó R (1947a) Action currents in volume conductors. In *Studies From The Rockefeller Institute for Medical Research*, Vol. 131, Rockefeller University Press, New York.

Lorente de Nó R (1947b) Action currents in volume conductors. In *Studies From The Rockefeller Institute for Medical Research*, Vol. 132, Rockefeller University Press, New York.

Lucas JH, Gross GW, Emery DG, and Gardner CR (1985) Neuronal survival or death after dendrite transection close to the perikaryon: Correlation with electrophysiologic, morphologic and ultrastructural changes. Central Nervous System Trauma 2, pp. 231–255.

McCaig CD (1986a) Electric fields contact guidance and the direction of nerve growth. J. Embryol. Exp. Morphol. 94:245–255.

McCaig CD (1986b) Dynamic aspects of amphibian neurite growth and the effects of an applied electric field. J. Physiol. 375:55–69.

McCaig CD (1987) Spinal neurite reabsorption and regrowth *in vitro* depend on the polarity of an applied electric field. Development 100:31–41.

McCaig CD (1989) On the mechanism of nerve galvanotropism. Biol. Bull. (in press).

McCaig CD and Dover PJ (1989) On the mechanism of oriented myoblast differentiation in an applied electric field. Biol. Bull. (in press).

McLaughlin S and Poo M-M (1981) The role of electro-osmosis in the electric field-induced movement of charged macromolecules on the surfaces of cells. Biophys. J. 34:85–93.

Mattson MP and Kater SB (1987) Calcium regulation of neurite elongation and growth cone motility. J. Neurosci. 7:4034–4043.

Meiri H, Spina ME, and Parnas I (1981) Membrane conductance and action potential of a regenerating axonal tip. Science 211:709–711.

Nieuwkoop PD and Faber J (1967) *Normal Table of Xenopus laevis (Daudin),* ed. 2, North Holland, Amsterdam.

Nitsch L and Wollman SH (1980a) Suspension culture of separated follicles consisting of differentiated thyroid epithelial cells. Proc. Natl. Acad. Sci. U.S.A. 77:472–476.

Nitsch L and Wollman SH (1980b) Ultrastructure of intermediate stages in polarity reversal of thyroid epithelium in follicles in suspension culture. J. Cell. Biol. 86:875–880.

Nordenstrom BEW (1983) *Biologically Closed Electric Circuits.* Nordic Medical Publications, Stockholm.

Nuccitelli R (1978) Ooplasmic segregation and secretion in the Pelvetia egg is accompanied by a membrane-generated electrical current. Dev. Biol. 62:13–33.

Nuccitelli R, Poo M-M, and Jaffe LF (1977) Relations between ameboid movement and membrane-controlled electrical currents. J. Gen. Physiol. 69:743–760.

Ochs S, Worth RM, and Chan S (1977) Calcium requirement for exoplasmic transport in mammalian nerve. Nature 270:748–750.

Orida N and Poo M-M (1978) Electrophoretic movement and localization of acetylcholine receptors in the embryonic muscle cell membrane. Nature 275:31–35.

Patel N and Poo M-M (1982) Orientation of neurite growth by extracellular electric fields. J. Neurosci. 2:483–496.

Poo M-M and Robinson KR (1977) Electrophoresis of concanavalin A receptors along embryonic cell membrane. Nature 265:602–605.

Porter KR (1976) Motility in cells. In Goldman R, Pollard T and Rosenbaum J (eds.): *Cell Motility,* Vol. 3. Cold Spring Harbor Laboratory, Cold Spring Harbor, NY pp. 1–28.

Purves D and Lichtman JW (1985) The molecular basis of neuronal recognition. In *Principles of Neural Development.* Sinauer Associates, Inc., Sunderland, Massachusettes, pp. 251–270.

Rajnicek AM (1988) An endogenous sodium current mediates wound healing in *Xenopus* neurulae. Dev. Biol. 128:290–299.

Ramón y Cajal S (1928) *Degeneration and Regeneration in the Nervous System.* Hoffner, New York.

Rees RP and Reese TS (1981) New structural features of freeze-substituted neuritic growth cones. Neuroscience 6:247–254.

Robinson KR (1983) Endogneous electrical current leaves the limb and prelimb region of the Xenopus embryo. Dev. Biol. 97:203–211.

Robinson KR (1985) The responses of cells to electrical fields, a review. J. Cell Biol. 101:2023–2027.

Robinson KR and Jaffe LF (1975) Polarizing fucoid eggs drive a calcium current through themselves. Science 187:70–72.

Robinson KR and Cone R (1980) Polarization of fucoid eggs by a calcium ionophore gradient. Science 207:77–78.

Robinson KR and Stump RF (1984) Self-generated electrical currents through *Xenopus* neurulae. J. Physiol. 352:339–352.

Robinson KR and Muncy L (1986) Neurite growth and response to electrical fields in calcium-free medium. In Nuccitelli R (ed.): *Ionic Currents in Development.* Alan R. Liss, Inc., New York, pp. 279–284.

Roederer E, Goldberg NH, and Cohen MJ (1983) Modification of retrograde degeneration in transected spinal axons of the lamprey by applied DC current. J. Neurosci. 3:153–160.

Roots BI (1983) Neurofilament accumulation induced in synapses by leupeptin. Science 221:971–972.

Ross WN, Arechiga H, and Nicholls JG (1988) Influence of substrate on the distribution of calcium channels in identified leech neurons in culture. Proc. Natl. Acad. Sci. U.S.A. 85:4075–4078.

Rovainen CM (1967) Physiological and anatomical studies on large neurons of central nervous system of the sea lamprey *(Petromyzon marinus)*. I. Muller and Mauthner cells. J. Neurophysiol. 30:1000–1023.

Sanes JR (1985) Laminin for axonal guidance? Nature 315:714–715.

Schackman RW, Eddy EM, and Shapiro BM (1978) The acrosome reaction of strongylocentrotus purpuratus sperm. Dev. Biol. 65:483.

Schlaepfer WW (1974) Calcium-induced degeneration of axoplasm in isolated segments of rat peripheral nerve. Brain Res. 69:203–215.

Schlaepfer WW (1977) Structural alterations of peripheral nerve induced by the calcium ionophore A23187. Brain Res. 136:1–9.

Schlaepfer WW (1983) Neurofilaments and axonal cytoskeleton as determinants of stability and growth in regenerating axons. In Kao CC, Bunge RP, and Reier PJ (eds.): *Spinal Cord Reconstruction*. Raven Press, New York, pp. 59–73.

Schlaepfer WW and Bunge RP (1973) Effects of calcium ion concentration on the degeneration of amputated axons in culture. J. Cell Biol. 59:456–470.

Spitzer N (1981) Development of memberane properties in vertebrates. In *Trends in Neurosciences*. Elsevier/North Holland Biomedical Press, Amsterdam, pp. 169–172.

Spitzer NC and Lamborghini JE (1976) The development of the action poentital mechanism of amphibian neurons isolated in culture. Proc. Natl. Acad. Sci. U.S.A. 73: 1641–1645.

Stump RF and Robinson KR (1983) Xenopus neural crest cell migration in an applied electrical field. J. Cell Biol. 97:1226–1249.

Thiery J-P, Tucker GC, and Hirohiko A (1985) Gangliogenesis in the avian embryo: Migration and adhesin properties of neural crest cells. In Edelman GM, Gall WE, and Cowan WM (eds.): *Molecular Bases of Neural Development*. John Wiley & Sons, New York pp. 181–212.

Thompson S and Coombs J (1988) Spatial distribution of calcium currents in molluscan neuron cell bodies and regional differences in the strength of inactivation. J. Neurosci. 8:1929–1939.

Tweedle C (1971) Transneuronal effects in amphibian limb regeneration. J. Exp. Zool. 177:13–30.

Varon S, Manthorpe M, David GE, Williams LR, and Skaper SD (1988) Growth factors. In Waxman SG (ed.): *Advances in Neurology, Vol. 47: Functional Recovery in Neurological Disease*. Raven Press, New York, pp. 493–521.

Veraa RP, Grafstein B, and Ross RA (1979) Cellular mechanisms in axonal growth. Exp. Neurol. 64:649–698.

Voss J, Kuriyama K, and Roberts E (1968) Electrophoretic mobilities of brain subcellular particles and binding of γ-aminobutyric acid, acetylcholine, norephinephrine, and 5-hydroxytryptomine. Brain Res. 9:224–234.

Warner AE (1985) Factors controlling the early development of the nervous system. In

Edelman GM, Gall WE, and Cowan WM (eds.): *Molecular Bases of Neural Development.* John Wiley & Sons, New York, pp. 11–34.

Watson WE (1976) Responses to injury. In *Cell Biology of Brain.* John Wiley & Sons, New York, pp. 201–272.

Wood MR and Cohen MJ (1979) Synaptic regeneration in identified neurons of the lamprey spinal cord. Science 206:344–347.

Wood MR and Cohen MJ (1981) Synaptic regeneration and glial reactions in the transected spinal cord of the lampry. J. Neurocytol. 10:57–79.

Woodruff RI and Telfer WH (1973) Polarized intercellular bridges in ovarian follicles of the *Cecropia* moth. J. Cell. Biol. 58:172–202.

Woodruff RI and Telfer WH (1980) Electrophoresis of proteins in intercellular bridges. Nature, 286:84–86.

Yawo H and Kuno M (1985) Calcium dependence of membrane sealing at the cut end of the cockroach giant axon. J. Neurosci. 5:1626–1645.

Zelena J (1969) Bidirectional shift of mitochondria in axons after injury. In Barondes S (ed.): *Cellular Dynamics of the Neuron.* Symp. Int. Soc. Cell Biol. Vol. 8, pp. 73–94.

Zelena J, Lubinska L, and Gutmann E (1968) Accumulation of organelles at the ends of interrupted axons. Z Zellforsch. Mikrosk. Anat. 91:200–219.

Electric Fields in Vertebrate Repair, pages 117–170

CHAPTER 4

Artificially Controlling Axonal Regeneration and Development by Applied Electric Fields

Richard B. Borgens

Center for Paralysis Research, Department of Anatomy, School of Veterinary Medicine, Purdue University, West Lafayette, Indiana 47907

INTRODUCTION

What cues initiate the growth or regeneration of neurons? What signals provide developing neurites information concerning their direction of growth and the specificity of their targets? These questions have been central to understanding not only the development of the nervous system in particular but of the developing embryo in general. In a previous chapter we have suggested that the development of certain tissues during organogenesis (those that form from condensations of migratory cells) may in part be determined by a subepidermal voltage gradient at precise locations in the embryo. This gradient may also provide a coarse guide for neurites that innervate these areas (Borgens, 1984). Circumstantial support for this last notion comes from studies of the responses of both neurons and other cells in culture to exogenously applied electric fields. As discussed in the preceding chapter, we do not suggest that the highly specific formation of embryonic neuron-target cell contacts are principally controlled by extracellular fields. Such specific contacts, made during development and regeneration, are probably mediated by several types of directional and recognitional cues that may include electrical ones. One theme carried throughout this volume is that by gaining an understanding of the developmental electrophysiology of injured tissues, one can perhaps develop novel ways to artificially modulate growth and regeneration through electrical means. These techniques may then become part of the clinical domain as they have in fracture healing (Brighton, 1981) and may in skin wound healing (Barker et al., 1982; Wheeler et al., 1971) (refer to Vanable, Chapter 5, McGinnis, Chapter 6, this volume). It is probable that this suggestion may be especially true with respect to the damaged nervous system. Here we will also explore the experimental basis for this notion.

EFFECTS OF APPLIED ELECTRIC FIELDS ON CULTURED GANGLIA

There have been roughly a dozen attempts to influence the growth of neurons in culture between 1920 (Sven Ingvar's first attempt) and 1979 (Jaffe and Poo's critical and unambiguous test). These latter authors provided a summary of the defects in the older literature that we will not include here in its entirety. It is worth mentioning, however, that these older investigations (and some of the modern ones!) were simply not carried out with an understanding of the fundamentals of current flow in extended conductive media. For example, in one experiment, electrical connection to the culture chamber was made with only one electrode. Another common complication that makes the early experiments (using explanted ganglia) difficult to interpret is that markers were not used in the substrate to index the movement of the explant itself. This is especially important if the cell mass *itself* moves in the applied field. When neurites stick to the substrate with their tips (facing the cathode, for example) and the cell mass moves toward the anode, the cathode-facing fibers may stretch—simulating growth toward this pole (and additionally simulating an anodal suppression of growth). The study of Marsh and Beams (1946) was in most ways the most rigorous of these early experiments. They were careful to use nonpolarizing electrodes in contact with their media, and they understood the geometry of the chamber, the character of the current flow, and fields imposed upon explants of chick medullary tissue (placed on plasma clots). They reported that at fields of about 100 mV/mm, neurite growth was suppressed at the anode and enhanced at the cathode side of the central cell mass. They also reported that neurite growth toward the cathode was accelerated in rate when compared with the anodal side of the explant. In addition, deflections of individual neurites toward the cathode were reported.

A more modern study of the influence of electric fields on nerve growth (Sisken and Smith, 1975) was flawed in design and thus could not unequivocally demonstrate an effect of the imposed fields on cultured chick trigeminal ganglia. Platinum electrodes were in direct contact with the culture medium, probably contaminating it with electrode products. Were observed responses the result of an electrical field or of chemical reactions associated with electrolysis? The imposed field was nonuniform, and its magnitude was not stated (nor could it be determined from the published data). Lastly, the movement of the cell mass was not recorded nor was the behavior of the culture substratum—a mat of ill-defined nonneuronal cells. One could reasonably suggest that the substrate cells may have moved in the field, carrying neurites attached to their surface. Since all of the prior studies were suspect, Jaffe and Poo (1979) performed a similar experiment employing substrate markers to more properly index the responses of chick dorsal root ganglion (DRG) to applied fields and separated the culture from possible electrode contamination with salt bridges. DRG explants were responsive to fields on the order of 30–100 mV/mm. At these field strengths the cell mass indeed moved toward the anode. However, when this movement was taken into account, it was still apparent that there was a marked growth asymmetry between neurites emerging at the cathode, when compared with the anode side of the culture chamber. At 70 mV/mm or higher, neurite growth was exaggerated toward the cathode. In one case, after an explant had exhibited a growth asymmetry in the presence of a field for 20 hours, the investigators reversed the polarity of the field. In the former condition, there was a marked enhancement in the rate of neurite expansion toward the cathode (when compared with the anode side of the cell mass). After field reversal, this enhancement in rate was transferred to the former anodal-facing neurites (now facing the cathode), while the former cathode-facing neurites (now facing the anodal side of the chamber) actualy *slowed down* (Fig. 1). Jaffe and Poo also demonstrated that growth asymmetry was not a result of formation of gradients of morphogens [such as nerve growth factor (NGF), used in all of the media to promote outgrowth], since alterations in the *viscosity* of the media had no effect on polarized growth responses. These authors did not, however, observe marked *deviations* in the direction of growth of single neurites (toward the cathode) as reported by Marsh and Beams (1946). As limited as these results were, they constitute the first unequivocal demonstration of an electrical field effect on growing neurites in culture.

Though the DRG explant experiments did not demonstrate a possible field effect on the *direction* of growth of neurites, subsequent reports have confirmed this earlier observation. These studies employed cultures of single disaggregated neurons that vastly improved the resolution of growth responses.

DISAGGREGATED NEURONS IN CULTURE

Kenneth Robinson's group was the first in modern times to employ disaggregated and developing single neuroblasts (taken from neurula stage

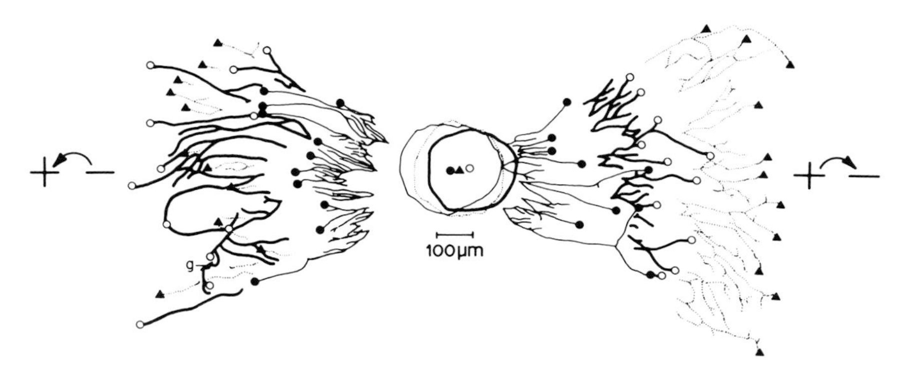

Fig. 1. The effects of reversing the electrical field on cultured dorsal root ganglia (DRG) (tracing of the central cell mass and neurites). The halo of neurites before (●) and after (○) a 20-hour exposure to a 70 mV/mm field with the cathode originally to the left, and then with the field reversed (cathode to the right) for another 20 hours (▲). During the first 20 hours in the field, the neurites facing the cathode grew 40% faster than the anode-facing side. After the field was reversed, the growth of the neurites facing the new cathode increased their rate of growth by 34% while the neurites facing the new anode retreated slowly. Shifts in the central cell mass are designated by the symbols (●,○,▲) within it. (Reproduced from Jaffe and Poo, 1979, with permission of the publisher.)

Xenopus embryos) to study the effects of an imposed electrical field. These individual cells did not develop well in culture; only a small percentage of explanted cells produced neurites (Hinkle et al., 1981). An initial, and crucial observation, was that the percentage of cells that began development was strikingly enhanced after the field was imposed across the culture chamber. This raises the possibility that an extracellular field may be a normal component of the natural environment of these cells. As discussed in the previous chapter, neuroblasts do experience an organized voltage gradient during their early development within the embryo. It is a satisfying observation that cells removed from this environment are viable in culture but do not develop neurites readily. Such behavior is stimulated when an external voltage gradient is reintroduced across their membranes.

This series of experiments substantiated the observations made in the explant studies of Marsh and Beams (1946) and Jaffe and Poo (1979) and greatly extended them. First, the *direction* of growth of single neurites was strongly influenced by the polarity of the field (Fig. 2). Neurites preferentially grew toward the cathode, sometimes bending through great arcs to do so. Neurites originating at the anodal side of the cell body were either resorbed, retarded in their growth, or made sweeping bends to project toward

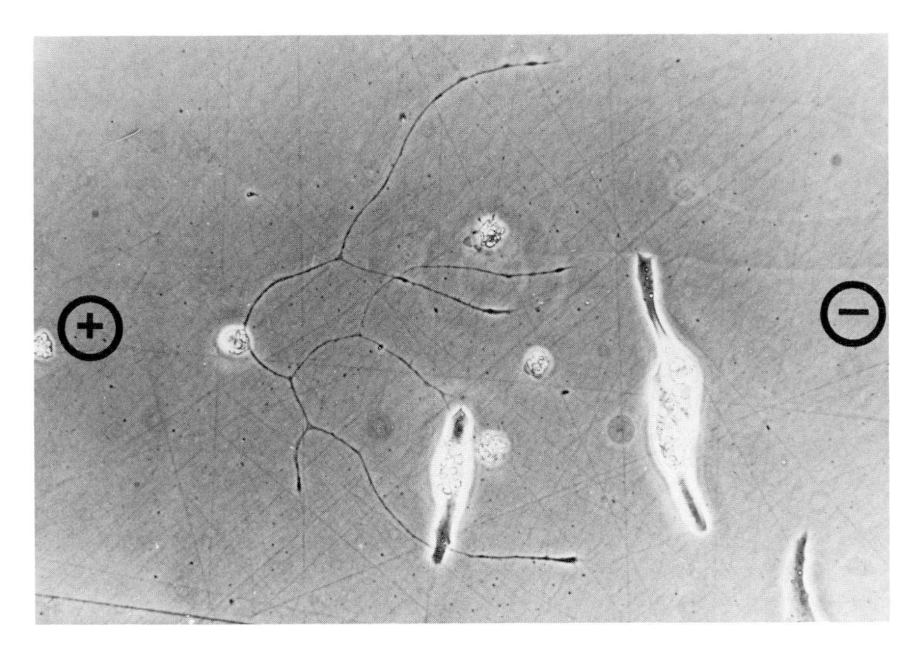

Fig. 2. The effects of an applied field on developing *Xenopus* neurons in culture. In this photomicrograph the cathode was immersed in the culture medium to the right, the anode to the left. The electrodes were several centimeters distant from the cells shown here. Cells were grown on tissue culture plastic and were experiencing fields of about 100 mV/mm. Note the alignment of the neurites *parallel* to the long axis of the field, orienting toward the cathode. Note also that myoblasts have developed their bipolar axis of symmetry *perpendicular* to the long axis of the electric field. This arrangement of developing nerve and muscle is similar to that in the embryo, where such cells reside in an *endogenous* electrical field of a similar polarity. Photo courtesy of Kenneth R. Robinson and Colin McCaig.

the cathode side of the chamber (Hinkle et al., 1981). The weakest magnitude of field where detectable responses could be observed was about 10 mV/mm. These striking observations were themselves extended by Patel and Poo (1982), who also used disaggregated and cultured single neurons under comparable conditions. In this study, a strong orientation was also observed in growing neurites—fibers grew toward the negative pole of the field. Patel and Poo reversed the polarity of the field (once a distinct trajectory had become established) and observed that fibers would *redirect* their growth in accord with the new polarity of the imposed field—always growing toward the cathodal side of the chamber. It was observed that if the field was turned off, there was a subsequent loss of neurite orientation. A greater number of neurites emanating from cell bodies within an applied field was also observed when compared with the control population of cells (cultured without an applied field). Overall, the average number of neurites and the average neurite length were found to be greater in cultures in which fields were imposed. Patel and Poo also performed another interesting experiment in which the lectin Con A was introduced to the cultures. This compound is well known to bind to receptors protruding from the plasma membrane. Binding of sufficient amounts of Con A inhibited the growth responses of neurons to the fields, suggesting that these field-induced responses may be mediated by molecular components protruding into the extracellular space from the plasma membrane.

Using a culture system similar to the above, McCaig (1986a,b) has measured a two- to threefold rate increase when neurites bend toward the cathode. Increases in the rate of growth (over control populations) were not detected in fibers aligned parallel to the field. Curiously, in neurites orienting toward the negative pole of the applied field there was also detected a two- to threefold increase in the length of the *rest periods* (Argiro et al., 1984) punctuating the growth of neurites. Such an increase in the duration of rest periods suggests that even in the presence of the field, growth rate *per* total distance of linear elongation may be under neuronal control. A periodic increase in the rate of growth in response to the field punctuated by increased periods of quiescence would produce similar overall rates of growth in control cells and cells exposed to a field—if the frequency of the rest periods were the same in electrically treated and control populations. We do not have measurements of this frequency, at the present time, however. McCaig (1986a) also noted that when contradictory directional cues were provided to neurons simultaneously by ''contact guidance'' mechanisms (scratches on the substrate) and an applied field, neurites chose to align themselves using cues provided by the field (Fig. 3). All of the above responses were observed when cells were exposed to a steady *direct current* (DC) electrical field. Recently, McCaig (1987) has studied the *resorption* of neurites facing the anode of an imposed field. Normally some degree of absorption was observed in control cultures. (About 3% of the neurites retracted while 7% were completely absorbed.) A distally positive field greatly enhanced this occurrence (67% retracted; 17% resorbed completely). Cathode-facing neurites were saved from retraction or resorption. Interestingly, a *reversal* of the field promoted an accelerated regrowth in about one-half of the neurites studied. These polarized growth-absorption effects were mirrored in the anatomical features of the growth cones. Anodal-facing growth cones had fewer filopo-

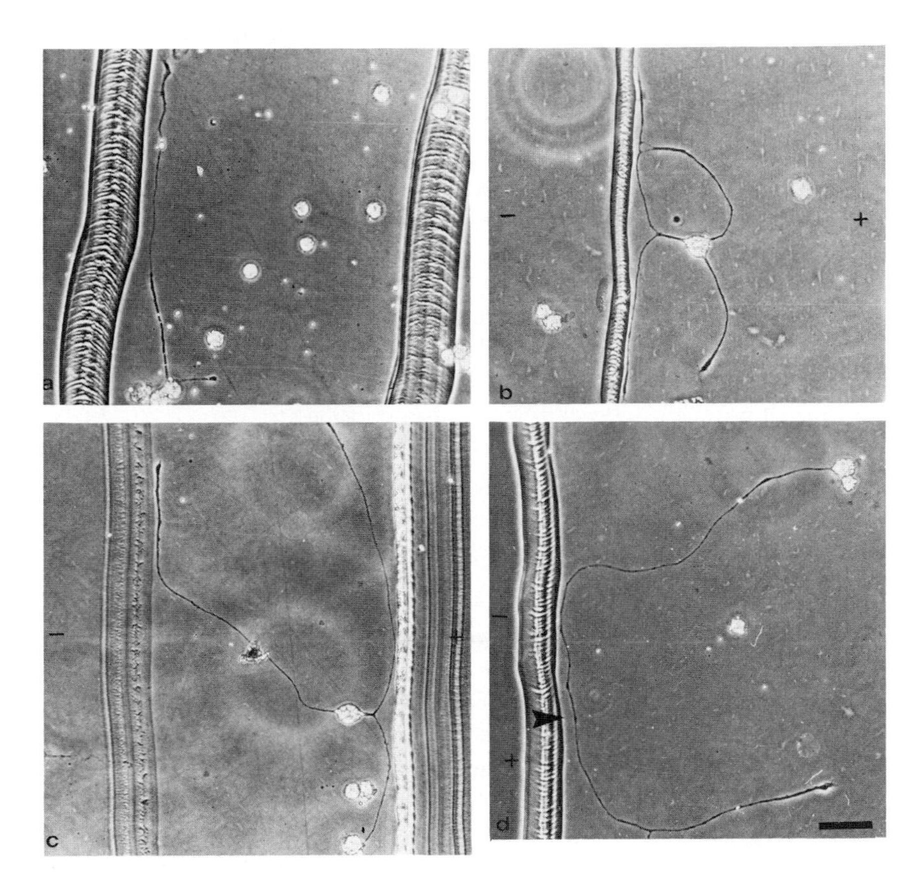

Fig. 3. **a:** Neurite in control culture growing by contact guidance along the edge of a score in the plastic substrate. **b:** Neurites turn to grow toward the cathode in a steady electric field. When confronted by a score on the cathodal-facing side, neurites turn to grow by contact guidance along the score. E = 120 mV/mm for 12 hours. Cathode at left. **c:** Two neurites at right turn away from contact guidance on their anodal sides. The other neurite (left) turns to grow by contact guidance along the score on the cathodal-facing side. E = 133 mV/mm for 12 hours. Cathode at left. **d:** Neurite growth before and after reversal of the polarity of the electric field. In the upper half of the figure, above the arrowhead, the neurite grew toward the score. When the score was contacted, a field was imposed (with the cathode to the left). At the arrowhead (2 hours later), the field polarity was reversed. The neurite then turned away from contact guidance on the new anodal-facing side, preferring to follow the electrical guidance cue by growing toward the new cathode (right); E = 153 mV/mm. During the next 2 hours, the cathode was at the left. The photograph was taken after 11 hours, with the anode at the left. Photos courtesy of Colin McCaig.

dia than cathodal-facing growth cones (emanating from the same cell body). A *reversal* of the field also reversed this morphology. The effects of uniformly applied *pulsed DC fields* and focally applied DC fields (that were both steady and pulsed) applied near cultured neurons with a micropipette have been tested as well as steady DC applications (Patel and Poo, 1984). In general, pulsed fields (intermittent DC fields) produced growth responses in neurites growing from single developing neuroblasts. The degree of the response depended on the frequency, amplitude, and duration of the pulse. It is reasonable to suggest that altogether neurons responded in a similar fashion to all such field applications as they had to steady DC fields. This finding is in direct contrast to the observation that perfectly *symmetrical alternating current* (AC) fields (of a magnitude comparable to DC applications) do not induce such growth responses (Patel and Poo, 1984).

We note one report in which explanted mammalian neurones fail to respond to an applied electric field (DeBoni and Anderchek, 1986). Although these authors have claimed to provide support for the occasional negative results of early investigators (Ingvar, 1947; Weiss, 1934), it is by no means clear that the general behavior of the neurite was even analyzed in this unrefereed report. The bulk of the data presented concerned *filopodial activity,* and there is no documentation of the overall responses of the explants themselves or the behavior of individual neurites in the applied field. It is probable that this lack of an observed response resulted from a failure to record the more pertinent aspects of nerve growth, and is probably a true case of "missing the forest for the trees."

It is important to stress that these neuronal responses to applied fields are unambiguous and are not related to effectors other than the electric field. As mentioned, Jaffe and Poo (1979) demonstrated that the growth responses of explanted ganglia could not be mediated by gradients of diffusible morphogens in the culture medium, since they tested the responses of the explants to fields in culture media of differing viscosity (using methyl cellulose). The use of cultures experiencing symmetrical alternating fields is also a convincing control. An old argument posited by Paul Weiss (1934) was that applied electric fields induced a certain ordering of the culture substratum, and it is this order that provides the basis for vectoral responses of neurites. Studies of electrically induced changes in birefringence in a variety of macromolecules (such as collagen) demonstrate that fields on the order of thousands of volts per centimeter are necessary (Jaffe and Nuccitelli, 1977) to affect the ordering of such molecules. Neurites respond to fields on the order of *tens of millivolts per millimeter*. Moreover, in some of the modern studies of disaggregated and cultured neurons exposed to a steady extracellular field, natural substances were not used as a substrate. Cells were grown on tissue culture plastic (Hinkle et al., 1981; McCaig, 1986a,b, 1987). Lastly, one cannot easily suggest that these growth responses are simply the result of an electrophoretic effect on the growth cone or neurite *itself*. To our knowledge, all known cells are moved toward the *anode* under physiological conditions when a sufficient voltage gradient is imposed on them (Ambrose, 1965). It is more probable that the effects of weak electric fields (as described here) operate through modulation of natural *electrophysiological* controls of growth.

RESPONSES OF NERVES IN SITU: NONMAMMALIAN VERTEBRATES

Responses of regenerating nerve fibers to applied fields have been observed in three "whole-animals" studies. An enhancement of peripheral nervous system (PNS) regeneration has been documented in two species of amphibians: *Xenopus laevis* (Borgens et al., 1979) and *Rana pipiens* (Borgens et al., 1977); and in the central nervous system (CNS) of the ammocoete larvae of the lamprey, *Petromyzon marinus* (Borgens et al., 1981).

In the frogs, the responses of nerves were observed as one tissue component of the electrically induced regeneration of limbs. *Xenopus* is a hypomorphic regenerator, producing only a cartilage spike covered with skin in response to amputation. Large adult frogs of the genus *Rana* completely heal limb amputations, resulting in a blunt stump (refer to a complete review of these experiments in Borgens, Chapter 2, this volume). The original intent of these studies was to induce a greater degree of limb regeneration in these species by applying an electrical field across the core tissues of the limb stump. The implanted stimulators used have been previously described; however, I wish to recall for the reader that a wick electrode (a hollow silastic tube filled with amphibian Ringer's solution) carried current to the target tissues in the forelimb stump while a platinum wire electrode served as the anode (located subdermally on the back of the animal). We are certain that these stimulating electrodes polarized quickly after implantation. Since only 100–200 nA of total current was driven through this circuit (prior to implantation), it is probable that the *chronically* imposed current (and its associated field) within the limb stump would have been *extremely minute*. Nevertheless, striking changes were induced in the limbs' responses to amputation. In *Rana* (Borgens et al., 1977), hypomorphic outgrowths (that bore little resemblance to normal limbs in external shape) were induced. These contained a striking degree of regenerated tissue representing most of the components of the limb, including muscle, bone, cartilage, and nerve.

The peripheral nervous tissue of the limb stump seemed to respond exceedingly well to the minute fields. In some limb regenerates, 20% of the newly regenerated tissue (as determined by morphometry) was nerve. Great whorls of nerve (appearing similar to neuromas) occupied much of the volume of the regenerated tissues.

Adult *Xenopus* responded differently to the applied field. In some of the experimental animals limb regenerates appeared remarkably like limbs in external anatomy (see Borgens, Chapter 2, this volume). Sham-treated controls developed a typical ''spike'' after amputation of the forelimb. Even though these ''pseudolimbs'' were limb-like in external appearance, they were still grossly hypomorphic in internal structure. No histogenesis of normal limb tissues was observed. Instead, the interior of the structures was an unjointed cartilage rod similar in most respects to the morphology of the ''spike'' that regenerated in sham-treated controls. There was, however, *one* important difference between the two groups. The cartilagenous core of the electrically treated limbs possessed large amounts of nerve that ramified throughout it to its tip. This nerve could be traced to originate from the large peripheral nerve trunks of the proximal limb stump. This abundance of nerve within the cartilage core of current-treated frogs (but not the cartilage core of the sham-treated animals) suggests that the field affected not only the external shape of the limb outgrowth and the amount of nerve regenerated, but also the *rate* of nerve regeneration as well, since chondrogenesis occurs *first* during the redifferentiation of limb components. Nerves are not known to penetrate cartilage once it has formed. Thus it seems probable that nerve of the forelimb stump regenerated at an accelerated rate in the presence of the field and invaded the regenerate prior to chondrification.

CNS tissue has been tested for its responsiveness to applied fields in the lamprey system (Borgens et al., 1981; Roederer et al., 1983). We have discussed the advantages of this model system in the previous chapter. Suffice it to repeat that the descending reticulospinal giant fibers provide an excellent means for studying the extent of axonal regeneration in the presence of a field, since these fibers are identifiable and easily manipulated, and much is known about their anatomy and regeneration (Rovainen, 1967; Wood and Cohen, 1979, 1981). After complete transection of the ammocoete spinal cord, reticulospinal axons regenerate across the lesion and form functional connections with either interneurons or other axons (Mackler and Selzer, 1985; Rovainen, 1967; Wood and Cohen, 1979, 1981). This functional regeneration occurs on the order of 150–200 days posttransection at 15°C, or about 100 days at room temperature when the cord is severed midway between the tip of the tail and brainstem.

Electric fields were imposed across completely severed lamprey spinal cords by surgically implanting wick electrodes on either side of the lesion, rostral and caudal to it. These electrodes were connected to an external voltage source, and the animal was left suspended in aerated pond water for about one week, after which time the electrodes were removed (Fig. 4). About 7–10 μA total current was driven through the circuit. Animals were maintained for another 40–45 days and then assayed for regeneration. Anatomical and electrophysiological techniques were used to determine the extent of regeneration compared with a sham-treated control group (Borgens et al., 1981).

Compound action potentials (APs) were evoked and recorded across the lesion using bipolar stimulating and recording electrodes. Stimulating electrodes were positioned behind the hindbrain, and recording electrodes were placed 1–3 mm caudal to the lesion. (This polarity of firing could easily be reversed without moving the electrodes themselves, to provide antidromic records.) An intracellular microelectrode was simultaneously inserted into identifiable giant fibers rostral to the lesion, allowing a comparison of whole cord (extracellular) and single cell AP records. When this regimen of orthodromic and antidromic stimulation and recording was ended for any one fiber, the individual axon was precisely identified by an iontophoretic injection of the intracellular marker Lucifer yellow (Stewart, 1978), the tip of the recording microelectrode being filled with the dye prior to impalement. These techniques provided clear identification of regenerating axons both anatomically and functionally. About 70% of the sham-treated animals failed to propagate APs across the lesion. (This was not surprising, since it usually takes axons about 180 days to reach the lesion. We assayed at about 50 days posttransection.) In the experimental group, 73% of the animals were found to propagate extracellularly recorded APs across the lesion both by orthograde and anterograde stimulation. These data suggested the presence of numerous axons traversing the lesion. Axons that propagated APs across the lesion were found to cross the plane of transection or to ramify throughout the scar after identification using the intracellular fluorescent dye. On the other hand, single axons (in the control group) that did not propagate APs across the lesion were found to terminate rostral to the scar (Fig. 4).

These recordings demonstrated that when one analyzed the correspondence between intracellular records, extracellular (whole cord) records, and the anatomy of single marked fibers, there was a clear relationship between physiology (the extent of AP propagation with respect to the lesion) and the precise anatomy of the individual regenerated axon. However, this approach was limited by the fact that few fibers per spinal cord could be analyzed in

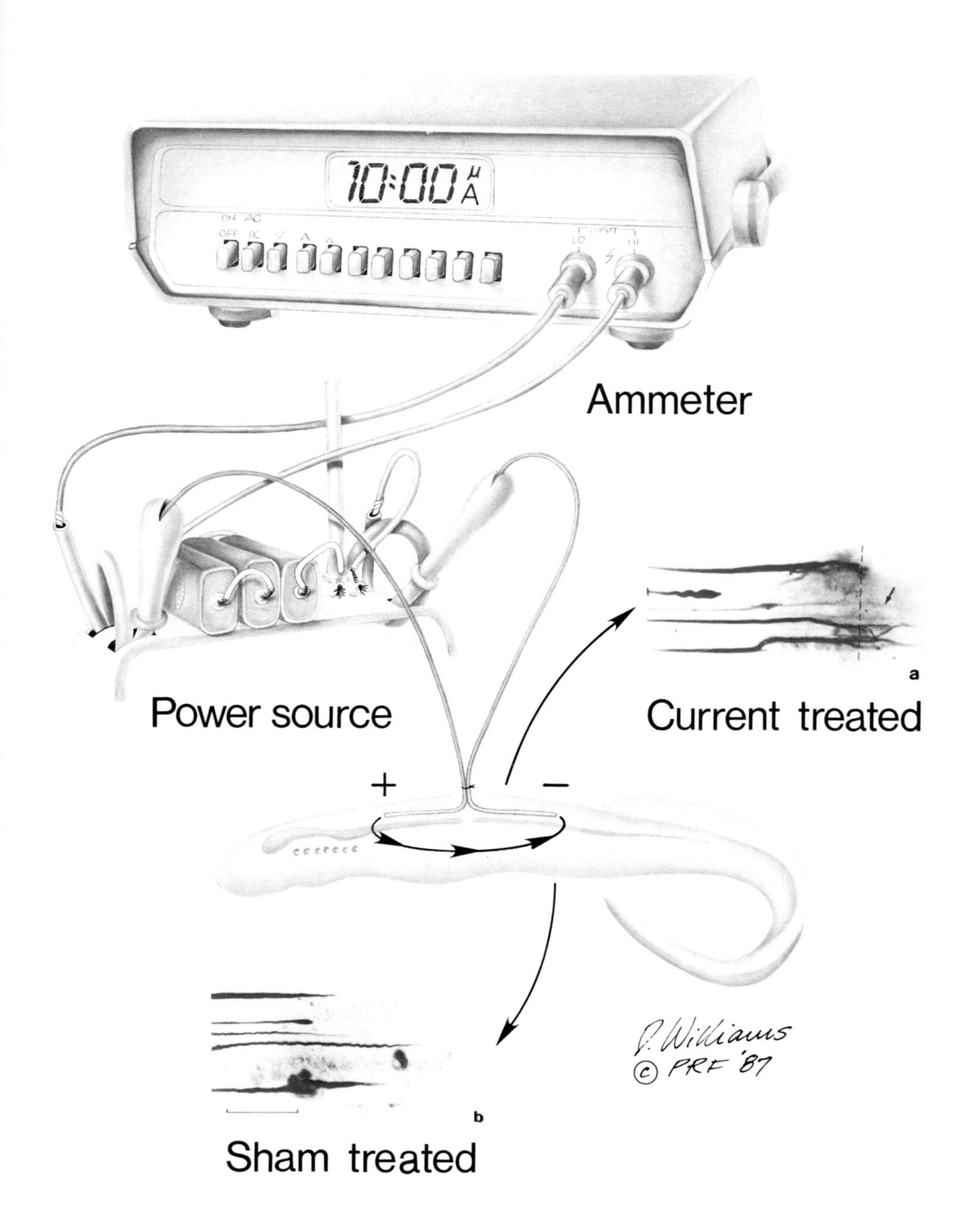
10:00 μA
Ammeter
Power source
Current treated
a
+
−
b
Sham treated
P. Williams
© PRF '87

this way. The extracellular records of AP propagation across the lesion suggested that large numbers of axons may have responded to the field, yet only a small sampling of these axons was permitted by intracellular injection (Fig. 4). We have also analyzed the anatomy of a large percentage of giant fibers in the cord by anterograde intracellular uptake of horseradish peroxidase (HRP). HRP was introduced to regenerated descending axons at the end of our experiments via a second and complete transection of the cord (about 2–3 mm caudal to the hindbrain). Thus many axons within the cord took up this intracellular marker and were identified. This procedure was very effective in displaying large numbers of axons within the cord and corroborated what the electrical records had suggested. Indeed, large numbers of axons either traversed the lesion or ramified throughout it in electrically treated animals. The sham-treated spinal cords did not show penetration of the scar by axons at 45–50 days posttransection. Occasionally an axon could be traced into the glial scar that developed around the plane of transection; however, this occurrence was very rare. Moreover, in control spinal cords individual fibers many times appeared blunt and in stasis—probably still in a degenerative phase prior to sprouting or regeneration. The terminal ends of fibers in the current-treated cords were indicative of actively regenerating axons, possessing typical growth cone morphology (Borgens et al., 1981).

Other responses to the applied field observed in this series of experiments were increased axonal branching, and perhaps an inhibition of retrograde axonal degeneration (Borgens et al., 1980, 1981; Borgens, 1982; see also Roederer et al., 1983). Recently we have obtained preliminary evidence that the field may also impose a direction on regenerating axons within the lamprey cord. In untreated cords, about 3–5% of large axons (over 30 μm

---

Fig. 4. Regeneration of ammocoete reticulospinal axons in an applied electrical field. An external voltage source supplied 10 μA total current to the lamprey larvae immersed in aquarium water. Indwelling wick electrodes filled with lamprey Ringer's were in continuity with reservoirs of lamprey Ringer's and AgAgCl contacts to the battery. Current output was monitored with an ammeter in series and controlled with a variable potentiometer. For about 6 days, a distally negative electrical field was applied across the completely severed spinal cord, since we were observing the responses of identifiable *descending* reticulospinal axons. The morphology of these axons was studied about 50 days posttransection by intracellular injection of Lucifer yellow; the micropipette also provided electrical records from these individual units. The insets are photomicrographs taken from Borgens et al., 1981. Current-treated cords displayed axons entering and traversing the glial scar and plane of transection (dashed line) (**a**). In sham-treated cords, regenerating axons had not yet projected into this region (**b**). In b, the plane of transection is to the right and out of the photographic field. (Reproduced from Borgens RB [1988]: Stimulation of neuronal regeneration and development by steady electrical fields, Adv in Neurology 47:547–564, with permission of Raven Press, New York.)

Fig. 5. Horseradish peroxidase (HRP) filling of giant fibers in the lamprey spinal cord. This photomicrograph of a spinal cord whole mount shows several giant axons filled by the uptake of HRP from an acute transection of the spinal cord through the hindbrain. The brain would lie to the left of the photographic field and an earlier transection of the cord (made 55 days previously) to the right and out of the photographic field. Note the large Müller axon (approximately 45 μm in diameter) that has reversed its direction of growth and has projected branches back toward the brain (arrow). Such reversed projections are not uncommon during the normal regeneration of lamprey spinal cord reticulospinal axons.

in diameter) show aberrant growth patterns—turning 180° and projecting back toward the brain (Wood and Cohen, 1979, 1981) (Fig. 5). We have noted that such reversed projections are extremely rare in current-treated animals. We are now attempting to test this notion further by imposing a distally positive field across transected cords. The expected result would be an increased number of aberrant reversals in axonal projection.

APPLIED FIELDS AND THE MAMMALIAN PERIPHERAL NERVOUS SYSTEM

The mammalian nervous system has posed a difficult problem for scientists wishing to test the effects of electrical fields on nerve regeneration in situ. In the periphery, axons already grow quite readily. The rate of growth (on the order of 1–2 mm/day) is substantial. Moreover, the anatomy is complex. Within any large peripheral nerve trunk are thousands of axons arranged in fascicles. The biologist has to compare regenerating nerves in control animals with regenerating nerves in experimental animals. This is not an easy task, considering the great individual variation in axon caliber and growth rates. Pomeranz began his studies (Pomeranz et al., 1984) by observing axonal sprouting (collateral sprouting) from *intact* peripheral nerve trunks in the presence of an applied field. Partial denervation of the rat hindpaw was accomplished (by sciatic nerve ligation) which induced sprouting of the *intact* saphenous nerve. The field was applied within the hindlimb by inserting a needle electrode (cathode) into the tip of the hindpaw, while the other electrode was inserted into the tail. The total current was 10 μA, applied for 30 minutes per day for 21 days following nerve ligation. Using a behavioral test (a withdrawal reflex test) and conventional silver impregnation histology to identify these cutaneous axons, a tenfold increase in nerve number (of collateral sprouts in cross section) was observed in experimental animals over control animals, and a significant increase in the rate of reflex return in current-treated animals was reported (Pomeranz et al., 1984). This group has recently applied this same approach to field application in studies of the *regeneration* of peripheral motor neuron axons, including an electromyographic analysis of function (Pomeranz, 1986). Their report suggests an enhancement in the rate of regeneration of axons within the severed (and reunited) sciatic nerve.

This preliminary data is difficult to evaluate. Current was applied for 30 days (30 minutes per day) using acupuncture needles in one group and wick electrodes in another. In the latter series, there was no description of the stimulator implant used. A 29% enhancement in the rate of functional recovery was claimed where the cathode was distal to the lesion; 16% where the anode was distal. (The anodal effect is difficult to square with the rest of the published literature of field effects on nervous tissue.) This data was presented in graphic form that did not include error bars or a clear indication of how many animals were even used in these groups. This preliminary report was succeeded by a full paper (McDevitt et al., 1987) in which some of these deficiencies were corrected for the needle-electrode study. Sizeable numbers of rats were used to test the effects of DC current on the regenerative response to crushed or severed sciatic nerve. Using electromyographic evaluation as an index of recovery, the data is consistent with the claim of early recovery in "cut and suture" groups, and a lack of an effect on crushed nerve preparations. Roman and colleagues (1987) also report a striking enhancement of peripheral nerve regeneration in adult rats. They separated the ends of a severed sciatic nerve within a silastic tube. A field was imposed within this tube (cathode distal), and the analysis of the response was anatomical. They reported a marked increase in the number of myelinated and unmyelinated axons within the "gap" when treated with electrical current (10 μA total current). This experiment is weakened, however, by the small numbers of animals studied (two to three rats) and the lack of another experimental group—one with the anode located distal to the lesion. Lastly, a recent report by Politis and colleagues suggests an enhancement of sciatic nerve regeneration based on counts of a fluorescent probe associated with neurofilament protein (Politis et al., 1988).

In summary, there is an indication that peripheral nerves in the mammal may respond to an applied current. Furthermore, this effect may be expected when nerves are cut and reconnected, but not when crushed. What is still needed is a complete investigation of this issue using thorough behavioral, anatomical, and physiological analysis of experimental and control animals. Until such a test occurs, the data thus far supporting an electrical field enhancement of mammalian peripheral nerve regeneration are suggestive, but not rigorous.

MAMMALIAN CENTRAL NERVOUS SYSTEM

It is probable that most neuroscientists who study neuronal regeneration or the controls of neuronal development have toyed with the notion of how their own particular studies may have applicability to one of the great riddles of neurobiology: the curious and unexplained lack of axonal regeneration in the adult mammalian CNS (Berry, 1979; Kiernan, 1979). As is well known, axons of the mammalian PNS regenerate—except when such axons enter the parenchyma of the spinal cord (for example, the central afferents of the dorsal root ganglia). Axons of first and second order motor neurons do not usually regenerate within the cord, but regenerate vigorously outside it (as components of peripheral nerve trunks). It is likewise certain that many nonmammalian vertebrates possess varying degrees of CNS regenerative power: they can sustain severe spinal injury and functionally recover from it. This functional recovery is in part predicated on a vigorous axonal regeneration. The list of these groups is large: ammocoete lamprey (Rovainen, 1967; Wood and Cohen, 1979, 1981); teleost fish and eels (Anderson and Waxman, 1981; Bernstein et al., 1978); sharks (Maron, 1963); salamanders and newts (Norlander and Singer, 1978; Stensaas, 1983); larval frogs and toads (Piatt, 1955); and the caudal cord of certain lizards (but not the thoracic cord) (Simpson, 1983).

In laboratory mammals, a recovery of function after subtotal lesions of the spinal cord suggests possible regenerative events; however, this view is controversial. It is still unresolved whether motor recovery in the laterally hemisected spinal cords of mammals is predicated on a local axonal regeneration (the formation of collateral sprouts innervating the affected area of the spinal cord from the contralateral side) or if function accompanies a physiological reorganization of the remaining or spared pathways in the cord (Eidelberg et al., 1981; Goldberger, 1980; Kucera and Wiesendanger, 1985; Lagasse et al., 1985; Robinson and Goldberger, 1985).

Recently there has been a resurgence of interest in spinal cord regeneration studies, probably because of a variety of reports suggesting that the CNS of vertebrates is a great deal more plastic than had been traditionally perceived. Our group (Borgens et al., 1986a, b) has attempted to influence axonal regeneration in the spinal cord of mammals (adult guinea pigs) using applied electric fields. The development of a proper model system was the first order of business, since it was our opinion that the lack of a rigorous effort in other laboratories probably resulted from the fact that traditional approaches to analyzing *axonal regeneration* in the damaged mammalian spinal cord have been ineffectual for the most part (Borgens et al., 1986a). Historically, the three most serious flaws have been: 1) a concentration on total transection of the cord or other gross inflicted trauma; 2) a failure to mark the exact plane of transection so that the degree of axonal ''dieback'' or the amount of regeneration (either axon number or the amount of linear elongation) could be indexed; and 3) a failure to develop an anatomical approach that would precisely identify the origins of axons and their trajectories. This would usually involve axonal tracts that include, for the most part, a single direction of projection. These efforts are necessary because a dense cicatrix—both fibrous (collagenous) and astroglial in composition—forms during the weeks to months after the original lesion. This scar may extend many millimeters rostral and caudal to the original lesion, disallowing a precise determination of the exact boundaries of the original transection. If one were to induce a millimeter or so of axonal regeneration in the cord, by what index would one measure it (or even be allowed to recognize it)? If one supposed there *were* responding axons and stained the cord with silver impregnation techniques (or the Luxol fast blue technique), they would in fact observe impressive numbers of axons, but in seeing everything in and around the lesion, one would probably learn very little. These techniques, for the most part, are unable to separate intact (surviving) fibers from originally injured fibers, extramedullary segmental fibers of a local origin to the lesion, and sympathetic fibers that enter the cord from interrupted capillaries and grow quite well in spinal cord parenchyma. In many cases, silver impregnation techniques stain reticular fibers and other nonneuronal processes as well as axonal processes. [(We have found that even those silver techniques that claim not to stain connective tissue in mammalian spinal cord or brain, in fact do) (Borgens et al., 1986a).] We first directed our efforts at solving these problems before turning our attention to developing an effective means of applying a field across the cord itself.

The solution to the problem of marking the boundaries of the lesion was to adopt Ann Foerster's device (Foerster, 1982), originally used in brain studies. This "staple"-shaped device is inserted into the lesion and left within it during the duration of the experiment (Fig. 6A). In our application, a dorsal hemisection of the spinal cord (past the level of the central canal) was made with a minute, electrolytically sharpened tungsten needle. The marker device is removed from the cord after fixation and prior to sectioning. The tiny holes left in the spinal cord tissues after thick sections (75–100 μm), along with the severed central canal, provide excellent markers for the lateral and ventral boundaries of the original lesion (Fig. 6B,C) (Borgens et al., 1986a).

We have adopted the use of the intracellular marker HRP from a previously discussed application in the lamprey spinal cord. We wished to fill many axons of one identifiable tract, containing axons of predominantly one projection. To accomplish this we concentrated on dorsal column axons (with a known projection and a well-understood anatomy). This dorsally located and shallow tract is especially accessible via a dorsal laminectomy. To *visualize* these fibers we made a shallow transverse transection of the spinal cord and packed this incision with crystals of HRP. Animals were allowed to survive to the next day to allow sufficient diffusion time. This fill point was usually on the order of two to three vertebral segments caudal to the original lesion. We only visualized long tract fibers (that spanned the cord at least two to three vertebral segments from the fill transection to the original dorsal column lesion) by this technique. Dorsal column axons entering the cord at higher (more rostral) segmental levels were not filled by this procedure, so we did not confuse these local axons with long tract axons. Since axons with origins more rostral or ventral to the point of the HRP fill were not observed, HRP-filled axons within the plane of the lesion and within the rostral segment of the spinal cord would *have to have regenerated there.* By combining the use of this HRP technique with the marking device, we were able to delineate with precision injured fibers from survivors at the level of the lesion (Borgens et al., 1986a).

We observed the responses to transection of dorsal column axons at 1 day, 10 days, 20 days, 50 days, and 90 days posttransection. The typical response was determined to be a vigorous retrograde degeneration of severed fibers for about 0.5–1.5 mm caudal to the lesion (Fig. 7A). At 10 days post-lesion, the terminal ends of these axons were swollen, and the general vicinity of the lesion was filled with dying axons, axonal debris, and free balls [after Ramón

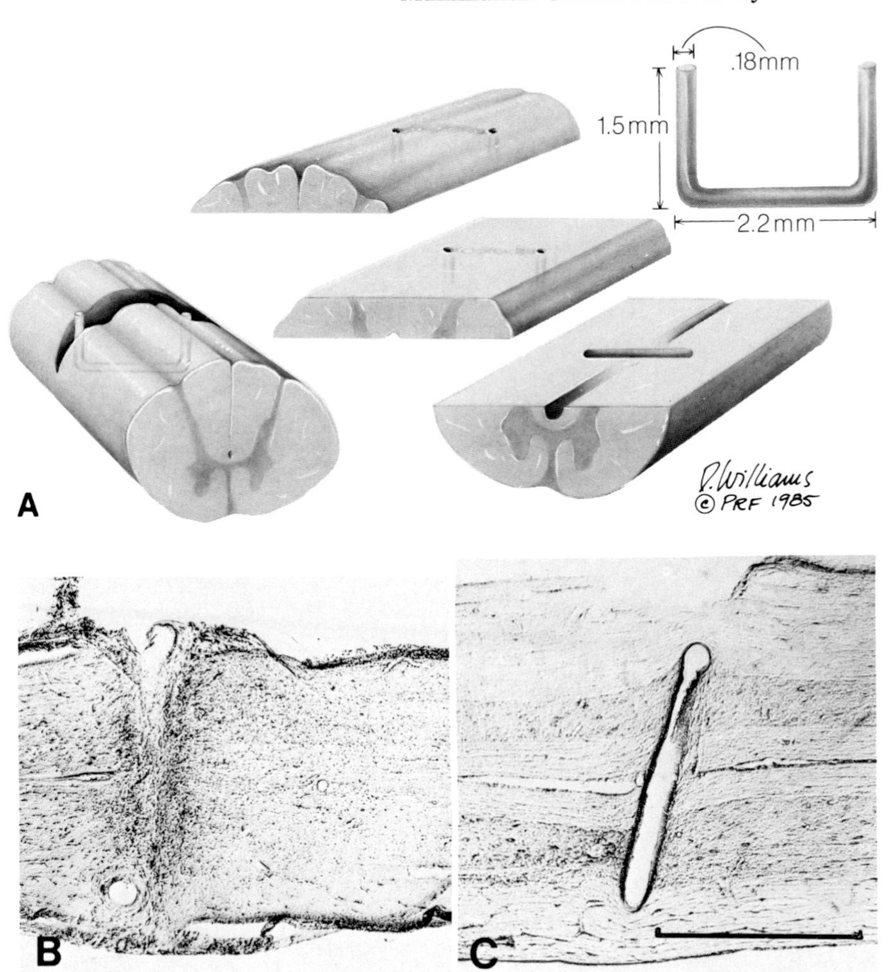

Fig. 6. **A:** This drawing demonstrates the size and placement of the marker device within a dorsal hemisection of the guinea pig spinal cord. Longitudinal-horizontal sections demonstrate the small ''holes'' left in the tissues when the marker is removed after perfusion-fixation. These holes mark the boundaries of the lesion and the exact plane of transection. **B:** The lateral boundaries of the lesion are shown in this 75-μm-thick section of spinal cord. **C:** The ventral boundary is depicted by the transverse hole. In B and C the scale bare = 2 mm. (Reproduced from Borgens et al., 1986a, with permission of the publisher.)

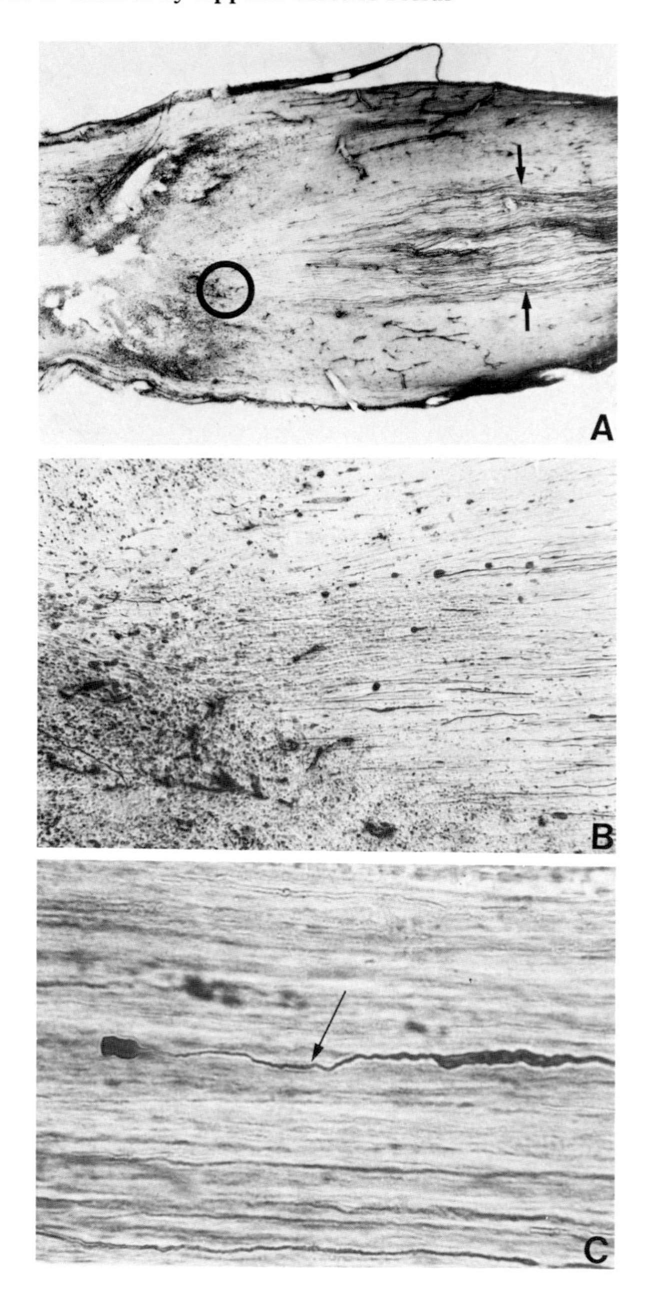
A
B
C

y Cajal (1928)] (Fig. 7B, C). Axons had regenerated for distances on the order of about 100 μm by 20 and 50 days post-lesion. This is a slightly more vigorous regenerative response than is traditionally reported (Gilson and Stensaas, 1974; Lampert and Cressman, 1964). Our ease in observing this endogenous regeneration was due to the clarity of *cleared* thick sections of spinal cord resulting in a high level of cytological detail in these large blocks of tissue (Figs. 7 and 8). In most respects the responses of severed dorsal column axons were similar to those described by others. In chronic preparations, the terminal ends of axons were well caudal to the caudal boundary of the glial scar. Axons simply did not project far enough to reach this level of the cord after "dieback." Only rarely have we observed fibers within the scar itself by 50 days posttransection (Borgens et al., 1986a).

Electrical Stimulation of the Mammalian Spinal Cord

We developed implantable stimulators that delivered between 1 and 10 μA total DC current. As in our other experiments (Borgens et al., 1977, 1980), we delivered this current to the target tissues via long wick electrodes, eliminating the possibility of electrode product contamination of the target tissues. These stimulators were placed within the peritoneal cavity of large (450–900-gram) guinea pigs (Borgens et al., 1986b). The abdomen was closed, and the wick electrodes were routed (beneath the skin) to the midthoracic region of the guinea pig, where three laminectomies were performed. The central one was the site of the dorsal hemisection. The two others were surgical approaches to the cord for the purpose of electrode placement (2 cm rostral and 2 cm caudal to the lesion) (Fig. 9). Current was applied for approximately two weeks. The control group was sham-treated (Borgens et al., 1986b); animals were implanted with batteries and electrodes

---

Fig. 7. **A:** Low-power photomicrograph of a caudal segment of guinea pig spinal cord 24 hours after transection. The plane of the transection is near the left-hand margin of the photomicrograph. The arrows point to the broad band of axons of the dorsal column anterogradely filled with HRP some 2 mm caudal to the right-hand margin of the photomicrograph. Note that most axons terminate about 1 mm caudal to the plane of transection. A typical area of axonal destruction and degeneration is circled and magnified in **B.** Note the terminal retraction clubs and the portions of degenerating axons. **C:** High-power photomicrograph showing the terminal end of a ~ 5-μm-diameter axon undergoing retrograde degeneration. Note that the terminal swollen ending (retraction club) is connected to an apparently healthy axon by a thin, thready portion of degenerating membrane (arrow). (A and C reproduced from Borgens et al., 1986a, with permission of the publisher.)

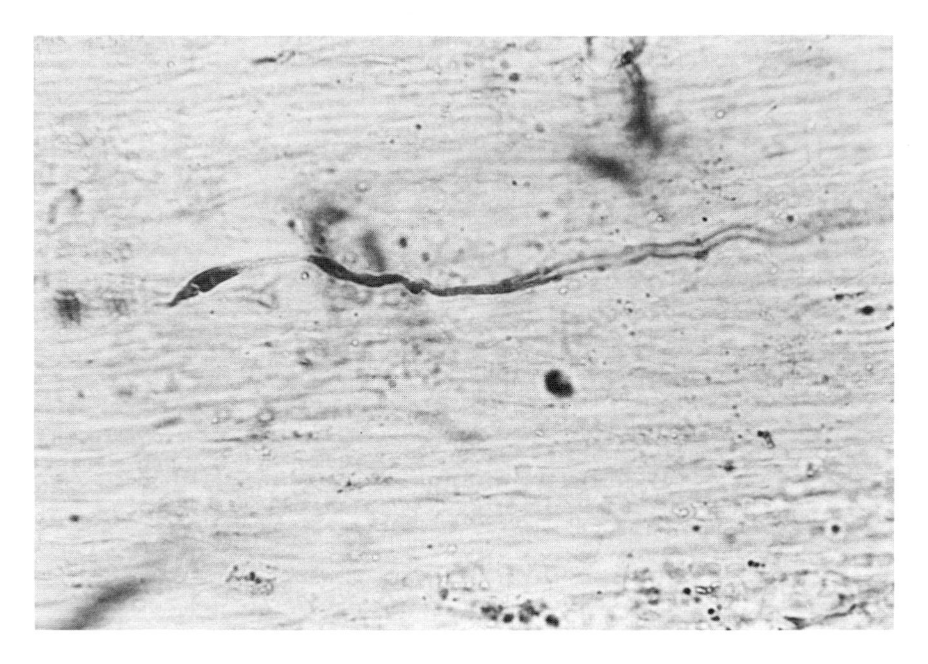

Fig. 8. A typical growth cone (filled with HRP) as observed in a cleared, 100-μm-thick section of guinea pig spinal cord. This large (5-μm-diameter) fiber was observed 50 days posttransection. (Reproduced from Borgens et al., 1986a, with permission of the publisher.)

except that the voltage sources were short-circuited (using silver conductive epoxy) and did not deliver current. In such sham-treated animals, the responses of dorsal column axons to spinal cord transection were altogether similar to the untreated animals previously described. The terminals of axons within the dorsal column lay well caudal to the glial scar. In only 2 animals out of 11 were fibers detected in the scar itself, and only at its caudal border (Borgens et al., 1986b). In the experimental group, the appearance of axons within the dorsal column was strikingly different. In the 10-μA series, axons grew to the boundaries of the glial scar and ramified throughout it. In 4 out of 11 spinal cords, axons deviated around the boundaries of the lesion and projected well into the rostral portion of the spinal cord. These regenerating fibers twisted and turned through this foreign terrain, sometimes turning 180° and projecting caudally. Typical growth cones (similar to Fig. 8) were found at the terminals of some fibers. In those fibers that circumnavigated the lesion to the rostral cord, the pathway chosen was determined by the axon's original position within the dorsal column. Lateral track fibers deviated laterally, skirting the edge of the glial scar, and adhering tightly to the dense glial

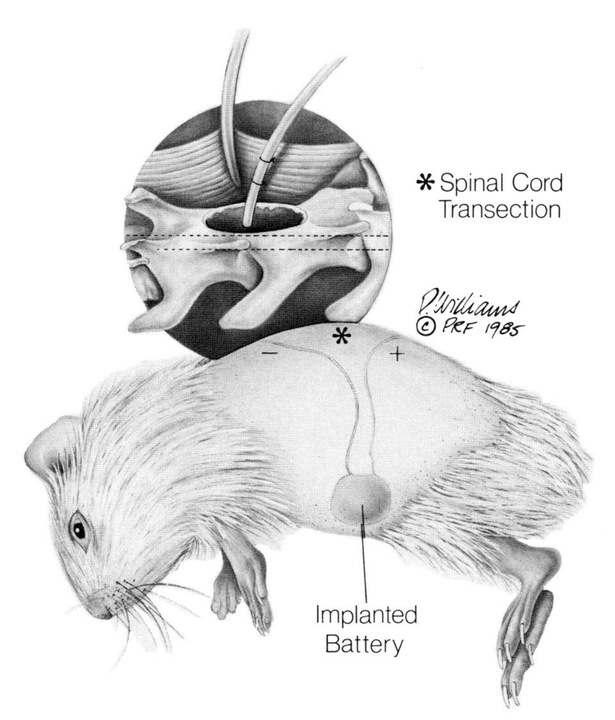

Fig. 9. This drawing shows the implantation of a stimulator unit within the peritoneal cavity of a guinea pig. The wick electrodes are routed out of the cavity, underneath the skin, and are surgically attached near the spinal cord (but not touching it) within the vertebral column. Since ascending dorsal column axons were to be induced to regenerate by this DC application (10 μA total current), the cathode was situated rostral to the dorsal hemisection of the spinal cord. (Reproduced from Borgens et al., 1986b, with permission of the publisher.)

capsule that had formed around the marker device (Fig. 10). Once entering the rostral cord, these fibers turned away from the glial scar and projected toward the brain. Another pathway around the lesion was followed by axons lying in the center of the dorsal column. These medial axons projected through the glial scar to the exact plane of transection. Here they apparently encountered a lamina (one that covers the face of the truncated cord after transection). This "membrane" may be a basal lamina (Bernstein et al., 1985) or the glial limitans (Reier et al., 1983). Medial dorsal column axons sharply descended this barrier, projecting first ventrally and then caudally, while circumnavigating the ends of the severed and swollen central canal and the ventral bar of the marker device. Axons then turned and ascended the face of the lesion (along the glia limitans) in the rostral segment. Once reaching a location approximate to their former position in the dorsal column

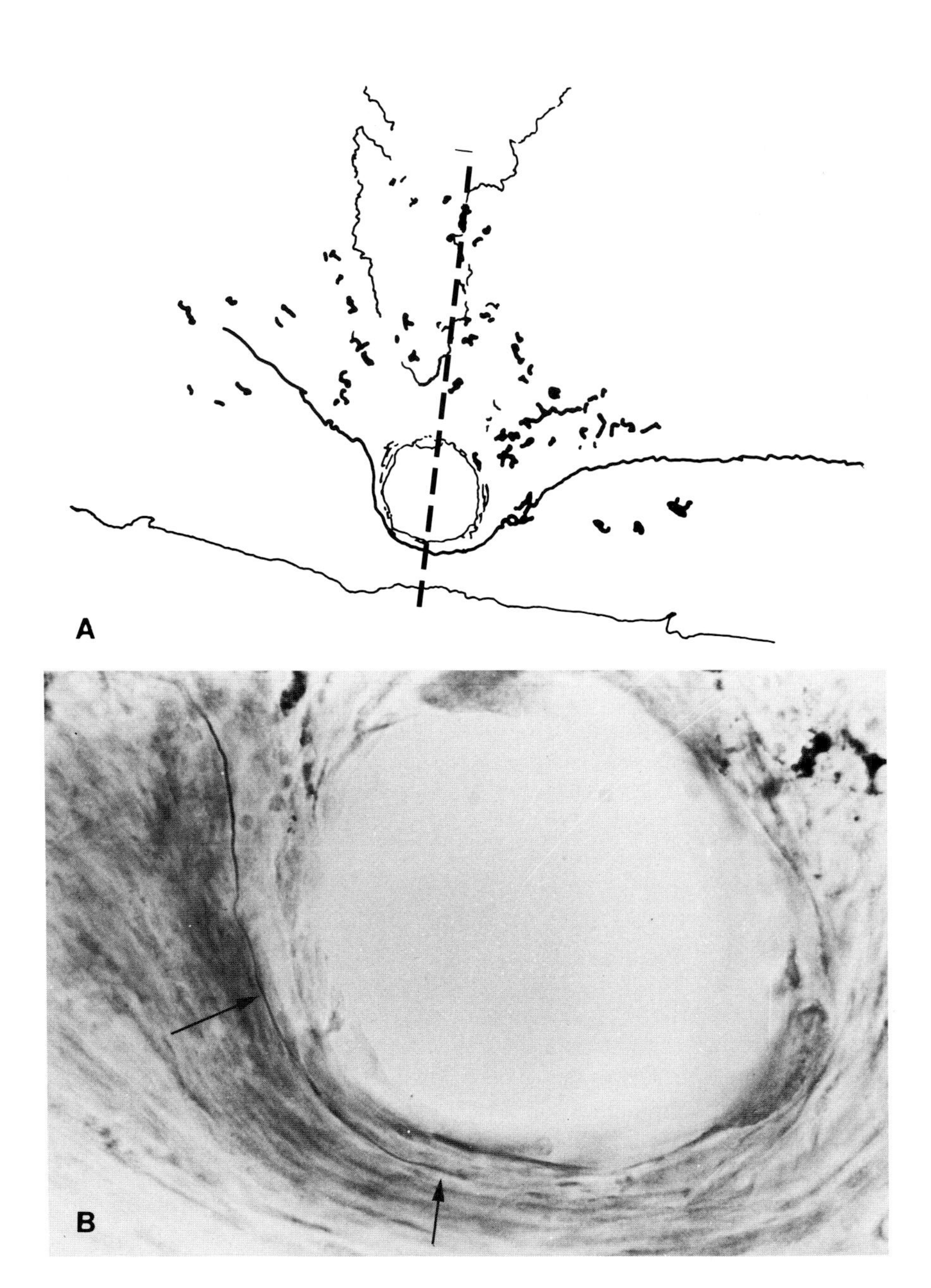
A
B

(but now in the rostral segment of the spinal cord), axons turned and projected toward the brain (Fig. 11). It is important to remember that these axons were filled with HRP two to three vertebral segments caudal to the lesion. The twisting and turning of individual axons, growth cone terminal endings (Figs. 8 and 12), the overall abnormal character of these axons when observed within the dense glial scar, and the axon's avoidance of certain obstacles (such as capillaries) were all observations that strongly suggested axonal *regeneration*. Some have suggested that these axons were *displaced* during surgery or postmortem. This is an easy snap judgment to make looking at the conformation of axons as described; however, this judgment does not square with the facts. If axons *were* displaced, similar anatomies should be present in controls as well as experimentals. They are not. Axons projecting around the scar and into the rostral cord are only found in rostrally negative electrical treatments, never in controls (Borgens et al., 1986b). Additionally, putative displaced axons should be observed 1 day, 2 days, or 10 days post-transection in all groups. They are not. Dorsally located axons could be traced *ventral* to a clearly severed central canal (Figs. 6 and 7, in Borgens et al., 1986b) (see also Fig. 11, this chapter); it appears impossible for this to have occurred in response to "pushing" or "pulling" axons out of place. Axons are not known to be capable of remaining intact given such physical and gross rearrangement. Moreover, we have observed terminal endings on ascending axons projecting well into the rostral segment of the spinal cord. Intact displaced fibers should not possess such terminals. Altogether, we think it probable that the electrical field did not simply induce a regenerative response, but, as well, awakened some more general developmental program—perhaps inducing even a measure of axonal pathfinding (Borgens et al., 1986b).

Fig. 10. Regenerative response of dorsal column axons to an applied electric field in the guinea pig spinal cord. **A:** This camera lucida reconstruction shows the pathway that *lateral* tract axons take during regeneration. The hatched line depicts the exact plane of transection (caudal segment of the cord to the left, rostral segment to the right). This fiber (filled two vertebral segments caudal to the transection) was traced to deviate laterally, along the edge of the glial scar, and to regenerate into the rostral portion of the cord for a distance of about 1 mm. **B:** A photomicrograph showing a portion of this same axon (arrows), visible in one focal plane as it deviated around the glial capsule surrounding the marker device. The orientation of the cord is similar to A. (Reproduced from Borgens RB [1988]: Stimulation of neuronal regeneration and development by steady electrical fields, Adv in Neurology 47:547–564, with permission of Raven Press, New York.)

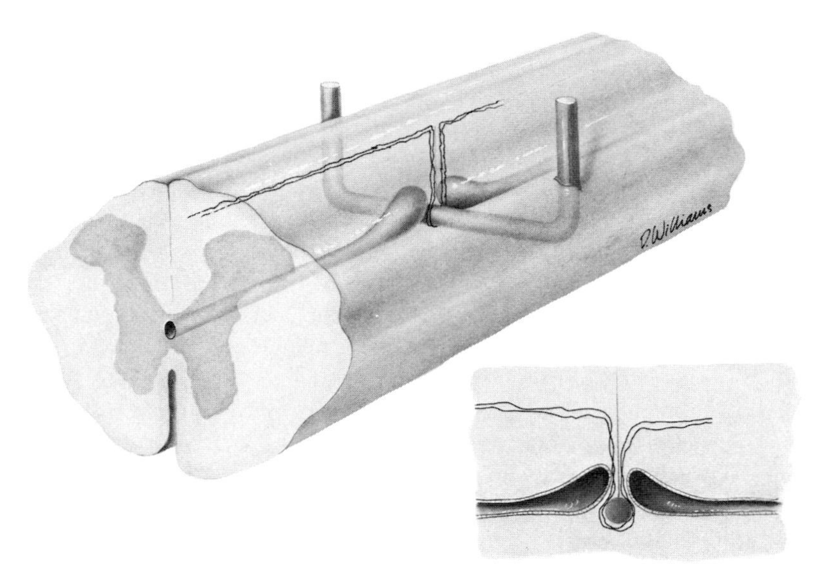

Fig. 11. This artist's reconstruction depicts the trajectory of regenerating medial dorsal column axons in an electrically treated guinea pig spinal cord. Note the abrupt dorsoventral trajectory of the axons at the plane of the lesion and their passage between the severed (and ununited) ends of the central canal and the horizontal bar of the marker device. The inset shows the path of axons around the marker between the ends of the severed central canal. The perspective in this drawing is left to right = caudal to rostral. This reconstruction was drawn from photomicrography presented in Borgens et al., 1986b.

Fig. 12. **A:** Effects of an applied electric field in a chronically injured spinal cord. This cord was transected (dorsal hemisection), and the stimulator system was implanted about 2 months later. Current was delivered to the cord (as in Fig. 9) for 2 more months after which the animal was sacrificed and prepared histologically. Note the encroachment of dorsal column axons to within 100 μm of the plane of transection (hatched line). Note the terminal ends of several axons (arrow), visualized because they are within the plane of focus in this thick section. **B:** A higher-power photomicrograph of these same axons showing their swollen and flattened growth cones. Some axons in this spinal cord regenerated in a dorsoventral trajectory (as in Fig. 11). In A and B, the rostral segment of the spinal cord is to the left of the photographic field.

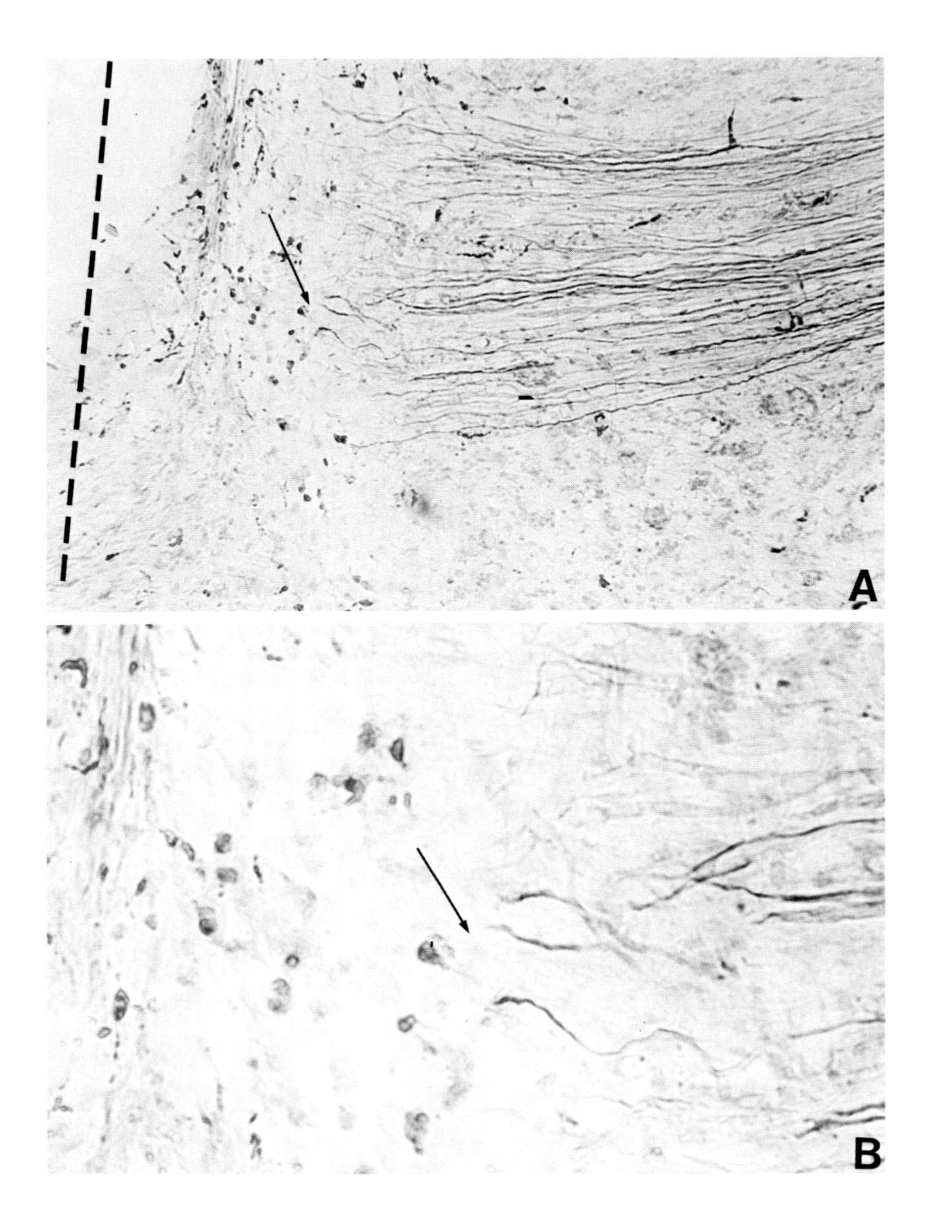
A
B

FUNCTIONAL REGENERATION

There are several fundamental difficulties with the analysis of behavioral recovery from spinal cord lesions, and with determining the relation between such recovery and axonal regeneration. First, there is the problem of separating behavioral components that are dependent on ascending or descending information from those behavioral components that are locally controlled or initiated. Second, in the case of subtotal lesions, there is the problem that spontaneous recovery of function may be based on modifications of surviving pathways rather than regenerative events. Third, there are intrinsic limitations on the ability to elicit and quantify repeatable behavior responses, particularly from the hindlimbs of a quadruped. To answer these problems we have adopted a behavioral model to determine the effects of electric fields on an otherwise *permanent* functional deficit. The defect we have chosen to study is a permanent areflexia in the cutaneous trunchi muscle (CTM) long tract reflex resulting from spinal cord hemisection (Figs. 13 and 14).

The CTM reflex is a behavioral function of cervical spinal cord motor units that depends on sensory input from lumbar and thoracic dorsal roots. The underlying circuitry of this reflex has been determined in some detail (Nixon et al., 1984; Blight, McGinnis and Borgens, unpublished studies). The behavior can be measured clearly and simply either visually or electromyographically to give quantitative information on both amplitude and spatial distribution of these sensory fields.

We demonstrate in Figure 13 the ability to determine the spatial distribution of the reflex response by marking the back skin of the animal and observing skin movements in response to light tactile stimulation. This figure also illustrates the chronic unilateral loss of responsiveness to ipsilateral stimulation below a vertical hemisection of the lower thoracic cord. We have found no evidence of recovery of this reflex on the lesioned side, at least during observation periods in excess of 2 years in guinea pigs. There appears to be no capacity for contralateral pathways to contribute to functional

---

Fig. 13. Photographs illustrating the cutaneous trunchi muscle (CTM) reflex in a guinea pig 6 months after thoracic right lateral hemisection of the spinal cord (arrow). **A:** The back skin of the guinea pig was shaved and marked with ink grid lines. **B:** Forceps lightly pinching flank skin produce a contraction of the CTM muscle (note drawing together of grid lines rostral to stimulus on the side of stimulation). **C:** Forceps poised for stimulation of the right side below the lesion to show the grid lines. **D:** Stimulation by pinching lightly below the lesion does not produce a CTM reflex. **E:** Stimulation above the level of the lesion on the right successfully elicits CTM contraction. **F:** Oscilloscope trace of electromyogram recorded from subcutaneous stainless steel wire electrodes. The lower trace indicates application of tactile stimulation. Scale: sweep duration = 1 second. Full scale = 5 mV.

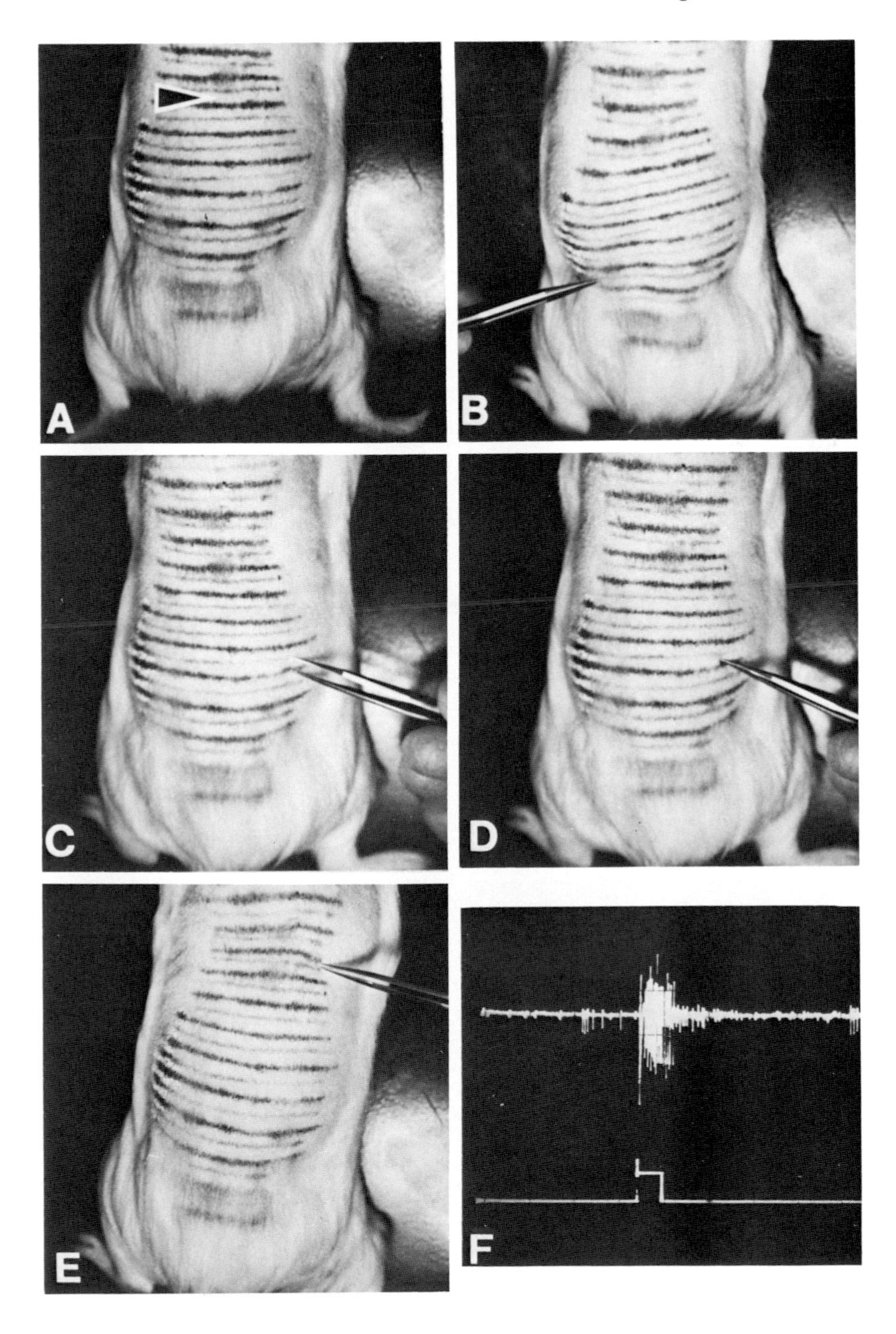
A
B
C
D
E
F

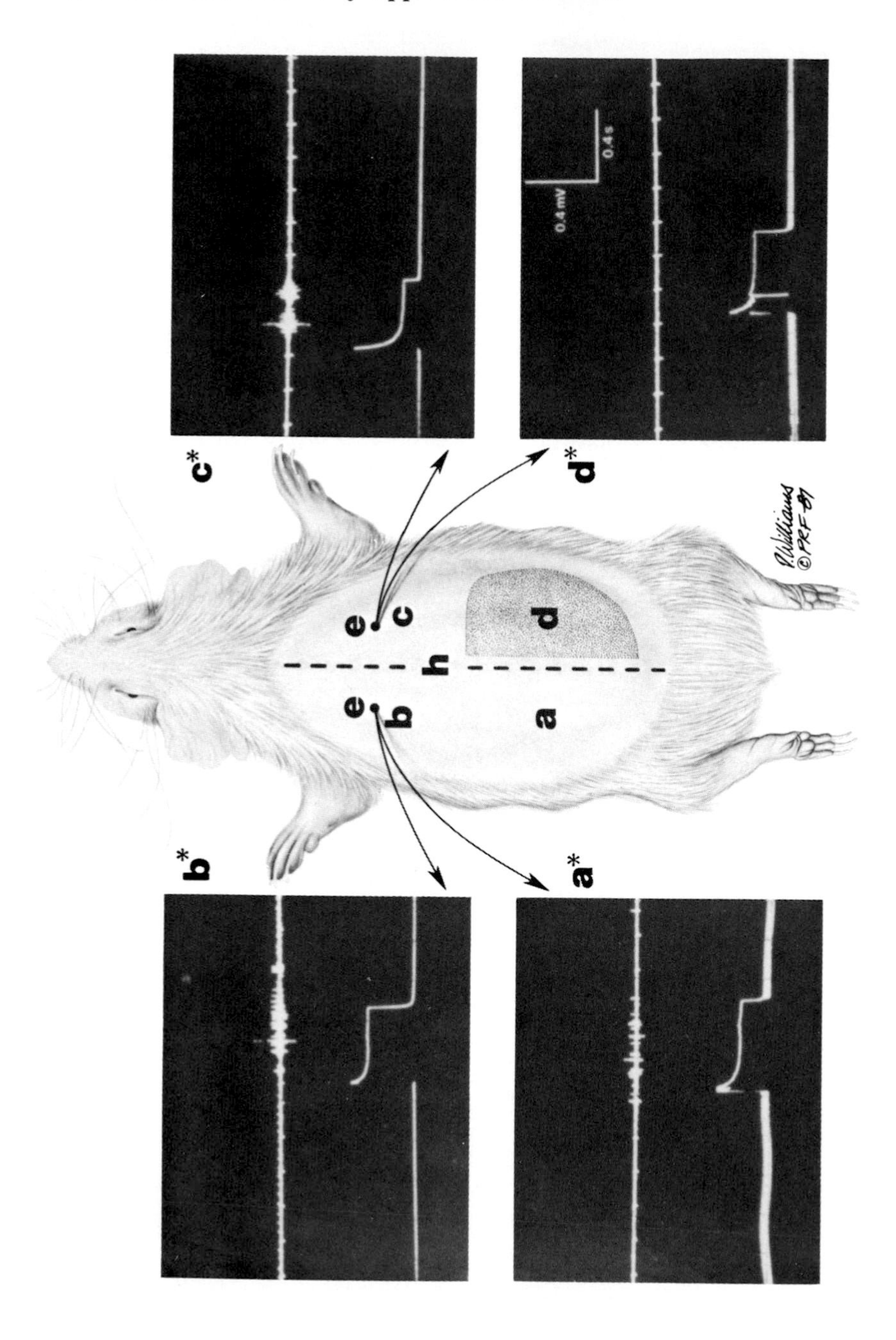
a*
b*
c*
d*
a
b
c
d
e
e
h
0.4 mV
0.4 s

restoration of the ipsilateral reflex. Therefore, chronic recovery of the reflex in conjunction with signs of regeneration of ipsilateral sensory pathways of the lateral tracts would be strong evidence indicative of functional recovery through reconnection of regenerated axons. We already have evidence of such functional recovery in a proportion of animals exposed to 35 μA total current (Fig. 15) (Borgens et al., 1987).

This application resulted in a 25% recovery of the CTM reflex when tactile stimulation was applied ipsilateral to—and below—the level of the spinal cord hemisection (Borgens et al., 1987) (Fig. 15). The defect was permanent in all sham-treated animals. Moreover, severance of the dorsal cutaneous nerves (the CTM sensory afferents) below the level of the lesion—and ipsilateral to it—produces a *loss of the recovery*. This demonstrates that the effect of the applied electrical field was upon the *central* projections of these nerves, and that changes in innervation of the CTM pathways induced by the applied field were *not* peripheral in character. In recent unpublished studies we have learned that recovery is dependent on the *polarity* of the application. Anodes located rostral to the lesion *do not* produce a recovery of the CTM reflex, which would be expected, since it is the *ascending* sensory afferents within the spinal cord that are the targets of the applied field.

Wallace et al. (1987) have also reported a beneficial affect of an applied field upon functional recovery in spinal-injured rats. The injury to the cord was produced by a "clip-compression" technique, and the behavioral analysis was based on the "inclined-plane" test for motor recovery. Here the animal is placed on a plane that is increased in angle from vertical until the subject falls off. The degree at which the animal fails to stay on the inclined plane is scored. This paper is flawed by the inconsistent claims reported both in the abstract and the Results section. It was suggested that there were no histological differences between the experimental and control groups to support the differences observed in behavioral tests. The significance of the latter differences were weakened by the questionable use of statistics to show

---

Fig. 14. Electromyogram (EMG) recordings of the cutaneous trunchi muscle reflex in the guinea pig. These oscilloscope records (upper trace in each recording) were obtained from subdermal electrodes (e) in the brachial region on either side of the midline. Stimulation was tactile, using forceps that were part of a circuit producing the stimulus artifact (lower trace in each recording). Traces **a***, **b***, **c***, and **d*** were obtained after stimulating regions a, b, c, and d respectfully. Note the complete absence of an EMG response to stimulation in d, which was below the right lateral hemisection of the spinal cord (h) and ipsilateral to it. This region of areflexia is a permanent consequence of transection of spinal ascending afferents in ventrolateral tracts. (Reproduced from Borgens et al., 1987, with permission of the publisher. Copyright 1987 by the AAAS.)

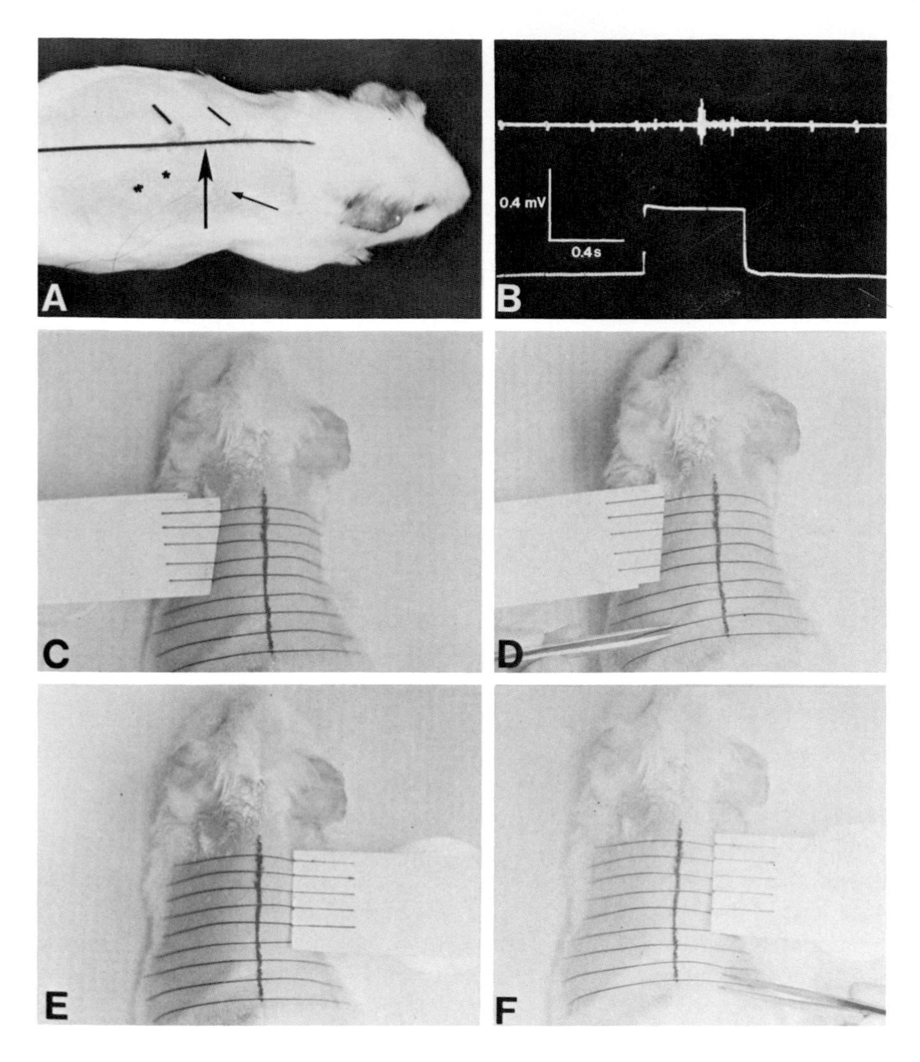

Fig. 15. Functional recovery in the cutaneous trunchi muscle (CTM) pathway in response to applied electrical fields. **A:** The large arrow points to the level of right lateral hemisection 65 days previous. The small arrow to the placement of right lateral electromyogram electrodes used to obtain recording shown in **B**. The asterisks mark the location of sites below the lesion and ipsilateral to it that elicited clear CTM responses. Drafting tape marks the location of the stimulating wick electrodes coming from the peritoneal stimulator implant, and also mark the midline. **C:** Lines on an index card align with lines drawn on the shaved back of this animal. Since the left-side CTM pathway is intact, note the drawing away of the grid lines as a result of skin contraction after tactile stimulation (**D**). **E:** Lines are similarly aligned prior to tactile stimulation of the lower right flank. **F:** Note the drawing away of the index lines drawn on the animal's back demonstrating a recovery of the CTM 65 days posthemisection. Such recoveries are only observed in current-treated experimental groups in which the cathode is located rostral to the lesion. Recoveries are never observed in experimental groups in which this polarity of application is *reversed* or in sham-treated groups. (Reproduced from Borgens et al., 1987, with permission of the publisher. Copyright 1987 by the AAAS.)

that the populations were different in behavioral scores. Only 2 weeks of the 14 weeks scored (and presented in a figure) clearly showed an improvement in inclined plane results.

Spinal Cord Regeneration: Other Techniques

There have been other techniques developed over the years aimed at inducing regeneration of CNS neurons, especially in the spinal cord of mammals. Some of the more important early approaches were the application of Pyromen (Matthews et al., 1979) and corticosteroids (Matthews et al., 1979; McMasters, 1962) and the surgical implantation of peripheral nerve bridges (Horvat, 1966; Ramón y Cajal, 1928; Sugar and Gerard, 1940). These early experiments were performed without the benefit of intracellular tracers, and without a focus on identifiable tracts within the cord. Thus the results offered in most of these early reports are not compelling. Aguayo, Richardson, and their collaborators (David and Aguayo, 1981; Richardson and Issa, 1984; Richardson et al., 1980, 1984) have recently reinvestigated the old technique beginning with Ramón y Cajal (1928) of placing a peripheral nerve "bridge" into CNS tissue. The "bridge" is actually a segment of peripheral nerve autographed to the CNS. PNS axons of such segments soon die, leaving the nonneuronal cells and the ensheathments of the nerve trunk. Additionally, these investigators have employed the intracellular marker HRP, thus unambiguously demonstrating that CNS axons that sprout within the boundary of the opening of the bridge will grow completely through it. These fibers sometimes reverse their direction of growth at the distal end of the bridge and retrace a route to the place where they had entered it. These experiments demonstrate conclusively that the environment of the PNS will support growth of CNS fibers. From a basic research perspective, this paradigm is very important. It provides a vehicle for testing what critical factors are provided by cells typical of the peripheral environment to support CNS axonal growth within the bridge. Although the numbers of axons traversing the "bridge" are indeed modest, the numbers can be strikingly increased by 100-fold (Richardson and Issa, 1984) by a prior "conditioning" lesion (McQuarrie, 1981) made to the bipolar neurons of the dorsal roots. It is these cells that contribute axons to bridges grafted to the dorsal spinal cord. All investigators attempting this technique have made one crucial observation: axons that grow into the bridge never leave it for more than a few hundred microns, i.e., central axons do not readily grow out of the peripheral nerve connective to reenter the environment of the cord or brain. One possible explanation of this finding is that when the bridge is inserted into brain tissue (for example), neurons whose axons are interrupted sprout within the boundaries of the open bore of the bridge. Sprouting and initial growth occur essentially in the early stages of this acute "injury." By the time fibers have grown completely through the bridge, it is probable that both ends may be sealed off by reactive gliosis, and/or by a basal lamina or a glial limitans. Thus fibers become trapped within the bridge and cannot easily reenter the cord or brain (Reier et al., 1983). Therefore at present these techniques appear to have little application to the issue of regrowth *within* the CNS parenchyma.

Another technique that shows promise as a possible means for affecting behavioral recovery after spinal lesions is the implantation of fetal tissue into defective cord areas. This approach has stemmed from studies of fetal transplants to the brain (Kolata, 1982). Here either genetic, or experimentally induced defects (resulting in very focused behavioral defects, such as genetic diabetes insipidus or hypogonadism in rodents) can be rescued. The behavioral defect is corrected by a graft of fetal brain tissue (from a suitable origin) into the lesioned area. Apparently the fetal nervous tissue grows well in these defective regions and supplies the necessary neurohormonal milieu to rescue the defect. There is, at present, no reason to suppose that adult neurons in the brain have been induced to grow by transplanted fetal material. In all cases that we are aware of, the restoration of defects in brain transplant studies has involved neurohormonal defects that only demand that the developing fetal material supply diffusible chemical transmitters or hormones in a more nonspecific way to reactivate those systems. In the case of *motor or sensory defects* caused by a lesion to the spinal cord, it is probable that such fetal growth (if it should occur in the cord environs) may be insufficient to restore function, since the control of these behaviors is predicated on a different physiology than the neuroendocrine control of various hormonally mediated diseases. Moreover, the adult mammalian spinal cord as a site for cell or tissue transplantation (or reinnervation by fetal nervous tissue) is somewhat distinct from brain [refer to a review of these technical matters by Das (1983)]. Attempts to reinnervate damaged spinal cords by such fetal grafts (Nornes et al., 1983) have not been very promising; however, this approach is still in its infancy. All of these techniques (singly or in combination) may eventually provide a laboratory model (perhaps a clinically meaningful one) for a rescue of behavioral defects caused by a spinal cord lesion in mammals.

Studies of the electrical induction of CNS axonal regeneration stand apart from these other techniques in several ways. First, the application of the electrical field is relatively noninvasive—no surgery is required to the cord itself (Borgens et al., 1986b). Second, there is no "gap" to cross, either across the cord lesion itself, or between the PNS environment interfaced to a CNS environment. In the presence of a field, regenerating axons within the lesioned spinal cord circumnavigate the lesion through remaining spinal cord parenchyma. In human injury as well as in animal models, the outer rim of the contused cord survives to a variable extent, and this tissue could feasibly provide the pathway for axonal regeneration. Complete severance of the cord is clinically rare, even in acutely fatal cases (Kakulas, 1984). Several problems would present themselves, however, if one wished to attempt such electrical applications in a clinical setting. For example, a clinician would desire to initiate the regeneration of *both* ascending and descending tracts within the damaged spinal cord. As discussed, a single polarity of applied electrical field (as used in animal studies) would probably enhance *growth* in one direction of axonal projection while enhancing retrograde *degeneration* in the other. A possible solution to this problem may be in utilizing the "window of effect," in which fields are applied to nervous tissue. Apparently growth induction is an immediate effect of field application, while resorption of neurites (or other degenerative effects) requires application times longer than 1 hour (McCaig, 1987). Perhaps an indwelling stimulator that oscillates the polarity of the applied field every 30 to 45 minutes may influence growth in one direction, but the opposite polarity of application would not be long enough to produce degenerative effects in axons projecting in the opposite direction. Weak applied currents are also known to produce capillary leakage (Nannmark et al., 1985) and may also contribute to the initial vascular insult produced by trauma. These techniques now offer intriguing possibilities for the clinic, being rather laboratory-bound at the present time. The character of behavioral responses that *may* occur in response to these laboratory techniques will ultimately dictate the value of these approaches as clinical tools. Their use will also need to be balanced with other clinical problems associated with spinal trauma, such as stabilization of the vertebral column, adequate control of cyst formation and other related vascular problems, and the special requirements of chronic, long-standing injuries. Although complicated, it is our opinion that these problems are now at least approachable, based on a new modern understanding of axonal regeneration and its control.

SUMMARY

1. Although at one time controversial, it is now clearly established that an applied electric field can profoundly affect the development and regeneration of nervous tissue. This fact has been established by modern, well-controlled culture experiments performed on single, disaggregated, and developing

neuroblasts as well as on cultured ganglia. Such neuronal responses to applied fields include an increase in the rate of growth of fibers facing the cathode; a decrease in growth rate of fibers facing the anode; resorption of newly formed neurites by the cell body when such fibers face the anode; an inhibition of axonal resorption in fibers facing the cathode; an increase in neurite outgrowth from cultured ganglia; vectored growth toward the cathode and away from the anode; an increase in the rate of growth in fibers that turn toward the cathode; an increase in periods of fiber quiescence during growth in the presence of the electrical field; and an increase in the number of growth cone filopodia and cytoplasmic spines in the presence of a distally negative applied field. It is also interesting that a weak electrical field can overcome and redirect neurite projections based on contact guidance cues.

2. Modern in vivo experiments include the imposition of electrical fields on: intact mammalian peripheral nerve, transected mammalian peripheral nerve, severed peripheral nerve trunks in amphibia, completely transected spinal cords in lamprey larvae, and dorsal hemisections in adult guinea pig spinal cords. The responses to these field applications in general mirror the responses of cultured neurites to fields including: an enhancement in the rate of regeneration of transected axons to distally negative fields, a decrease in retrograde degeneration in axons facing the cathode, an increase in axonal degeneration in fibers facing the anode, an increase in branching in fibers facing the cathode, and an overall increase in the regenerative response of both nonmammalian and mammalian nervous tissue to injury.

3. The responses of neurites growing or regenerating in an artificially applied field reinforces the notion that a critical control of *their development* may be the presence of an endogenous and persistent extracellular voltage gradient imposed on developing neuroblasts by currents driven through the embryo.

4. Electrical field effects on nerve regeneration may have important clinical relevance. In the adult mammal, applied electrical fields can induce regeneration of spinal cord axons. Furthermore, this regeneration is associated with a functional recovery.

ADDENDUM: IN VIVO STIMULATION OF MAMMALIAN NERVOUS TISSUE—MATERIALS AND METHODS[1]

The purpose of this discussion is to outline some of the practical aspects of delivering DC currents to nervous tissue of experimental animals. In

[1]This section written by Richard B. Borgens and Michael E. McGinnis.

designing stimulator units we have worked under the following constraints for the following reasons: small size so that small, easily available laboratory animals can be used; low cost so that large numbers of animals can be used for experiments; easy availability of parts so there are no limits because of custom manufacturing; easily modified circuits so current, frequency, etc. can be easily changed by the investigator; and reliability, because of the difficulty in monitoring the currents and the expense of animal research.

A survey of the literature reveals at least five different stimulator designs that have been used for applying direct currents to either peripheral or central nerves. (Descriptions of similar devices can be found in the bone literature.) None, however, have been described in enough detail for others to construct themselves. Some of these are unregulated, relying on either a limiting resistor (Kerns et al., 1987; Politis et al., 1988) or a manually operated potentiometer (Roman et al., 1987) to maintain constant current. Besides our earlier work (Borgens et al., 1986b, 1987), one other group has used wick electrodes (Pomeranz, 1986). Wallace et al. (1987) used a device (obtained through a commercial supplier) that appears similar to the one discussed in this paper, but it is not described in detail.

Here we provide a full description of a device that supplies constant regulated current, allows for easy monitoring of the current flow, is easily and inexpensively constructed, and can be easily modified to a variety of configurations. Variations of this design have been used to stimulate spinal cord, peripheral nerves, and bones in both guinea pigs and dogs in a variety of implant sites.

Electrode Choice

There are several advantages of using platinum (Pt) electrodes over wick electrodes (which were used in most of the studies discussed in this chapter and Chapter 2), and few disadvantages. The main advantage is reliability. Wick electrodes tend to become blocked with air bubbles or electrode products. The tube may also be kinked or compressed, which also leads to high resistance. The only comparable failure that occurs with Pt electrodes is fatigue of the lead wire at the solder joint with the Pt coil (refer to Fig. 16). In our experience this only occurs when the coil is implanted in the periphery and is subject to much motion. By extending the extent of the silicone rubber adhesive that is placed over the solder joint, an effective strain relief is created that eliminates the problem.

Wick electrodes have inherently high resistances that require substantial voltage if more than a few microamperes of current is desired. Platinum electrodes, on the other hand, never exceed 3 V of electrode potential when delivering physiological levels of current (100 μA or less). Since the LM 334 (constant-current source) drops about 1 V, the total voltage necessary to drive the DC stimulator under normal circumstances is 4 V.

Another disadvantage of wick electrodes is that Ag/AgCl, which is quite cytotoxic, is used as the metal electrode at the anodal pole. In such cases, one can expect silver ion to be deposited at the target tissue with time (at the cathode, silver ions would be captured and would not be expected to traverse the wick electrode). Other metals cannot be used because they produce gas bubbles as current is passed. Pt is well tolerated by the tissues and, with the current densities we use (relatively large surface area and low current levels), produces negligible tissue reaction as a result of electrode product formation. The final problem with wick electrodes is that they are difficult to manufacture, store, and sterilize, the main difficulty being the maintenance of a continuous fluid reservoir in the wick and electrode cavity.

The *only advantage of wick electrodes,* and *the reason for their use,* is the ability to distinguish between *true electrical effects* in the tissue and effects resulting from alteration of the local chemical environment. Given, however, that true electrical effects have been demonstrated, that electrodes can be moved a reasonable distance from the target tissues, and that there is little tissue reaction to Pt, it is reasonable to abandon wick electrodes as the method of choice except in new pioneering studies in which the nature of an effector (electrical or otherwise) is unknown.

Our initial experiments, in fact, employed *wick electrodes* to conduct currents from a remote electrode to the tissue to be stimulated (Borgens et al., 1986b, 1987). In this way, the direct effects of electric fields on the spinal cord could be separated from an indirect effect attributable to the presence of electrode products. As mentioned above, the wicks proved in the long run to be a rather unsatisfactory design, being both difficult to construct and prone to technical failings. The use of metal electrodes placed directly in the tissues avoided many of these problems, with, however, the attendant problem of the presence of electrode products in the tissues. A way to avoid the possibly deleterious effects of electrode products on the spinal cord is to move the electrodes away from the cord, leaving a "buffer" zone of muscle to absorb the effects of the local altered chemical environment around the electrodes. The drawback to this method is that the electric field strength in the cord may be drastically reduced by moving the electrodes farther away. To test whether this positioning of the electrodes would produce significant changes in the field strength within the cord, we performed a series of field strength measurements with different electrode arrangements. We also made a histological assessment of cellular damage in

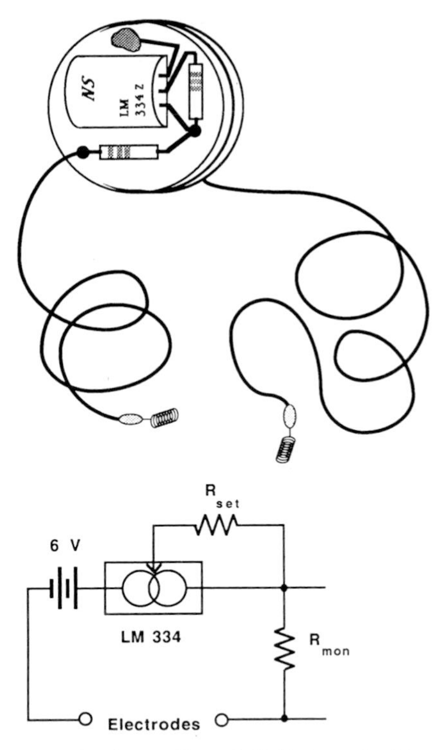

Fig. 16. DC constant current stimulator. Two 3 V lithium dioxide cells and a LM 334 constant current source provide regulated current to two Pt coil electrodes. Current magnitude is determined by the value of R_{set} and the current is monitored by measuring the voltage drop across R_{mon}.

tissue adjacent to the electrodes and in the spinal cord directly under the electrodes. We will describe below the stimulator used and the results of these tests.

DC Stimulator—Design and Fabrication

The stimulator that we have designed (Fig. 16) uses two 3-volt lithium dioxide cells (Ray-O-Vac #BR 1225 or equivalent) connected in series. Silver epoxy is used to attach leads to the cells or the cells to each other if stacked. These small crimp-style cells are not capable of withstanding normal soldering temperatures. (Although cells can be purchased with solder tabs, we have not found them useful.) The stainless steel case of the cells should be roughened or scratched to enhance the adhesion of the epoxy. A constant-current source (LM 334, National Semiconductor) and two resistors are soldered together and attached to the battery with silver epoxy. One resistor, R_{set}, is used to set the current level (I) to any value between 1 μA and 10 mA. The proper value for the resistor is given by the equation:

$$R_{set} = \frac{70.4 \text{ mV}}{I} \text{ at } 37°C$$

This resistor is connected between pins 2 and 3 of the LM 334. One can also place a 1,000-ohm resistor, R_m, in series with the output, allowing the current to be monitored by measuring the voltage drop across the resistor. (One microampere of current will produce one millivolt of potential.) Monitoring leads are soldered on either side of R_m. The monitoring leads, as well as the electrode leads, are made of a highly flexible silicone insulated multistrand wire (AS-155-36, Cooner Electronics, Chatsworth, CA). The monitoring leads can either be left long and exteriorized percutaneously or cut short and left inside the body. When left in the body, the cut ends are capped with medical-grade silicone rubber (Medical Adhesive Silicone Type A, Cat. #891, Dow Corning, Midland, MI). These ends are placed subcutaneously and are later exposed through a skin incision and the ends stripped for voltage measurements. The electrodes consist of coils of platinum-iridium (90%/10%; 0.178 mm in diameter; Engelhard, Carteret, NJ) soldered to the silicone insulated wire with the solder joint covered by silicone rubber. The electrode leads are then either soldered or epoxied to the rest of the circuit. The unit is then suspended by the electrode leads (and monitoring leads if used) and lowered into melted beeswax. Several dippings produce a 1–2-mm-thick coating of wax that inhibits rusting or corrosion of the electronics. The unit is then dipped in a silicone elastomer. We have used Silastic 382 by Dow Corning, but since it has recently been discontinued by the manufacturer we are trying Silastic 3110, which, although not labeled as medical-grade or biocompatible, appears to be well tolerated by guinea pigs. The unit is then gas-sterilized (ethylene oxide) and allowed to degas for 1 week before implantation. We have also had success with simply soaking the units in a chemical disinfecting solution (zepharin chloride) before implantation. At 1988 prices, the cost of materials for a stimulator unit is $9.00 (Table 1). The electronics can be fashioned in several shapes to meet the experimental needs. For subcutaneous implants we use a flat arrangement, while for peritoneal implants we use a stacked arrangement. The length of the leads is tailored to the situation.

Table 1. DC Battery Design Costs for 35 μA Unit

	Unit	Total
1. Battery: Two 3 volt lithium cells	0.40	2.80
Ray-O-Vac BR1225—35 mAh		
Newark Electronics, Stock No. 44F6692*		
2. Constant Current Source:		
One National Semiconductor LM-334Z		0.82
Hamilton Avnet		
3. Resistors:		
R_{set}—One 1910 Ohm 1/8 Watt,	0.21	0.21
Dale RN55D T–1%		
Newark Electronics, Stock No. 58F001		
R_{mon}—One 1000 Ohm 1/8 Watt T–5%	0.23	0.23
Allen-Bradley RC05GF1025		
Newark Electronics, Stock No. 10F300		
4. Electrode Wire: Anode-10 cm, Cathode-5 cm		3.00
Platinum-Iridium 10%,	$605/100 ft.	
0.007 In. Dia. Annealed	(0.20/cm)	
Engelhard, Stock No. PT10IR		
5. Hook up Wire: 30 cm—	1.25/Ft.	1.25
Stranded Silver Plated Copper		
AS155-28 Silicone Rubber Insulated		
Cooner Wire		
6. Beeswax—8 g	1.80/64 g block	0.23
Local hardware store		
7. Elastomer: Silastic #382 Dow Corning Medical	No longer available	
Grade Elastomer and Catalyst		
3110 RTV Silicone Rubber and F Catalyst	19.17/lb.	0.20
Dow Corning		
Brownell Electro		
Silastic Brand Medical Adhesive Silicone	17.80/tube	0.10
Type A (2 oz. tube)		
General Medical		
8. Silver Epoxy:	39.62/kit	0.04
#3021 E-Solder Conductive Adhesive		
ACME Chemical and Insulation Co.		
	TOTAL	$ 8.88

*Addresses for supply companies are listed in the appendix following the references.

Electrical Performance

The lithium dioxide cells that we use have a rated capacity of 35 mA hours, which will provide 20 μA of direct current for 73 days. This calculation is based on the manufacturer's data from tests conducted with discharge rates of 1 mA. Since the currents we use are normally one to two orders of magnitude smaller than this, it is our experience that somewhat longer life can be expected from these units (although we have not accumulated data on this point).

In a recent experiment, 32 stimulators set at 20 μA were used for 21 days. Of the 32 implants, there was one failure because of corrosion at a solder joint (the result of poor construction). The other 31 were delivering an average of 20.26 ± 0.07 μA (X ± SEM). The current ranged from a high of 20.83 μA to a low of 19.48 μA, which is a 6.7% variation, most of which is accounted for by the 5% tolerance of the set resistor used in this case. In any one animal, the day-to-day variation was always less than 1 μA.

In Vivo Field Measurements: Methods

The measurements were made on three deeply anesthetized guinea pigs (64 mg/kg ketamine, 0.64 mg/kg, acepromazine, and 14 mg/kg xylazine; IM). The differential voltage between two Ag/AgCl recording electrodes placed 1.3–3 mm apart in the cord parenchyma was amplified and displayed on a storage oscilloscope (Model 5113, Tektronix, Inc., Beaverton, OR). High-input impedance voltage followers were placed in the circuit ahead of the oscilloscope amplifier inputs. The electrodes were Teflon-insulated solid silver wire with AgCl electrolytically deposited on the exposed flush surface (0.25 mm in diameter; Medwire Corp., Mt. Vernon, NY). The two electrodes were sutured in place after positioning them through a laminectomy into the cord at T9-T10. Two other laminectomies were made, each two vertebral segments on either side of the recording site. Coils of platinum-iridium wire (10 cm in length, 0.178 mm in diameter; coil diameter 3 mm, coil length 3 mm) were then inserted in the two outside laminectomy sites to a position near the cord *but not touching it.* A switch was placed in series with a 50-μA constant-current source connected to these two current delivery coils. By activating the switch, current was allowed to flow through the circuit momentarily. The change in the differential voltage between the recording electrodes because of the current flow was easily measured on the storage scope. This voltage was divided by the distance between the recording electrodes and the magnitude of the current to yield a specific field strength per unit of applied current. This configuration, with the stimulating electrodes in laminectomies two segments rostral and caudal to the injury site (or in this case the recording site) and near the cord, but not touching it, is similar to what we have used previously (Borgens et al., 1986b, 1987), and for this discussion will be referred to as the *standard electrode arrangement.*

As long as one electrode remained within 12 mm (approximately two vertebral segments) of the recording site, the convergence of current to the electrode dominated the current distribution at the recording site. As both electrodes were moved away from the recording site, the field strength decreased. Moving either one or both of the electrodes closer than two vertebral segments from the recording site resulted in higher current density at the recording site. However, because of *nonuniformity* resulting from the *geometry of the field* close to the current delivery electrodes, we chose to restrict our studies to electrodes placed at least two segments away.

Moving the anode from within a laminectomy to a site on the muscle dorsal to the same area resulted in only a 10% drop in field strength, as does the converse of moving the cathode to a more superficial position while leaving the anode in the laminectomy. Even moving the anode to a position on the same side of the recording site as the cathode only results in a 43% reduction in field strength *while maintaining the original polarity at the measuring site*. It is interesting that even with removal of both electrodes to extremely distant sites, (the ear and foot), the field strength is still one-fourth of that found with the electrodes in the standard positions. Therefore, an increase in current by a factor of 4 would result in similar current densities at the cord without the need for critically placed surgical implants.

The field strengths that we measured in the guinea pig spinal cord agree amazingly well with calculations by Coburn and Sin (1985) of field strengths in the human spinal cord stimulated with extradural electrodes. In their model the electrodes were placed 20 mm apart, and the steady-state fields and current densities resulting from passing 10 mA of current were calculated. At a point midway between the two electrodes they calculated a field strength of 0.81 V/cm along the long axis of the cord in the region of the dorsal columns. This finding is equivalent to a field of 8.1 (μV/mm)/μA, which is quite similar to the 8.4 (μV/mm)/μA we recorded with the stimulating electrodes in the standard position (also approximately 20 mm apart). The nearly perfect agreement between Coburn and Sin's calculations and our measurements is surely coincidental, given the size difference between the thorax of a guinea pig and a human and differences in electrode placement, but the similarity does give confidence in the measurements. The gross appearance of the electrodes and surrounding tissue after several weeks of current application was very consistent. The cathode coil becomes coated with a granular white deposit. There is little or no tissue reaction. The anode coil has a golden red acellular layer attached to it, with a ring of cellular necrosis extending at most 100 μm from the Pt surface. Sometimes this electrode is surrounded by a fibrous capsule. Sham electrodes (coils that do not pass any current) appear bright and shiny, with no evidence of tissue reaction. We have also examined sections of spinal cord directly underneath the electrodes with silver impregnation, hematoxylin and eosin, and neutral red. No signs of necrosis or other changes were encountered when these areas were compared with control areas of the cord.

Oscillating Field Stimulator

How does one induce regeneration of axons projecting in one direction with a steady electrical field while not causing damage to fibers projecting in the opposite direction? How can one induce regeneration in both directions? This is an especially important issue if applied electrical fields are ever to be used as a component of a treatment for spinal cord injuries. We have discussed the rationale for a possible solution to this paradox: the oscillating field. When a field is reversed every 30 minutes, we believe growth responses are immediately initiated facing the cathode, while the destructive effects of the anodal field (imposed on fibers of the opposite projection) takes much longer. Prior to the occurrence of the latter, the field is reversed. We are involved in testing these notions in the guinea pig—and have started clinical trials in dog spinal cord injuries—using such oscillating field stimulators. We include below a description of their design and fabrication and the results of various other tests. We hope the following descriptions are useful; however, they necessitate more experience in electronics fabrication than the simpler, single-polarity stimulators previously discussed.

Design and fabrication. The oscillating field stimulator we have designed is capable of delivering a constant current that reverses polarity after virtually any time period desired. The circuit is shown in Figure 17, and parts are listed in Table 2. The timing circuit is based on a CMOS (complementary metal oxide semiconductor) 14-stage ripple-carry binary counter (CD 4060) that contains an oscillator. The frequency of the oscillator is set by a resistor (R_f) and capacitor (C_f) and can be varied from 500 kHz to less than 1 Hz. The output of the oscillator can then be divided by up to 14 binary stages to achieve very low-frequency oscillations (as low as one cycle every 5 hours). For example, we set the oscillator frequency at 4.5 Hz and divide by 14 stages to produce a frequency of 1 cycle/hour (equivalent to reversing the polarity every 30 minutes). To achieve very precise timing, a miniature potentiometer (trimpot, Spectrol model 64X) can be substituted for R_f and adjusted until the desired frequency is obtained. The output is taken from the binary stage desired (stage 14 in our case) as a 0–6-V square pulse (railed at the supply voltage). This voltage is applied to the inverting input of a low-power CMOS operational amplifier (ICL 7611), and a 3-V signal is

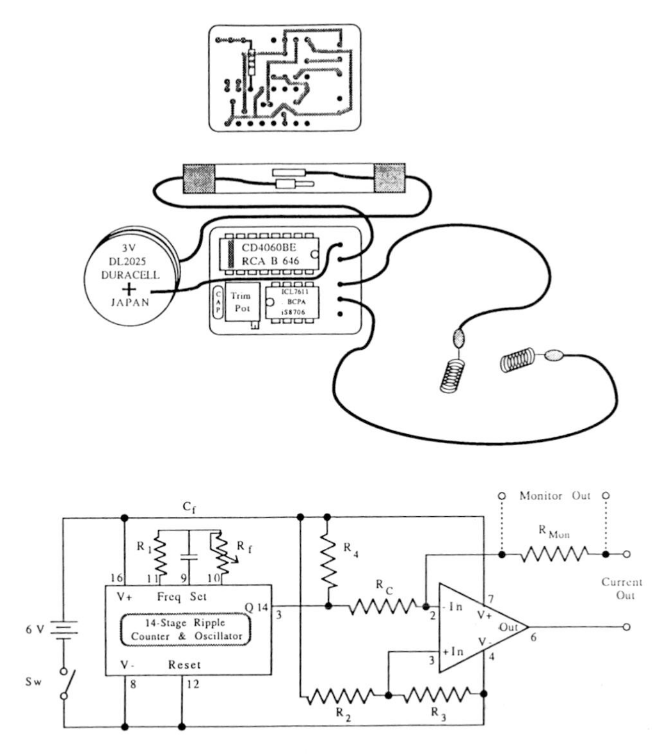

Fig. 17. Oscillating Field Stimulator. The circuit has been built on a circuit board, both the top and bottom of which are shown schematically. Most of the resistors are placed underneath the semiconductors. Current output is set by the value of R_c and the frequency of polarity reversal by R_f. The switch consists of a male and female connector encapsulated in a silicone tube. The tube can be stretched to insert the male into the female. The outside of the female connector is covered with heat shrinkable Teflon to prevent contact before it is desired.

applied to the noninverting input (obtained by dividing the supply voltage between R_2 and R_3). This then gives a net ±3V at the op-amp input that drives current through the electrodes. The magnitude of the current is determined by the value of R_c by the equation

$$I = \frac{3\ V}{R_c}$$

For currents less than 100 μA, the quiescent current of the op-amp is set at 1 μA by strapping pin 8 to the positive supply voltage. To deliver more

Table 2. OFS Battery Design Costs for 200 μA Unit*

	Unit	Total
1. Battery:		
Two 3 volt lithium cells	1.70	3.40
Duracell DL2025B-170mAh		
Newark Electronics, Stock No. 81F143[†]		
2. Low Power Op. Amp:		
One Intersil ICL 7611BCPA	1.86	1.86
Newark Electronics		
3. 14 Stage Counter/Divider and Oscillator:		
RCA CD4060BE	0.97	0.97
Newark Electronics		
4. Resistors:		
R_2,R_3,R_4—Three 1 Mega Ohm 1/8 Watt T-5%	0.23	0.69
Allen-Bradley RC05GF105J		
Newark Electronics, Stock No. 10F300		
R_{mon}—One 1000 Ohm 1/8 Watt T-5%	0.23	0.23
Allen-Bradley RC05GF102J		
Newark Electronics, Stock No. 10F300		
R_1—One 6.2 Mega Ohm 1/8 Watt T-5%	0.23	0.23
Allen-Bradley RC05GF625J		
Newark Electronics, Stock No. 10F300		
R_c—One 15000 Ohm 1/8 Watt T-1%	0.21	0.21
Dale RN55D		
Newark Electronics, Stock No. 10F300		
R_f—One 2 Mega Ohm Trimming Potentiometer	1.95	1.95
Spectral Type 64X		
Newark Electronics, Stock No. 67F5792		
5. Capacitor:		
One 0.047 uF 50 Volt	0.11	0.11
Sprague Z5U Mini Size Ceramic		
Newark Electronics, Stock No. 29F013		
6. Printed Circuit Board:	$20/20 unit board	1.00
Size 0.95 x 0.80 Inches		
In-House Fabrication		
7. Electrode Wire:		
Two 20 cm Wires	$605/100 ft.	8.00
Platinum-Iridium 10%, 0.007 Inches dia.	(0.20/cu)	
Annealed		
Engelhard, Stock No. PT10IR		
8. Hook Up Wire:		
48cm – Stranded Silver Plated Copper	125/Ft.	2.00
AS 155–28 Silicone Rubber Insulated		
Cooner Wire		

Continued next page.

Table 2. OFS Battery Design Costs for 200 μA Unit (*continued*)

9. Beeswax – 10 g	1.80/64 g block	0.28
Local hardware store		
10. Mechanical Switch:		
3.5 cm Silastic Medical Grade	31.99/50 ft.	0.08
Silicone Tubing		
(0.132 in. I.D. x 0.183 in. O.D.)		
Dow Corning Cat. No T5715–11		
American Scientific Products Cat. No. 75715–11		
One set – Mini Gold Pin Connector and Socket	0.24	0.24
Amphenol 220–S02–100		
Newark Electronics		
11. Elastomer:		
Silastic #382 Dow Corning Medical	No longer available	
Grade Elastomer and Catalyst		
3110 RTV Silicone Rubber and F Catalyst	19.17/1b.	0.20
Dow Corning		
Brownell Electro		
Silastic Brand Medical Adhesive Silicone	17.80/tube	0.10
Type A (2 oz tube)		
General Medical		
12. Silver Epoxy:		
#3021 E–Solder Conductive Adhesive	39.62/kit	0.04
ACME Chemical and Insulation Co.		
	TOTAL	$21.59

*Cost design for dog study. For a guinea pig study, electrode wire and hook up wire costs would be reduced to make a battery design total of $15.09.

†Addresses for supply companies are listed in the appendix following the references.

than 100 μA, this pin should be left floating, which programs the quiescent current to 10 μA. The device draws approximately 65 μA from the battery to supply 20 μA to the tissues. Using the small (12-mm-diameter) BR 1225 lithium dioxide cells (35 mA hours) allows a minimum of 22 days of continuous operation. For longer times or higher currents, we use the larger BR 2025 cells, which have a 120-MA hour capacity.

To facilitate construction of the stimulators, small circuit boards were produced. We have used these oscillating field units in preliminary clinical trials, testing their effects on chronic and acute dog paraplegia. For the canine studies, we place the two cells and the circuit board together in one package dipped in beeswax and coated with a silicone elastomer. These packages are too large for rodent studies. In guinea pigs the circuit board and batteries are encapsulated separately and connected by flexible wires, which can then be mounted subdermally, one on either side of the animal. An additional concern is that as soon as the battery is connected to the circuit, the timer begins to run even though no current is being drawn from the electrodes. This consumes approximately 45 μA. To prevent this current drain, the power must not be connected to the circuit until the time of implantation, which can be done by a variety of techniques, including:

1. Having a normally closed magnetic reed switch incorporated in the circuit with a magnet taped to the finished unit (thus holding the switch open) until the circuit is to be used.
2. Using a latching magnetic read switch to connect and disconnect power as needed.
3. Using a sealed mechanical switch. We have developed our own design encased in a flexible transparent tube. We use these in our canine studies, which are too expensive to rely on a magnetic switch.
4. Soldering a joint just prior to insertion and sealing with light-activated dental adhesive or silicone adhesive squirted inside a piece of silicone tubing slipped over the solder joint (Loeb and Gans, 1986).

Electrical performance. Implants into guinea pigs have proven the stimulators to be reliable and stable, with less than 5% variation in current levels or frequency. The main problem encountered was that the unit began to erode through the skin of the animals after about 7 days. This was especially apparent in smaller animals weighing less than 450 grams. We will attempt to miniaturize these units further for peritoneal implantation in the rodent. The implants into dogs have been well tolerated for periods up to 6 months in eight dogs.

pH measurements. A second possible advantage of an oscillating field, besides being able to induce growth in both projections of axons, is that there should be a reduction in the net production of electrode products. As a direct (or indirect) measure of electrode product formation at the electrodes, we chose to follow changes in pH of the local electrode environment during oscillating field stimulation and to compare this change with the type of pH change observed during steady field stimulation.

The formation of metal salts and other interfacial electrode reaction products (peroxidase formation, hydrogen gas formation, etc.) is associated with characteristic alteration in the pH of the local environment of the electrode studied, which can be demonstrated by measuring the pH at the electrodes. Each platinum electrode was placed in a well with 4 ml of unbuffered physiological saline. (We used unbuffered media in order to determine the actual magnitude of pH change exposed to—and buffered by—target tissues and fluids at the electrode site.) The wells were then electrically connected to each other through a Ringer's agar bridge. The pH of the solution in each well was monitored by a combination pH electrode. Before applying current, the pH in each well was approximately 7.0. Experiments were done with both direct currents and oscillating currents, using 35 μA supplied by the two stimulators previously described.

After 24 hours of constant 35 μA, the cathode well was at a pH of 9.8, and the anode well was 3.4. By 48 hours, the anodal pH was 2.4. This experiment was repeated with an oscillator in series with the voltage source that reversed the polarity of the field every 45 minutes. During the 60 hours the test was run, the largest pH change seen at either electrode was 0.4 pH units.

Summary

The in vivo field strength measurements demonstrate that the field strength within the cord at the site of the lesion depends on the location of the current delivery electrodes. The convergence of current to an electrode produces high current density and hence higher field strength near each electrode. The closer one electrode is to the lesion site, the less critical is the placement of the other to maintain high field strengths. However, as the current delivery electrode approaches the lesion, the less uniform is the current direction. At a lesion exactly half-way between two electrodes placed on the midline, the current will all be oriented along the long axis of the animal. As one electrode is moved closer to the lesion, there will be a larger vertical (dorsoventral) component of the current at the lesion (assuming that the electrodes remain a few millimeters dorsal to the target tissue). As a compromise between uniform current direction and maximum field strength in our spinal cord studies, we chose to position the current delivery electrodes two vertebral segments on either side of the lesion. It appears that the critical distance to be within the convergence zone of an electrode (i.e., that area in which the current convergence to the electrode so dominates the field strength that the position of the other electrode is relatively inconsequential) is approximately 1 cm in our guinea pigs. Therefore, by placing one electrode 1 cm from the lesion, the position of the other becomes inconsequential and is a matter of experimental convenience. Since the anode is potentially the most cytotoxic of the electrodes, we routinely place it farthest away from the area of interest.

REFERENCES

Ambrose EJ (1965) *Cell Electrophoresis*. Little, Brown and Co., Boston.

Anderson MJ and Waxman SG (1981) Morphology of regenerated spinal cord in *Stenarchus albifrons*. Cell Tissue Res. 219:1–8.

Argiro V, Bunge MB, and Johnson MI (1984) Correlation between growth cone form and movement and their dependence on neuronal age. J. Neurosci. 4:3051–3089.

Barker AT, Jaffe LF, and Vanable JW Jr. (1982) The glabrous epidermis of cavies contains a powerful battery. Am. J. Physiol 242:R358–R366.

Bernstein JJ, Wells MR, and Bernstein ME (1978) Spinal cord regeneration: Synaptic renewal and neurochemistry. In Cotman CW (ed): *Neuronal Plasticity*. Raven Press, New York, pp. 49–71.

Bernstein JJ, Getz R, Jefferson M, and Kelemen M (1985) Astrocytes secrete basal lamina after hemisection of rat spinal cord. Brain Res 327:135–156.

Berry M (1979) Regeneration in the central nervous system. In Smith WT and Cavanaugh JB (eds.): *Recent Advances in Neuropathology*. Churchill Livingstone, New York, pp. 67–111.

Borgens RB (1982) What is the role of naturally produced electric current in vertebrate regeneration and healing? Int. Rev. Cytol. 76:245–298.

Borgens RB (1984) Are limb regeneration and limb development both initiated by an integumentary wounding? Differentiation 28:87–93.

Borgens RB, Vanable JW Jr., and Jaffe LF (1977) Bioelectricity and regeneration. I. Initiation of frog limb regeneration by minute currents. J. Exp. Zool. 200:403–416.

Borgens RB, Vanable JW Jr., and Jaffe LF (1979) Small artificial currents enhance *Xenopus* limb regeneration. J. Exp. Zool. 207:217–225.

Borgens RB, Cohen MJ, and Jaffe LF (1980) Large and persistent electrical currents enter the transected lamprey spinal cord. Proc. Natl. Acad. Sci. U.S.A. 77:1209–1231.

Borgens RB, Roederer E, and Cohen MJ (1981) Enhanced spinal cord regeneration in lamprey by applied electric fields. Science 213:611–617.

Borgens RB, Blight AR, and Murphy DJ (1986a) Axonal regeneration in spinal cord injury: A perspective and new technique, J. Comp. Neurol. 250:157–167.

Borgens RB, Blight AR, Murphy DJ, and Stewart L (1986b) Transected dorsal column axons within the guinea pig spinal cord regenerate in the presence of an applied electric field, J. Comp. Neurol. 250:168–180.

Borgens RB, Blight AR, and McGinnis ME (1987) Behavioral recovery induced by applied electric fields after spinal cord hemisection in guinea pig. Science 238:366–369.

Brighton CT (1981) Current concepts review: The treatment of non-unions with electricity. J. Bone Joint Surg. 63-A:847–859.

Coburn B and Sin WK (1985) A theoretical study of epidural electrical stimulation of the spinal cord—Part I. Finite element analysis of stimulus fields. IEEE Trans. Biomed. Eng. BME-32:971–977.

Das GP (1983) Neural transplantation in the spinal cord of the adult mammal. In Kao CC, Bunge RP, and Reier PJ (eds.): *Spinal Cord Reconstruction*. Raven Press, New York, pp. 367–396.

David S and Aguayo AJ (1981) Axonal elongation into peripheral nervous system "bridges" after central nervous system injury in adult rats. Science 214:931–933.

DeBoni U and Anderchek KE (1986) Quantitative analysis of filapodial activity of mammalian neuronal growth cones, in exogenous electrical fields. In Nuccitelli R (ed.): *Ionic*

Currents in Development. Alan R. Liss, Inc., New York, pp. 285–293.

Eidelberg E, Walden JG, and Nguyen LH (1981) Locomotor control in macaque monkey. Brain 104:647–665.

Foerster AP (1982) Spontaneous regeneration of cut axons in adult rat brain. J. Comp. Neurol. 210:335–356.

Gilson BC and Stensaas LJ (1974) Early axonal changes following lesions of the dorsal columns in rats. Cell Tissue Res. 149:1–20.

Goldberger ME (1980) Motor recovery after lesions. Trends Neurosci. 3:288–298.

Hinkle L, McCaig CD, and Robinson KR (1981) The direction of growth of differentiating neurons and myoblasts from frog embryos in an applied electric field. J. Physiol. 314:121–131.

Horvat JC (1966) Comparison des réactions régénératives provoquées dans le cerveau et dans le cervelet de la souris par des greffes tissulaires intravaciales. CR Assoc. Anat. 51:487–499.

Ingvar S (1920) Reaction of cells to galvanic current in tissue culture. Proc. Soc. Exp. Biol. Med. 17:198–199.

Ingvar D (1947) Experiments on the influence of electric current upon growing nerve cell processes *in vitro*. Acta Physiol. Scand. [Suppl] 13:150.

Jaffe LF and Nuccitelli R (1977) Electrical controls of development. Annu. Rev. Biophys. Bioeng. 6:445–476.

Jaffe LF and Poo M-M (1979) Neurites grow faster toward the cathode than the anode in a steady field. J. Exp. Zool. 209:115–127.

Kakulas BA (1984) Pathology of spinal cord injuries. Cent. Nervous System Trauma 1:117–129.

Kerns JM, Pavkovic IM, Fakhouri AJ, Wickersham KL, and Freeman JA (1987) An experimental implant for applying a DC electrical field to peripheral nerve. Journal of Methods, 19:217–223.

Kiernan JA (1979) Hypotheses concerned with axonal regeneration in the mammalian nervous system. Biol. Rev. 54:153–197.

Kolata G (1982) Grafts correct brain damage. Science 101:342–344.

Kucera P and Wiesendanger M (1985) Do ipsilateral corticospinal fibers participate in the functional recovery following unilateral pyramidal lesions in monkeys? Brain Res 348:297–303.

Lagasse P, Campney HK, Steinberg GS, and Kroll W (1985) Patterned electric stimulation: Treatment of spastic hemiparesis. Arch. Phys. Med. Rehabil. 66:550.

Lampert PW and Cressman M (1964) Axonal regeneration in the dorsal columns of the spinal cord of adult rats. Lab. Invest. 13:825–839.

Loeb GE and Gans C (eds.) (1986) *Electromyography for Experimentalists.* The University of Chicago Press, Chicago, IL, pp. 105–106.

Mackler SA and Selzer ME (1985) Regeneration of functional synapses between individual recognizable neurons in the lamprey spinal cord. Science 229:774–776.

Maron K (1963) Regeneration of the spinal cord in shark embryos (scyliorhinus canicula). Folia Biol. (Krakow) 11:269–275.

Marsh G and Beams HW (1946) *In vitro* control of growing chick nerve fibers by applied electric currents. J. Cell Comp. Physiol. 27:139–157.

Matthews MA, St. Onge MF, Faciano CL, and Gelderd HB (1979) Spinal cord transection: A quantitative analysis of elements of the connective tissue matrix formed within the site of lesion following administration of piromen, cytoxan or trypsin. Neuropathol. Appl.

Neurobiol. 5:161–180.

McCaig CD (1986a) Electric fields contact guidance and the direction of nerve growth. J. Embryol. Exp. Morphol. 94:245–255.

McCaig CD (1986b) Dynamic aspects of amphibian neurite growth and the effects of an applied electric field. J. Physiol. 375:55–69.

McCaig CD (1987) Spinal neurite reabsorption and regrowth *in vitro* depend on the polarity of an applied electric field. Development 100:31–41.

McDevitt L, Fortner P, and Pomeranz B (1987) Application of weak electric field to the hindpaw enhances sciatic motor nerve regeneration in the adult rat. Brain Research 416:308–314.

McMasters RE (1962) Regeneration of the spinal cord in the rat: Effects of piromen and ACTH upon the regenerative capacity. J. Comp. Neurol. 119:113–125.

McQuarrie IG (1981) Acceleration of axonal regeneration rat somatic motoneurons by using a conditioning lesion. In Gorio A, Millesi H, and Mingrino S (eds.): *Posttraumatic Peripheral Nerve Regeneration: Experimental Basis and Clinical Implications.* Raven Press, New York, pp. 49–58.

Nannmark U, Buch F, and Albrektsson T (1985) Vascular reactions during electrical stimulation. Acta. Orthop. Scand 56:52–62.

Nixon BJ, Doucette R, Jackson PC, and Diamond J (1984) Impulse activity evokes precocious sprouting of nociceptive nerves into denervated skin. Somatosensory Research 2: 97–126.

Norlander RH and Singer M (1978) The role of epidyma in regeneration of the spinal cord in the urodele amphibian tail. J. Comp. Neurol. 180:349–374.

Nornes H, Bjorklund A, and Stenevi U (1983) Reinnervation of the denervated adult spinal cord of rats by intraspinal transplants of embryonic brainstem neurons. Cell Tissue Res. 230:15–35.

Patel NB and Poo MM (1982) Orientation of neurite growth by extracellular electric fields. J. Neurosci. 2:483–496.

Patel NB and Poo MM (1984) Perturbation of the direction of neurite growth by pulsed and focal electric fields. J. Neuroscin 4:2939–2947.

Piatt J (1955) Regeneration in the central nervous system of amphibia. In Windle WF (ed.): *Regeneration in the Central Nervous System.* CC Thomas, Springfield, IL, pp. 20–46.

Politis MJ, Zanakis MF, and Albala BJ (1988) Facilitate regeneration in the rat peripheral nervous system using applied electric fields. Journal of Trauma, 28:1–6.

Pomeranz B (1986) Effects of applied DC fields on sensory nerve sprouting and motor-nerve regeneration in adult rats. In Nuccitelli R (ed.): *Ionic Currents in Development.* Alan R. Liss, Inc., New York, pp. 251–260.

Pomeranz B, Mullen M, and Markus H (1984) Effect of applied electrical fields on sprouting of intact saphenous nerve in adult rat. Brain Res. 303:331–336.

Ramón y Cajal S (1928) *Degeneration and Regeneration in the Nervous System.* Hoffner, New York.

Reier PJ, Stensaas LJ, and Guth L (1983) The astrocytic scar as an impediment to regeneration in the central nervous system. In Kao CC, Bunge RP, and Reier PJ (eds.): *Spinal Cord Reconstruction.* Raven Press, New York, pp. 163–195.

Richardson PM and Issa VMK (1984) Peripheral injury enhances central regeneration of primary sensory neurones. Nature 309:791–793.

Richardson PM, McGuinness UM, and Aguayo AJ (1980) Axons from CNS neurones regenerate into PNS grafts. Nature 284:264–286.

Richardson PM, Issa VM, and Aguayo AJ (1984) Regeneration of long spinal axons in the rat. J. Neurocytol. 13:165–182.

Robinson GA and Goldberger ME (1985) Interfering with inhibition may improve motor function. Brain Res. 346:400–421.

Roederer E, Goldberg NH, and Cohen MJ (1983) Modification of retrograde degeneration in transected spinal axons of the lamprey by applied DC current. J. Neurosci. 3:153–160.

Roman GC, Rowley BA, Strahlendorf HK, and Coates PW (1987) Stimulation of sciatic nerve regeneration in the adult rat by low intensity electric current. Exp. Neurol. 98:222–232.

Rovainen CM (1967) Physiological and anatomical studies on large neurons of central nervous system of the sea lamprey: (Petromyzon marinus), I Muller and Mauthner cells. J. Neurophysiol. 30:1000–1023.

Simpson SB (1983) Fasiculation and guidance of regenerating central axons by the ependyma. In Kao CC, Bunge RP, and Reier PJ (eds.): *Spinal Cord Reconstruction.* Raven Press, New York, pp. 151–163.

Sisken BF and Smith SD (1975) The effect of minute directed electrical currents on cultured chick trigeminal ganglia. J. Embryol. Morphol. 33:29–41.

Stensaas LJ (1983) Regeneration in the spinal cord of the newt, *Notophthalmus (triturus) pyrrhogaster.* In Kao CC, Bunge RP, and Reier RJ (eds.): *Spinal Cord Reconstruction.* Raven Press, New York, pp. 121–149.

Stewart WW (1978) Functional connections between cells as revealed by dye-coupling with a highly fluorescent naphthalimide tracer. Cell 14:741–761.

Sugar O and Gerard RW (1940) Spinal cord regeneration in the rat. J. Neurophysiol. 3:1–19.

Wallace MC, Tator CH, and Piper I (1987) Recovery of spinal cord function induced by direct current stimulation of the injured rat spinal cord. Neurosurgery 20:878–884.

Weiss P (1934) In-vitro experiments on the factors determining the course of the outgrowing nerve fiber. J. Exp. Zool. 68:373–448.

Wheeler PC, Wolcott LE, Morris JL, and Spangler RM (1971) Neural consideration in the healing of ulcerated tissue by clinical electrotherapeutic application of weak direct current: Findings and theory. In Reynolds DV and Sjoberg AE (eds.): *Neuroelectric Research.* CC Thomas, Springfield, IL, pp. 83–99.

Wood MR and Cohen MJ (1979) Synaptic regeneration in identified neurons of the lamprey spinal cord. Science 206:344–347.

Wood MR and Cohen MJ (1981) Synaptic regeneration and glial reactions in the transected spinal cord of the lamprey. J. Neurocytol. 10:57–79.

APPENDIX: SUPPLY COMPANIES

1. ACME Chemical and Insulation Company, Division of Allied Corporation, P.O. Box 1404, New Haven, CT 06505, (203) 562–2171

2. American Scientific Products, 1210 Waukegan Road, McGaw Park, IL 60085, (800) 323–4515

3. Brownell Electro, 565 Busse Road, Elk Grove, IL 60007

4. Cooner Wire, 9186 Independence, Chatsworth, CA 91311, (818) 882–8311

5. Engelhard, 700 Blair Road, Carteret, NJ 07008, (201) 321–5722

6. General Medical, 1850 W. 15th St., Indianapolis, IN 46202, (317) 634–8560

7. Hamilton Avnet, 485 Gradle Drive, Carmel, IN 46032, (800) 692–6025

8. Newark Electronics, 1676 Viewpond S.E., Grand Rapids, MI 49508, (800) 253–9221

Electric Fields in Vertebrate Repair, pages 171–224

CHAPTER 5

Integumentary Potentials and Wound Healing

Joseph W. Vanable, Jr.

Department of Biological Sciences, Purdue University, West Lafayette, Indiana 47907

INTRODUCTION

Compared with most aspects of development, the healing of wounds in the skin seems to be a relatively simple process. It is, nonetheless, a process whose repeated effective execution during the lifetime of most organisms is of no small consequence to their well-being (Needham, 1964). Furthermore, there are numerous circumstances in which wound healing does not occur properly. For instance, paraplegics and other persons confined to bed or to wheelchairs frequently develop decubitus ulcers, severe and difficult-to-heal skin lesions at points where the prolonged pressure of the skeleton produces an ischemia that leads to serious breakdown of the skin (Wolcott et al., 1969; Sather et al., 1977; Thiyagarajan and Silver, 1984). Wounds in the skin of persons with certain diseases, such as diabetes (Goodson et al., 1980), epidermolysis bullosa (Carter and Lin, 1988), chronic venous stasis, peripheral arteriosclerosis, and sickle cell anemia (Wolcott et al., 1969), do not heal easily. It is worthwhile, therefore, to examine the process of wound healing, not only for the intrinsic value of understanding an important biological process, but also to identify ways to promote this process in individuals in which it is not occurring properly.

The promotion of wound healing, to be sure, is a topic that has received considerable attention (e.g., Alvarez and Biozes, 1984; Eaglstein, 1984b; Eckersley and Dudley, 1988); to review all facets of this subject would be impractical. However, a focus on just one aspect of this area, the role of epidermally generated electrical potentials in the healing of vertebrate skin, is both feasible and worthwhile. There are indications that electrical potentials can promote the healing of wounds, although the evidence is by no means clear. Moreover, the rationale for why there should be an electrical component to wound healing is only in the earliest stages of its development. I therefore propose to review briefly the process of wound healing, focusing particularly on those aspects that might be affected by steady electrical fields, and then consider what indications there are that these fields play a role in promoting wound healing. In doing this, I will attempt to develop a rationale for how fields might act in this process.

THE PROCESS OF WOUND HEALING: A BRIEF REVIEW

There seem to be almost as many systems of dissecting wound healing into its component parts as there are descriptions of the process. A useful scheme is outlined by Pollack (1984), who identifies six aspects of vertebrate wound healing: inflammatory response, epithelization, fibroplasia, neoangiogenesis, wound contraction, and collagen remodeling. It would be well to review these aspects (several of which occur nearly simultaneously), to provide a common ground for considering which ones might be affected by electric fields. Most of the information available on wound healing concerns mammals; the description that follows draws heavily on several recent reviews (Lambert et al., 1984a,b; Peacock, 1984; Pollack, 1984; Winter, 1972). There are, however, several notable studies of healing in amphibians (Lash, 1955; Derby, 1978; Radice, 1980a,b; Donaldson and Dunlap, 1981; Donaldson et al., 1982, 1985; Mahan and Donaldson, 1986; Atnip et al., 1987) that provide valuable insight into certain processes of wound healing, particularly the migration of epidermal cells to cover the wound; these will be briefly considered, where appropriate.

Inflammatory Response

When a wound bleeds, the cavity produced by the wound becomes filled with both cellular and dissolved elements of the blood, several of which contribute to providing a substratum hospitable to the adhesion of cells migrating into the wound. Platelets release thrombospondin, a protein that helps in this capacity (Tuszynski et al., 1987). Among the dissolved elements is fibrinogen (Brown et al., 1988), which, when it clots, produces a substratum for future epithelial and fibroblast cell migration (Pollack, 1984). Also found in clotted blood is fibronectin, which combines with fibrin and collagen to make an especially attractive substratum for cell migration (Clark et al., 1976; Grinnell, 1984; Pollack, 1984; Donaldson et al., 1985), although the nature of this interaction in amphibians is unclear (Atnip et al., 1987). The fibronectin brought in with the blood is later augmented by its production by the fibroblasts that later migrate to the wound (Lambert et al., 1984a).

Polymorphonuclear leukocytes (neutrophils) appear several hours after wounding, migrating from capillaries that, although intact, have become "leaky." They serve as the earliest debriders of the wound, ingesting debris created by the wounding and bacteria contaminating the wound (Pollack, 1984). There is evidence (Deuel et al., 1981, 1982; Pierce et al., 1988) that this immigration is stimulated by the chemotactic activity of platelet factor 4 and platelet-derived growth factor (PDGF) released by platelets deposited in the wound cavity as a result of the wound's bleeding. There is also evidence that PDGF (Pierce et al., 1988) and another growth factor found in platelets, transforming growth factor-β (TGF-β) (Mustoe et al., 1987), accelerate wound healing at least in part by promoting the migration of monocytes and fibroblasts to the wound.

When a wound is exposed to the air, the neutrophils are soon dried up in the wound crust. However, if an occlusive dressing is used, the drying is prevented, and these cells and the other leukocytes that follow them accumulate between the film used for the dressing and the wound bed. Short or long-lived, the neutrophil appears to be a dispensable part of wound healing, provided that the wound is not badly infected (Ross, 1980; Mertz and Eaglstein, 1984): Simpson and Ross (1972) found that when wounds are depleted of neutrophils by antineutrophil serum, wound healing still occurs normally.

One to two days after wounding (in mammals), monocytes arrive at the wound site (Fishel et al., 1987) and transform into macrophages (Ross and Benditt, 1961), which continue the process of phagocytosis: necrotic tissue, foreign material, and other debris are ingested and recycled. There is evidence that in addition to performing this crucial function, macrophages promote the migration of fibroblasts to the wound, as well as their subsequent proliferation upon arrival (Leibovich and Ross, 1975) (see pp. 177–178).

Epithelization

While the inflammatory phase of wound healing is progressing, epidermal cells begin migrating to cover the wound. In shallow wounds in mammalian skin, the vast majority of these cells are derived from the appendages of the skin, such as hair follicles and sweat glands (Winter, 1972). In deep wounds, this source of cells is removed, leaving only the edge of the wound to supply the cells. In each case, cells begin migrating to cover the wound at a rate of about 7 μm/hr as early as 8 hours after wounding. It is clear that superficial wounds can be covered more quickly than deep ones, because such wounds leave hair follicles intact, and they provide many loci of epithelial cell sources scattered in these wounds. For example, in a 6.25-mm^2 shallow wound in pig skin, there are 160 hair follicles to supply cells to cover the wound (Winter, 1972). Epithelization can be promoted by a factor or factors produced by the epidermal cells themselves. Eisinger et al. (1988) have found that extracts and supernatant fluids of cultured epidermal cells (epidermal cell-derived factors [EDF]) promote epithelization of wounds made in the skin of domestic swine by stimulating keratinocyte migration and proliferation.

In a wound that has been allowed to dry, epithelial cell migration occurs within the dermis that is situated just below the dried surface of the wound. In order to make their way through this mat of collagen fibers, these migrating cells secrete collagenase (Grillo and Gross, 1967). If the drying of the wound is excessive, this burrowing process occurs at even deeper levels, and epithelization is retarded further (Lambert et al., 1984b; Eaglstein, 1984a). If the wound is properly covered, however (that is, within 2 hours of wounding and maintained for at least 24 hours [Eaglstein et al., 1988]), epidermal cell migration occurs on top of the dermal layer, not through it (Winter, 1972; Eaglstein, 1984a). Such wounds are covered more quickly than wounds that have been allowed to dry (Winter, 1972; Eaglstein, 1984a; Hinman and Maibach, 1963; Fisher and Maibach, 1972; Chvapil et al., 1986).

In order to migrate, epidermal cells must detach from their subjacent basement membrane. To do this they lose hemidesmosomal junctions, which become fewer and less substantial (Martinez, 1972). They also increase their content of myosin (Gabbiani et al., 1978) and develop actin microfilaments (Krawczyk, 1971; Gabbiani and Ryan, 1974; Reibel et al., 1978; Andersen, 1980), whose integrity is understandably crucial for the cell migration that produces epithelization. In the newt, if microfilament structure is broken down by cytochalasin B treatment, epithelial migration to cover the wound is inhibited (Donaldson and Dunlap, 1981).

In the mammalian literature, there is disagreement about the way in which epidermal cells progress to cover a wound, particularly concerning the extent to which epithelial cell migration and cell division are involved. Many investigators (Viziam et al., 1964; Odland and Ross, 1968; Croft and Tarin, 1970; Martinez, 1972; Lambert et al., 1984b) describe the movement of a sheet of epithelial cells across the connective tissue bed remaining after wounding, not unlike that seen with great clarity in amphibian wound healing (Radice, 1980a,b; see below). However, two investigators of mammalian wound healing, Krawczyk (1971) and Winter (1972), have proposed a rolling leapfrog mechanism, whereby the first cells to migrate progress only two or three cell diameters over the wound surface, providing a substratum on which the next cells migrate. These cells, in turn, provide a substratum for the next cells, and so on.

Winter (1964, 1972), studying wound healing in 15–20-week-old pigs, proposes that as this process continues, the cells for covering the wound are not derived from undamaged cells at the wound's edge, but from cell division that occurs in the epidermal cells that initially migrate, and their progeny. This suggestion is at variance with observations made by Krawczyk (1971), who found that the rate of epithelization of blister wounds made in the skin of 2-day-old mice was unaffected by colchicine or vinblastine sulfate treatment. (The effectiveness of this treatment in blocking mitosis was assessed by observation of increasing numbers of arrested mitoses in epithelial root sheath cells of hair follicles 12, 18, and 24 hours after treatment.) Also arguing against a crucial role for mitosis in epithelization are the observations of Fisher and Maibach (1972): they found that biopsy punch wounds in human skin healed significantly faster in wounds occluded with tape or with plastic film than did uncovered wounds, with no significant differences in the mitotic indices estimated in these groups.

Although it may not be critical for the initial covering of the wound, cell division is certainly required for the restoration of the full thickness of the new epidermis. This cell division usually begins in basal cells 24 hours after they have stopped migrating and have settled down on the wound surface (Krawczyk, 1971). It is promoted by epidermal growth factor (EGF) and, even more effectively, by transforming growth factor-α and vaccinia growth factor, which are homologous to EGF (Schultz et al., 1987). The effect of EGF can be important for proper healing, at least of deep corneal wounds: Tsutsumi et al. (1988) have identified an EGF-like substance in mouse tears. In mice whose tears had been depleted of this EGF-like substance by prior submandibular gland removal, deep corneal wounds did not heal normally unless exogenous EGF was applied.

Epithelization can be seen much more clearly in amphibians than it can in mammals, and there is a strong consensus that the migration of epidermal cells from the edge of the wound is the only process by which epithelization is achieved. Lash (1955) has traced this cell movement in the epithelization of 1-mm-diameter wounds made in larval *Ambystoma* skin by following cells marked with carmine particles. Radice (1980a), taking advantage of the optical clarity of *Xenopus* tadpole tail fin, directly observed the migration of epidermal cells to cover 200-μm wounds made in the epidermis, using Nomarski differential interference microscopy. To complement these elegant in vivo studies, Radice (1980b) also examined the cytological details of this migration, which confirmed the in vivo observation that in this simple epithelium composed of two cell layers, it is the cells of the lower layer that actively migrate over the dermal substratum as a loosely connected sheet to cover the wound, pulling the upper cells along with them. As is the case with newt wound healing (Donaldson and Dunlap, 1981; Mahan and Donaldson, 1986), lamellipodia of the cells at the leading edge of the sheet appear to provide the motive force for this migration.

The rate of progress of these loose sheets is much more rapid than is typical for healing of wounds in mammalian skin. Wound healing in mammalian skin occurs over days or even weeks, with epithelial migration rates ranging from 7 to 20 μm/hr [(Winter, 1964); the rates at the high end of the range are for wounds that are covered, and therefore not allowed to dry]. On the other hand, amphibian skin wounds heal within hours, with migration rates ranging from 60 to more than 600 μm/hr (Lash, 1955; Winter, 1964; Donaldson and Dunlap, 1981; Donaldson et al., 1982; Radice, 1980a,b; L.R. Robinson, 1985). Similarly, the delay between wounding and the beginning of epithelial cell migration is far less in amphibians than in mammals: Radice (1980a) could detect the beginning of lamellipodium extension within seconds after wounding; in mammalian skin wounds, it is more typical to see the earliest beginning of epithelial cell migration 8 hours after wounding (Winter, 1972).

Epithelization of wounds made in two unusual areas of the mammalian integument, the epithelium of the oral cavity and the cornea, may provide examples that perhaps bridge the gap between wound healing in amphibian and mammalian skin. In linear incision wounds of rat gingiva (Martinez, 1972), epithelial cell migration begins between 3 and 6 hours after wounding, and epithelization is complete by 12 hours, as opposed to the usual 72 hours in comparable skin wounds. A direct comparison of the healing of comparable wounds made in the oral mucosa and back skin of the same rats directly confirms that the former is distinctly more rapid than the latter (Sciubba et al., 1978): initiation of epithelization, completion of epithelization, and repair of supporting connective tissue were all more rapid in the healing of the oral mucosa wounds than in the skin wounds. Phagocytosis by macrophages was more active at 72 hours in oral mucosa wounds than in the skin wounds, but there was no appreciable difference in either the time of onset of this activity or the activity at 96 hours postwounding in mucosa and skin wounds.

Corneal wound healing is also more rapid than the healing of skin wounds: rates of epithelial migration of 15 to 20 μm/hr, characteristic of rapidly healing covered skin wounds, are typical (Winter, 1964), and rates as high as 67 μm/hr have been reported (Ubels et al., 1982). The delay between wounding and the beginning of cell migration is also short, compared with typical skin wounds; response to a needle prick wound occurs between 1 and 3 hours (Friedenwald and Buschke, 1944; Maumenee, 1964). A common denominator of these rapid rates is that in each case, including skin wounds that are occluded, the wound is kept moist during healing. As will be seen later in the discussion (see pp. 185–187), this fact may have significance with respect to the role of electric fields in wound healing.

Fibroplasia

To heal a wound made in the skin, it is not sufficient merely to clear away bacteria and the debris caused by the damage and to cover the defect with an epithelium. In both deep and shallow wounds (Alvarez et al., 1983b) the skin's connective tissue must be repaired, also. This is the function of the fibroblast.

The source of the fibroblasts that repair the connective tissue during wound healing has been debated in the past (see Ross and Odlund, 1968; Ross, 1968; Chapter 4 in Peacock, 1984, for discussion). Particularly persistent has been the notion that cells from the blood, principally mononuclear leukocytes, may transform into fibroblasts. However, the weight of evidence now is that this process does not occur, but rather that wound fibroblasts originate from the connective tissue surrounding the wound (Ross, 1968).

During the first several days after wounding, therefore, fibroblasts migrate into the wounded area, and begin to proliferate and mature into cells that synthesize collagen and other glycoproteins that form the fabric of the skin's dermis (Ross, 1968; Odlund and Ross, 1968). This fibroblast migration occurs while polymorphonuclear leukocytes are debriding the wound, at about the same time that lymphocytes are moving in and transforming into macrophages to further ingest debris, and while epithelial cells are covering the wound. There is strong evidence that macrophages promote both fibroblast migration to the wound site, and their subsequent proliferation there. When the macrophage population of a wound is reduced to very low levels by combined administration of cortisone and antimacrophage serum, the arrival of fibroblasts is delayed, and their cell division rate depressed for several days (Leibovich and Ross, 1975). A considerable amount of effort has been expended on further characterizing this effect (Leibovich and Ross, 1975; DeLustro et al., 1980; Martin et al., 1981; Bitterman et al., 1982; Deuel et al., 1981; Glenn and Ross, 1981; Martin et al., 1981, 1984;

Diegelmann et al., 1982; Wharton et al., 1982; Dohlman et al., 1984; Leslie et al., 1984; Baird et al., 1985; reviewed recently by Shimokado et al. 1985; by Martinet et al., 1986, and by Turck et al., 1987). It would appear that macrophages produce several distinct growth factors, and that at least some of the macrophage-derived growth factors (MDGF) that have been isolated and partially purified in these studies are closely related, or even identical to other, rather well-characterized growth and chemotactic factors. Baird et al. (1985) present evidence that the MDGF they studied is indistinguishable from fibroblast growth factor (FGF) studied and characterized by Gospadarowicz et al. (1978). There is also evidence that part of MDGF's activity can be attributed to interleukin-1 (IL-1) (see Dinarello's review, 1984, although the MDGF studied by Dohlman et al., 1984, was apparently free of IL-1 activity). Leslie et al. (1984) and Martin et al. (1984) present evidence that activated macrophages can produce arachadonic acid metabolities other than prostaglandin, including leukotriene B_4, which act as chemotactic agents for monocytes. Finally, it is probable that still another type of MDGF consists of one or more forms of PDGF (Shimokado et al., 1985; Martinet et al., 1986; Turck et al., 1987), which is also a chemotactic agent for monocytes (Deuel et al., 1982; Ross et al., 1986). It is apparent that the molecular nature of MDGF activity is complex, and far from being settled. However, there is ample evidence that, in addition to their debriding function, wound macrophages produce factors that attract monocytes and fibroblasts to the wound, and promote their proliferation there.

Neoangiogenesis

Just before fibroblasts begin to synthesize collagen and other connective tissue components actively, buds develop from capillary tubes in the region surrounding the wound (Ross and Benditt, 1961). These buds grow into the wound, making numerous anastamoses and developing a rich network of vasculature (Schoefl and Majno, 1964) that supplies oxygen and nutrients to the healing wound, and carries away carbon dioxide and other wastes. Oxygen is critical for many processes, but it is especially so for collagen synthesis. According to Niinikoski (1980), collagen synthesis is impaired at PO_2s lower than 20 mm Hg; it is therefore understandable that the vascularization of a wound precedes collagen synthesis by the fibroblasts that have migrated into the wound region.

Oxygen also promotes rapid epithelization. Silver (1972) and Winter (1972), studying the effect of occlusive dressings on the rate of epithelization, both found a direct correlation between the oxygen permeability of the film used to occlude wounds and the rate of epidermal cell migration to cover them.[1] Indeed, locally applied hyperbaric oxygen has been a successful strategy in speeding the epithelization of second-degree burn wounds (Miller, 1980). [Administration of oxygen via inhalation often has little effect, or can even inhibit wound healing (Kulonen and Niinikoski, 1968), because abnormally high blood P_{O_2} usually results in vasoconstriction of peripheral vessles (Silver, 1980).]

Vascular sprouts make their way into the wound region via a multiple-step process (Garner, 1986). The endothelial cells that are about to form a sprout enlarge, and their intracellular junctions break down, so that migration can occur. The basement membrane of the vessel that is about to sprout is destroyed locally by collagenolysis so that the endothelial cells can bleb through and migrate away. As they migrate, they proliferate in a zone just proximal to the leading tip. While all this is going on, a lumen forms in the sprout, beginning at the region closest to the vessel from which it originated.

Wound fluids promote angiogenesis (Banda et al., 1982) via a multitude of its components (see Garner, 1986, and Folkman and Klagsbrun, 1987, for recent reviews.) Among the possibilities are several prostaglandins, macrophage pluripotential growth factor, immunocompetent T lymphocytes, polymorphonuclear neutrophil leukocytes, specific angiogenic factors, and relatively nonspecific growth factors, such as platelet-derived growth factor, epidermal growth factor, and fibroblast growth factor (FGF) (Garner, 1986; Folkman and Klagsbrun, 1987). It appears as though capillary endothelial cells themselves can promote their own growth, by producing FGF (Schweigerer et al., 1987). Another candidate for stimulation of angiogenesis in wound healing is, again, a factor or factors released by macrophages. Clark et al. (1976) have briefly presented data that argue for a role of macrophages (particularly activated macrophages) in the stimulation of neovascularization, as have Polverini et al. (1977) and Polverini and Leibovich (1984), in somewhat more detail. Angiogenic factors, therefore, are not restricted just to tumors, where they first were studied so elegantly (Folkman et al., 1971), but have also been reported in activated macrophages (Wall et al., 1978; Greenberg and Hunt, 1978; Thakral et al., 1979; Martin et al., 1981; Knighton et al., 1983) and in skin (Wolf and Harrison, 1973). The FGF (Gospodarowicz et al., 1978) that is produced by activated macrophages

[1]Alvarez et al. (1983a), using two quite different dressings, DuoDERM (oxygen-impermeable) and Op-Site (oxygen-permeable), found no advantage of the latter over the former dressing. It is possible that these wounds were so shallow that the circulation provided the major source of oxygen supply (Eaglstein, 1984a), or that other favorable properties of DuoDERM outweighted any disadvantage it might have in its lack of oxygen permeability.

(Baird et al., 1985) is able to stimulate cell division in vascular endothelial cells (Baird et al., 1985; Ross, 1986). Knighton et al. (1983) provide convincing evidence that PO_2 controls production of angiogenic factor by macrophages; macrophages cultured in 20% O_2 produced no angiogenic factor, while macrophages cultured in 2% O_2 secreted a material into the culture medium, which, when concentrated into a pellet that was then implanted into rabbit cornea, strikingly promoted vascularization of the cornea. When the macrophages cultured in 2% O_2 were returned to 20% O_2, they stopped producing this angiogenic factor. It is of interest that removal of angiogenic factor leads to regression of blood vessels newly formed in test corneas (Ausprunk et al., 1978). It could well be that the reduction in marked vascularity of wounds that is commonly seen as the synthesis of collagen tapers off is attributable to this PO_2 control of macrophage angiogenesis factor production (Knighton et al., 1983).

If neoangiogenesis is promoted by a factor or factors produced by macrophages, then it would be expected that depriving an animal of monocytes would considerably inhibit the growth of capillaries into a wound. Just such a result is reported by Fromer and Klintworth (1975). Using irradiation with 1,500 rads or local injections of methyl prednisolone acetate to render their rats leukopenic, they found that without leukocyte infiltration of silver nitrate wounds in the cornea, no vascularization occurred. When lower doses of X-irradiation were used (1,100–1,300 rads), leukocytes invaded the wound, and neoangiogenesis occurred that was indistinguishable from that occurring in nonirradiated rats, or in rats whose heads had been irradiated with 1,500 rads while the rest of the body was shielded. In these experiments, the highest leukocyte count observed in animals perceived to have failed to initiate neoangiogenesis was 1% of normal, while at the lower doses (which permitted normal angiogenesis), the leukocyte count ranged from 2.5% to 15% of normal.

However, a caveat about the role of macrophages in angiogenesis is provided by the work of Sholley and his colleagues (Sholley and Cotran, 1978; Sholley et al., 1978). In one series of experiments (Sholley and Cotran, 1978), they rendered rats leukopenic with 800 rads of X-irradiation to the whole body, with only the area to be wounded shielded. This dose was less debilitating than the one used by Fromer and Klintworth (1975). Two days after irradiation, at the time when thermal wounds were made in the shielded area, circulating monocytes had been reduced to 3.6% of preirradiation levels, and by 5 days, to 1.7%. (By 5 days postirradiation, neutrophils and lymphocytes were also drastically depleted, to 3.1% and 0.9%, respectively, of preirradiation values.) Despite this drastic reduction in monocytes, and therefore, presumably, of activated macrophages, there was, if anything, a slightly *higher* ^{3}H-thymidine labeling index in the vascular endothelium of wounds of irradiated animals than in wounds made in nonirradiated control animals ($0.025 < P < 0.05$, control wounds vs. wounds in animals irradiated 7 days prior to sampling). In wounds made in irradiated animals, the connective tissue was highly vascularized, and epithelization was advanced, even though monocytes were still only at 1.3% of preirradiation values. In a second series of experiments (Sholley et al., 1978), rats were made more severely leukopenic by combined irradiation with 800 rads and injections with an antibody produced in response to a peritoneal exudate cell population consisting of 75% neutrophils and ~25% monocytes ("antineutrophil serum"). With this regimen, the total circulating white blood count was reduced to very low values, and there was no detectable monocyte or neutrophil infiltration of the wounds. Nevertheless, angiogenesis still occurred, albeit to a reduced extent: vascular length was 67% of control values on day 3, and 33% on day 4 after wounding with silver nitrate.

Similar results were seen by Eliason (1978) in observations made on burn wounds made in rabbit cornea. Blood vessels invaded the corneas of animals made leukopenic by X-irradiation, but to a lesser extent than they did in controls; at 2 days postcautery, leukopenic capillary invasion was approximately 85% of the extent seen in controls; at 3 days, 83%; at 4 days, 71%, and at 5 days, 75%. The degree of leukopenia was comparable to that achieved in the studies discussed above: at the time of cautery, the peripheral white blood count was 2.4% of normal, and this dropped to 0.5% of normal by day 5 after cautery.

There is one other study that could be of relevance to this issue, that of Leibovich and Ross (1975) with wound healing in guinea pigs made nearly completely devoid of macrophages via combined steroid and antimacrophage serum (see pp. 177–178). Although no clear statement is made about angiogenesis in these wounds, a delay in vascularization was seen in this study (Ross, personal communication). In these experiments, as with the study of Fromer and Klintworth (1975), monocytes were almost completely eliminated.

The conflict raised by these studies is not readily resolved. Sholley et al. (1978) quite reasonably attribute the difference between their results and those of Fromer and Klintworth (1975) to two factors: 1) the X-ray dose used in the latter experiments was well over the LD_{50} for their rats, and so survival times after wounding were short (2 days as opposed to their 4 days), and the surviving animals were severely debilitated; and 2) vascularization was studied without prior perfusion of colloidal carbon to increase the contrast of the newly formed blood vessels. However, in their discussion they do not address the local prednisolone injection series by Fromer and Klintworth (1975), in which no vascular infiltration into the cornea was observed during the course of a 2-week observation period.

It must be concluded that, although there is strong evidence that macrophages produce a substance or substances that can greatly stimulate neoangiogenesis, it is not yet clear that this stimulation is essential for the occurrence of this process during wound healing.

Wound Contraction

The size of a wound can decrease markedly via wound contraction, a process that occurs in certain areas of human skin, and more widely in other animals. Wound contraction occurs independently of epithelization and is caused by a special type of fibroblast, the contractile myofibroblast (Ross, 1968; Chapter 4 in Peacock, 1984). Myofibroblasts appear to be derived from fibroblasts; they contain actin and resemble smooth muscle cells in structure and function (Chapter 4 in Peacock, 1984; Lambert et al., 1984a). The centripetal pull of these cells serves to draw the edges of the wound together, particularly when the wound is made in skin that is only loosely connected to underlying tissues. This process can be quite rapid, ~700 μm a day, and so it can be a significant factor in healing (Pollack, 1984). Wound contraction can greatly decrease the area of the scar that ultimately forms (Zitelli, 1984), but it also can seriously distort the end result (Peacock, 1984, Chapter 6).

Collagen Remodeling

The collagen that is first synthesized by the fibroblasts that migrate into the wound is in the form of tropocollagen fibrils, which are thin and randomly oriented. Some of this newly deposited collagen is digested by collagenase and replaced by newly synthesized collagen having a high degree of cross-linking and a primary orientation that is parallel to lines of stress, providing strength to the healed wound (Lambert et al., 1984a; Pollack, 1984). Collagen remodeling begins during the period of fibroplasia, and continues for some time, even for more than a year (Lambert et al., 1984a). Typically, this process is well controlled, and usually results in an improvement of the appearance of the healed wound. However, under certain not very well understood circumstances, this new collagen synthesis occurs to excess, producing a hypertrophic scar or in extreme cases a keloid, a benign but usually grotesque fibrous tumor that invades areas beyond the original boundaries of the lesion (Garcia-Velasco, 1972; Murray et al., 1984).

ELECTRICAL FIELDS AND HEALING OF SKIN WOUNDS

Epidermis of the Skin Acts as a Battery

To understand how electrical fields might contribute to wound healing, it first is essential to discuss several aspects of the electrical characteristics of vertebrate skin that are relevant to this issue. Clarification of the electrical characteristics of skin that are relevant to wound healing began many years ago, but only recently has real progress been made in this area. The existence of wound currents has been recognized for more than 200 years; in early experiments, about 1 μA of current was found to leave a wound in human skin immersed in saline (see Barker et al., 1982 and Jaffe and Vanable, 1984 for brief reviews).[2] More recently, Illingworth and Barker (1980), measured currents with densities of from 10–30 $\mu A/cm^2$ leaving the stump surface of children's fingers whose tips had been accidentally amputated. Subsequently, in collaboration with Barker, Jaffe and Vanable (Barker et al., 1982) studied in some detail the similar current-generating capacity of guinea pig (cavy) skin, and how it might relate to wound healing. Work has also begun with another system, amphibian skin (L.R. Robinson, 1985; Stump and Robinson, 1986; McGinnis and Vanable, 1986a,b; Rajnicek et al., 1988), in which, as mentioned above, wound healing is more rapid and perhaps simpler to study than mammalian wound healing.

The electromotive force (EMF) driving currents from wounds made in skin is the transepithelial potential (TEP) that the epidermis actively maintains across itself. This epidermal battery is best understood in the amphibian, which has been intensively studied for many years (recently reviewed by Kirschner, 1983; Van Driessche and Zeiske, 1985; Sariban-Sohraby and Benos, 1986). Embedded in the outer membrane of the outer living layer of epidermal cells are Na^+ channels that are radically different from the Na^+ channels usually found in plasma membranes of most other cell types. In contrast to the latter, the Na^+ channels of the apical membrane of the skin's mucosal surface allow Na^+ to diffuse from outside to inside the cell (Fig. 1). This diffusion occurs because these cells maintain very low internal concentrations of Na^+ by the active outward transport of Na^+ by their basal and

[2]The standard convention is to consider current a flow of positive charge. It is important to recall that while in metallic conductors current is the flow of electrons, in aqueous solutions current is the flow of charged ions.

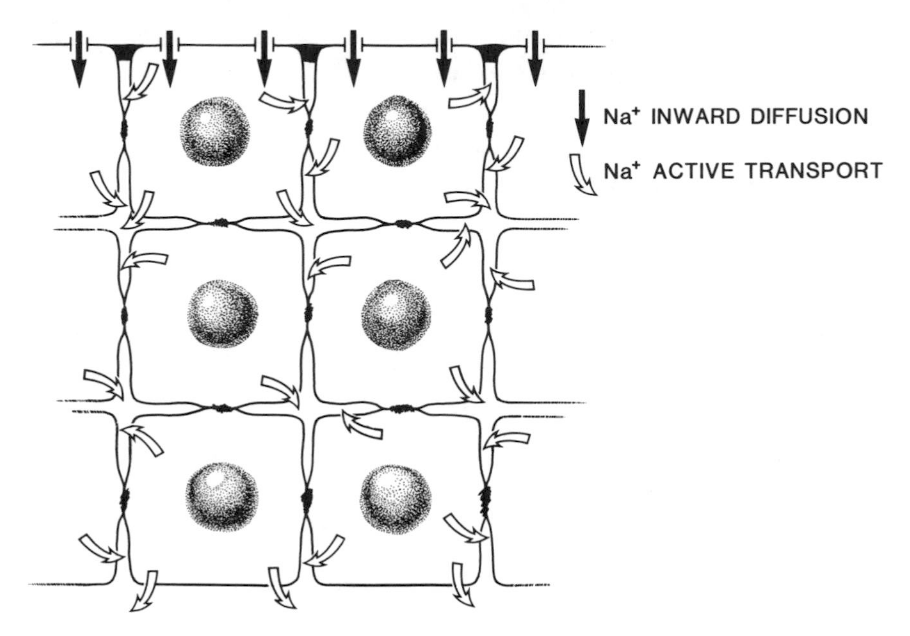

Fig. 1. Diagram of a Na^+-transporting syncytial epithelium. Sodium ions enter the outer (in this diagram, upper) cells of the epithelium via specific channels in the outer membrane of these cells, migrating along a steep electrochemical gradient. Once in the cell, they can diffuse to other cells of the epithelium via the gap junctions that join the cells of the epithelium, and they can be actively transported from all of these cells via electrogenic ''pumps'' that are in all of the plasma membranes of the epithelium except the outer membrane that has the Na^+ channels. This results in a transport of Na^+ from the water bathing the epithelium to the internal body fluids of the animal, and the generation of a potential of the order of 50 mV across the epithelium. Tight junctions between cells at the outer edge of the epithelium minimize the escape of the positive charge transported inwards, and therefore minimize the collapse of the potential generated by the epithelium's electrogenic Na^+ transport. See text for details and references.

lateral membranes, and by the entire membrane of the other epidermal cells below the mucosal layer (with which the surface cells interconnect via gap junctions [Farquhar and Palade, 1965]). This outward active transport keeps the internal concentration of Na^+ in the epidermal cells low, providing a chemical gradient down which the external Na^+ diffuses into the epidermal cells (Farquhar and Palade, 1964; Mills et al., 1977; Rick et al., 1978; DiBona and Mills, 1979; Robinson and Mills, 1987), resulting in a net transport of Na^+ from the water in which the animal is immersed into its internal body fluids, and the generation of an electrical potential across the epidermis, with the body fluids positive with respect to the outside of the skin. Tight junctions (Farquhar and Palade, 1964, 1965) at the outer regions of the outer cells minimize leaks of ions out of the body fluids back into the water in which the animal is immersed; these, therefore, minimize local current loops that would short-circuit the epidermal battery and collapse its potential.

The potential across amphibian skin varies with the species and depends a good deal on the composition of the water in which the animal is immersed. For example, for *Rana pipiens* immersed in tap water (0.25 mM Na^+), the TEP is about 25 mV; in water that is 111 mM Na^+, the TEP is almost 70 mV (Bentley, 1975). Skin of the red-spotted newt, *Notophthalmus viridescens,* immersed in water that is 1.5 mM Na^+, has a TEP of about 60 mV (McGinnis and Vanable, 1986a,b).

In the cavy, the skin's TEP varies considerably, ranging from 4 mV (inside positive) in regions of skin with a good deal of hair, to as much as 80 mV in regions relatively free of hair, such as the postotic bald spot and the footpad (Barker et al., 1982). The TEP of human skin has also been measured (Barker et al., 1982; Foulds and Barker, 1983), with values ranging from about 10 mV to almost 60 mV, depending on the region measured (but not the age [range, 22–70 years] or sex of the person [Foulds and Barker, 1983]).

The sodium channels of the outer-facing membranes of the outer living layer of cells can be blocked by amiloride and some of its analogues (Eltinge et al., 1986), depriving the epidermal cells of the major cation that they transport (Sariban-Sohraby and Benos, 1986; Eltinge et al., 1986). When this is done, the transepithelial potential across the skin is drastically reduced (McGinnis and Vanable, 1986b), and wound currents are reduced, abolished, or even reversed to low-density incurrents that probably are attributable to a reversed TEP resulting from Cl^- transport (L.R. Robinson, 1985; Eltinge et al., 1986; McGinnis and Vanable, 1986b). The ionic basis for generating the mammalian epidermal transepithelial potential has not been studied extensively. However, Barker et al. (1982) found that application of amiloride to a slit made in cavy tarsal pad skin reduced the transepithelial potential at the slit to about one-half its original value, suggesting that mammalian skin has a Na^+-dependent battery similar to that found in amphibians.

Lateral Fields in the Vicinity of Wounds

When a wound is made in the skin, an electrical leak is produced (Barker et al., 1982) that short-circuits the epidermal battery at that point (Fig. 2), allowing current to flow out of the wound (as long as the wound is not

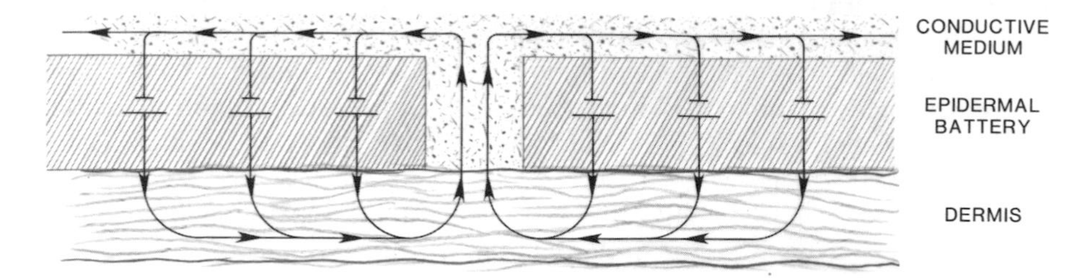

Fig. 2. Current path when an epithelium is wounded. When a wound is made in an epithelium, the potential across the epithelium drives a current (considered to be a flow of positive charge) through the subepidermal region and out of the wound. This current returns to the battery via some conductive path outside the epithelium. In amphibians and, possibly, mammalian cornea and oral epithelium, and skin that is kept moist by an occlusive dressing, this is simply the fluid bathing the epithelium. In mammalian skin, this conductive path normally is between the living and dead layer of the epidermis.

allowed to dry). At the wound, the potential drop from outside the skin to inside is relatively low. To either side of the wound, as the distance from the wound increases, the potential across the skin is found to be greater and greater, until a point is reached, a few millimeters from the wound in the cavy (Barker et al., 1982), at which the potential across the skin is the full value normally found in unwounded regions of that skin. In the cavy, most of this increase occurs in the first 0.5 mm of skin bordering the wound (Barker et al., 1982), so that, in skin with a TEP of 40–80 mV, there is quite a steep lateral voltage gradient in the vicinity of the wound. For six series of measurements, the average lateral voltage gradient was 140 mV/mm (Jaffe and Vanable, 1984) (Fig. 3). If the wound is allowed to dry, the wound resistance increases and blocks the flow of current so that there is no lateral potential drop at the edge of the wound (Barker et al., 1982).

Direct measurements of TEPs in the vicinity of small wounds made in the skin of *N. viridescens* have not yet been done, but they recently have been made in the vicinity of the relatively large wound that is made when the forelimb is amputated, as it is for studies of limb regeneration (McGinnis and Vanable, 1986b). Here again, quite substantial lateral voltage gradients exist in the fraction of a millimeter at the edge of the wound, this time in the region underlying the skin. Immediately after amputation, over the first 125 μm in from the edge of the wound, the field strength averaged about 60 mV/mm. As the wounds healed by migration of cells from the epidermis, the currents escaping through them were reduced because of the resistance created by the

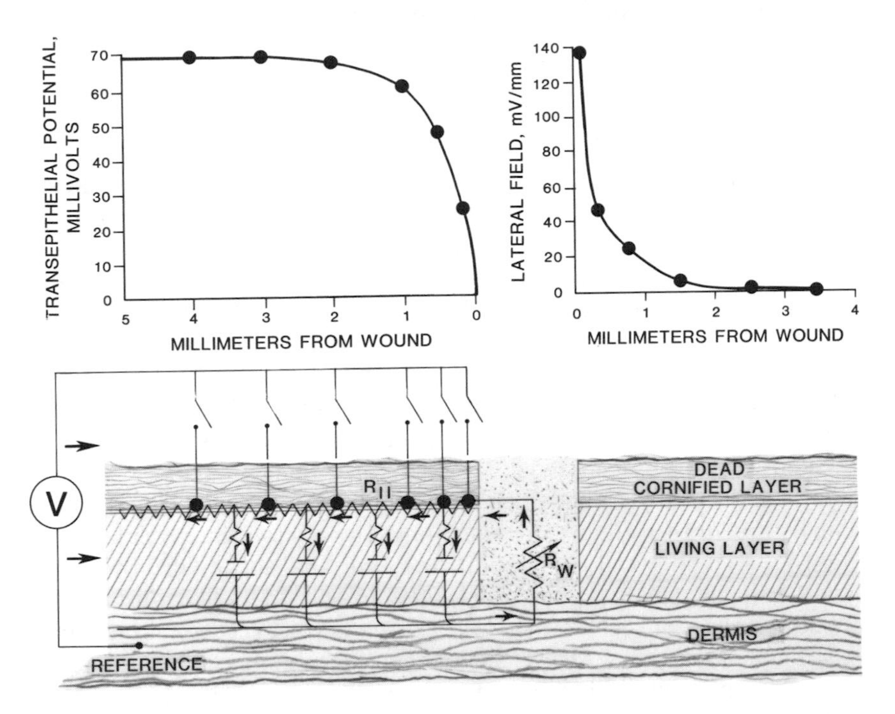

Fig. 3. Lateral potential in the vicinity of a wound made in mammalian skin. Left: schematic illustration of the circuit diagram of the skin, the electrodes for measuring transepithelial potentials in the vicinity of the wound, and (above), representative data obtained from such determinations of transepithelial potentials. (For actual data and more detailed information, see Barker et al., 1982. $R_{\|}$, resistance of the return path of the current to the epidermal battery, along which the lateral potential drop occurs near the edge of the wound; R_w, resistance of the wound. When R_w is low, as when the wound is kept moist, current driven by the epidermal TEP flows out the wound and returns to the battery via $R_{\|}$. When R_w is high, little or no current flows in the circuit, and little or no lateral field exists at the edge of the wound. Right, below: diagrammatic representation of the relevant layers of mammalian skin corresponding to the circuit on the left; above: the lateral field in the vicinity of the wound, plotted as a function of distance from the wound edge (based on the hypothetical data plotted on the left).

epithelium that forms, and with this, the lateral voltage gradients in the vicinity of the wound also diminished (McGinnis and Vanable, 1986a,b); by 6 hours after amputation, gradients were reduced to 15 mV/mm, and by 12 hours to about 5 mV/mm. The decrease in wound current that occurs with healing provides a convenient indirect way of monitoring healing as it occurs in amphibian skin.

In Vitro Effect of Electrical Fields on Cells

It is clear, then, that there are fields in epidermis close to the edge of wounds made in mammalian and in amphibian skin. With mammalian wounds, the fields exceeded 100 mV/mm. In amphibians, they are somewhat more modest, approximately 50 mV/mm. It is important to consider whether electrical fields of such magnitude could affect the behavior of cells in any way. It seems clear that they can. K.R. Robinson (1985) has recently reviewed the cases in which it is well established that the migration and/or orientation of cells can be controlled by imposing relatively modest electrical fields across them in vitro. Included in the total list of cells that can be affected are epithelial cells, fibroblasts, macrophages, and leukocytes.

Erickson and Nuccitelli (Nuccitelli and Erickson, 1983; Erickson and Nuccitelli, 1984) found that embryonic quail somite fibroblasts migrate to the cathode in fields with quite modest strengths. At field strengths as low as 1 mV/mm, they could just barely detect this directed movement; at higher field strengths (50–100 mV/mm), they saw a more clearly directed movement toward the cathode, and at 100–150 mV/mm and above, the fibroblasts aligned perpendicularly to the field, in addition to moving toward the cathode. Cooper and Schliwa (1985) have found that epidermal cells from fish scales migrate toward the cathode when placed in fields of from 50 to 1,500 mV/mm. Epidermal cells from *Xenopus* tail fin may be more sensitive than this: Muncy and K.R. Robinson (K.R. Robinson, 1985) have detected their migration toward the cathode in fields with strengths as low as 8 mV/mm.

In contrast to the cases just considered, the behavior of macrophages that are subjected to imposed fields is complex and incompletely understood. Orida and Feldman (1982) report that macrophages (from mouse peritoneum) migrate preferentially toward the anode when placed for 30 minutes in fields of relatively high strengths, ranging from 390 to 1,170 mV/mm. Cells were judged to be positive (migrating toward the anode) if the pseudopod axis was directed toward the anode within a sector 45° above or below the axis of the electric field; negative, if the pseudopod was directed outside this sector; or indeterminate, if no axis could be discerned. At 150 mV/mm, with this method of scoring and for the relatively short time of observation that was used, the movement of these cells was not significantly different from that of control cells with no field imposed. This finding might suggest that these macrophages are somewhat less responsive to electric fields than other cells that have been studied, and that the direction of their migration in an imposed field is opposite that normally seen in studies of other cell types. It is interesting, however, that osteoclasts, which may be derived from macrophages (Sminia and Dijkstra, 1986), migrate toward the anode (Ferrier et al., 1986). Studies with lower field strengths over a longer duration of time should be conducted before definite conclusions can be drawn about the direction that macrophages migrate in an electrical field.

Earlier studies of leukocyte and macrophage galvanotaxis have also been carried out, by Dineur (1892), by Monguió, (1933), and by Fukushima et al. (1953). Monguió's studies must be interpreted with caution; he subjected cells (probably activated macrophages, since he introduced a mixture of quartz sand and carmine powder into the lymph sac of his frogs before isolating the cells from the lymph sac) to current densities and field strengths that may have been very high. His currents were delivered through micropipettes whose diameter ranged from 60 to 100 μm, kept approximately 100 μm apart. At the orifices of these electrodes, the density of the delivered currents ranged from 600 to 6,000 mA/cm^2 (not $\mu A/cm^2$). If one accepts his assertion that the cells between these electrodes experienced a 200 to 400 times lower current density than existed at the electrode orifices, then (using the 400-fold attenuation) these figures become 1.5 mA/cm^2 to 15 mA/cm^2. At the low end of the range of currents used, cells moved toward the cathode. It can be estimated that at 1.5 mA/cm^2, assuming a ρ of 100 Ω-cm, the field strength would have been a quite reasonable 15 mV/mm. At higher current densities (estimated, with his 400-fold diminution assumption, to be 15 mA/cm^2), Monguió's cells moved toward the anode. What may be questioned here, however, is whether there could have been a 400-fold diminution of current density between such closely spaced current delivery pipettes. It therefore is not clear whether these observations have any physiological significance, since these field strengths could have been as high as 60 V/mm. Fukushima et al. (1953) carefully studied normal human leukocytes subjected to quite reasonable fields. (Assuming a ρ of 100 Ω-cm for their physiological saline and a 0.2 mm depth of fluid under their 22 × 22-mm coverslip, with 60 μA of current, the field strength used can be estimated to be ~ 10–20 mV/mm.) In such fields, the leukocytes migrated towards the anode. This finding agrees well with the much earlier but quite interesting studies of Dineur (1892). Dineur placed electrodes into the peritoneal cavity of frogs and mice and observed the aggregation of cells to these electrodes after passing current from a single Daniel cell ($Zn\text{-}ZnSO_4|Cu\text{-}CuSO_4$; develops ~ 1 volt [Getman and Daniels, 1947]). In normal frogs and mice, many more *globules blancs* aggregated around the anode than the cathode. However, when an inflammatory reaction was elicited, the cells (probably activated macrophages) aggregated preferentially to the cathode, rather than the anode. It therefore may well be that whether macrophages migrate to the anode or to the cathode depends both on the field strength and on whether the macrophages are activated or not. It would appear that activated macrophages migrate to the cathode at low field strengths.

Effect of Electric Fields on Capillary Permeability[3]

Nannmark et al. (1985) have reported that capillary permeability to macromolecules and leukocytes can be increased by electric fields. Using direct observation by fluorescence microscopy, they found that when 5, 20, or 50 μA of DC or 20 μA of AC current was passed between electrodes placed in contact with an everted hamster cheek pouch, FITC-dextran (MW 150,000) leaked out of the cheek pouch capillaries after a lag time of from 30 to 160 minutes. Histological observations of these preparations showed that in addition to the increased permeability to macromolecules, there was a marked extravasation of white blood cells from the capillaries. Neither macromolecular leakage nor white blood cell extravasation was observed in nonstimulated controls. Similar results were seen in capillaries growing into special chambers inserted into the tibiae of rabbits. Nannmark et al. (1985) found that the increase in capillary permeability was spread throughout the field, without any preference for the anode or cathode. No changes in blood flow rate were seen in these experiments, which might be surprising in view of observations of Sawyer and others (Sawyer et al., 1960) that small currents (20 μA) occasionally produce thromboses in small arteries (15–80 μm) and veins (40–200 μm) in the rat mesentery. It would appear that at least a partial explanation of this apparent discrepancy is found in the fact that in practice, to produce electrical hemostasis, tens of milliamperes of current must be used, with the anode directly contacting the bleeding vessels (Sawyer and Wesolowski, 1964).

The current densities used in experiments of Nannmark et al. (1985) were not reported, but it seems likely that even at the low end of the range used, they would be somewhat greater than those that would occur normally, especially below the wounded epidermis, where the relevant capillaries would be found. However, it would be desirable to establish a minimum value for the current needed to produce this effect. If anything, there was a greater delay between stimulation and onset of macromolecular leakage with the 50-μA than with the 5-μA current, so it is possible that the lowest current used, 5 μA, is still above the threshold needed for an effect. It may well be,

[3]I am indebted to M.E. McGinnis for calling my attention to this material.

therefore, that fields of low strength found in vivo could be effective in promoting the inflammatory response (see pp. 172–173). Be this as it may, these studies are quite likely to have relevance to results seen when such currents are delivered to skin wounds from an external source, since they are typically well above even 50 μA (see pp. 198–199).

What Aspects of Wound Healing Could Be Affected by Electric Fields?

On the basis of the in vitro studies on the effect of imposed electric fields on the movement of cells, it is apparent that the lateral electric fields that exist in the epidermis bordering wounds could, in theory, foster the cell movement that must occur during wound healing. There is good evidence that, in vitro, the migration of cells relevant to wound healing (epidermal cells, leukocytes, macrophages, and fibroblasts) can be promoted by the imposition of electric fields of the same magnitude that exists in epidermis at the margin of wounds made in guinea pig skin. Of these, the macrophage is the least certain candidate, in view of the somewhat unsettled state of information available on the effect of fields on these cells (see pp. 188–190). However, it seems possible that, in vivo, lateral fields at the edge of wounds could directly promote the inflammatory response, epithelization, and fibrogenesis associated with wound healing.

In deciding how probable such actions of the lateral fields at wound margins might be, several matters must be considered. First, it is important to recall that epithelization sets up a resistive barrier to the wound currents that must flow in order to have lateral electrical fields in the epidermis (McGinnis and Vanable, 1986a, b). Therefore only events of wound healing that precede complete epithelization can be seriously considered as processes susceptible to a direct influence by epidermally generated lateral electrical fields. On these grounds, collagen remodeling can safely be ignored. It also is unlikely that wound contraction per se could be directly affected by these fields, although to the extent that fibroblast immigration might be fostered by lateral electrical fields, this process could be indirectly affected. This leaves as remaining possibilities the inflammatory response, epithelization and fibrogenesis (which were considered plausible on the basis of in vitro studies), and angiogenesis, since in mammalian skin all of these processes are at least initiated before epithelization is complete (Ross and Benditt, 1961).

Although the field strength in mammalian skin is equal to or even exceeds field strengths found to affect cell migration in vitro, it is important to consider how the polarity of the fields that are set up at the edge of a wound compares with the polarity of the effect seen on the relevant cells in vitro.

Here the question becomes more complex, both from the standpoint of the polarity of the lateral field at the edge of the wound, and from the standpoint of the polarity of the response of a particular cell type.

The lateral fields we measured in cavy epidermis are those that exist in the narrow space between the upper living layer of the epidermis and the outer dead, cornified layer of the skin (Barker et al., 1982). The polarity of these fields is such that the regions close to the wound edge are more positive (or, less negative) than those further away from the wound (Fig. 4A), since these fields exist in the return path of the wound current from the wound to the outer surface of the epidermal battery. The fields measured in the amphibian epidermis (McGinnis and Vanable, 1986a, b) are of the opposite polarity, since these are on another leg of this circuit, where the wound current flows from the inner surface of the epidermal battery to the wound (Fig. 4B). A similar subepidermal field must also exist in mammalian skin but this has not yet been measured. In amphibians, if the skin is exposed to air and allowed to dry partially, there would be a proximodistal lateral potential along the outer surface of the skin battery in the vicinity of a wound. Such potentials have been measured by several investigators (Monroy, 1941; Becker, 1960; Rose and Rose, 1974; Lassalle, 1979). However, as a practical matter, in water, the resistance of this path is too low for there to be an appreciable potential drop there. Therefore, in speaking of the polarity of the lateral field at the edge of a wound, it is necessary to distinguish between the mammal and the amphibian, and, in the case of the mammal, between the outer (measured) or inner (at present unstudied) lateral field. It is interesting to note that the circuitry of mammalian corneal and oral epithelium may be different from that of skin; the keratinized layer of these epithelia is thin and normally moist. Therefore the location of the significant lateral wound field might well resemble that of amphibian skin more than it does mammalian skin. It is also possible that when mammalian skin is kept moist by an occlusive dressing, it too might have a circuitry that is more amphibian than mammalian (see next section).

From the foregoing, it is evident that in dry mammalian skin, epidermal cells at the edge of a wound (especially those close to the stratum corneum) are subjected to a rather strong field, and if their movement is to be promoted by this field, they would have to be so constituted that they migrate toward the positive end of a field. The epidermal cells in the lower layers would possibly also be exposed to a field, but one that is likely to be weaker (since the resistance of this region is likely to be less than in the very narrow region between living and dead layers of the skin). It also would have a polarity opposite to the upper field. On the other hand, in cornea and oral epithelium

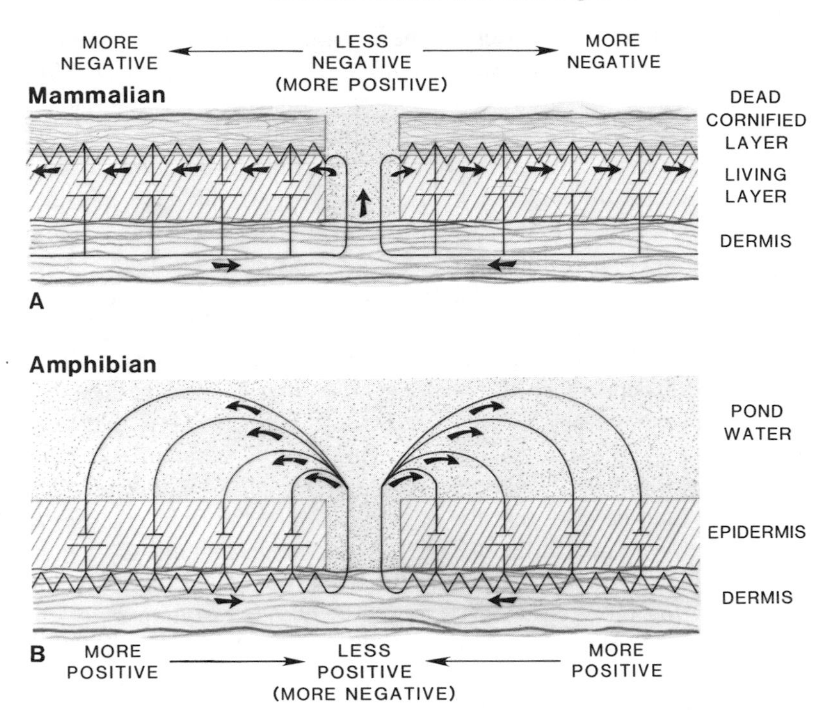

Fig. 4. Comparison of the locus of the major lateral potential in mammalian and amphibian skin. **A:** Mammalian skin. A steep lateral potential is developed in the high-resistance space between the living and dead cornified layers of the epidermis (see Fig. 3). This space is on the return path of the current from the wound to the epidermal battery, in which the current (positive charge) is moving, as it must, from more positive to less positive regions. In this path, then, points close to the wound are more positive than points away from the wound. **B:** Amphibian skin (and possibly mammalian cornea and oral epithelium, and skin that is kept moist by an occlusive dressing). The dead cornified layer is thin and moist (and is ignored in the diagram), and is not a significant barrier to the movement of Na^+ into the epidermal cells. The water bathing the outer surface of the epithelium provides a very low-resistance path for the return of the wound current to the epidermal battery. Here the significant locus of lateral potential drop is subepidermal, where the current flows from the epidermal battery towards the wound, again from more positive to less positive. In this path, then, points close to the wound are less positive (more negative) than points away from the wound.

(and perhaps moist skin), the significant lateral field may well be subepidermal.

So far, as discussed above (pp. 188–190), fields have been imposed in vitro on teleost (Cooper and Schliwa, 1985) and amphibian (K.R. Robinson,

1985) epidermal cells, and each of these migrates toward the negative, or cathodal end of the imposed field. (Although K.R. Robinson [1985] feels there may be a subpopulation of *Xenopus* epidermal cells that migrate toward the anode.) In amphibians, the polarity of the field at the edge of the wound would, then, be correct for promoting epidermal cell migration to cover the wound, at least for the majority of these cells. It would be of considerable interest to study the behavior of cultured mammalian epidermal cells from skin, cornea, and oral epithelium in an imposed field, to see whether or not they behave as predicted from the circuitry of each of these epithelia.

Whether the movement of mammalian leukocytes and macrophages into the wound would be promoted during normal wound healing by epidermally generated fields depends on the outcome of further studies on these cells in vitro, and on whether they become "activated" before or after arriving at the wound site. To the extent that such a field would exist in subepidermal regions of the skin, where these cells would be located, it will be recalled that its polarity would be such that the negative end of the field is closer to the wound (Fig. 4A). It presently appears that monocytes and macrophages must be activated in order to migrate toward the negative pole, or cathode, in fields of low strength. Fibroblasts, at least those of quail somites, do migrate toward the cathode in modest electrical fields, and so their immigration to the wound site could be promoted by subepidermal fields in the mammal, should they be of sufficient magnitude. Finally, although apparently there are no data on the movement of capillary endothelial cells in imposed fields, the possibility exists that capillary permeability could be increased by in vivo fields, and this would certainly increase the supply of the leukocytes of various sorts that are so crucial for successful wound healing.

To summarize, it would appear that in mammalian skin, cells of the upper epidermal layer(s) at the edge of a wound would be subjected to rather strong electrical fields, and if these cells are so constituted that they migrate toward the anode, it would seem likely that lateral fields at the wound's edge would be a positive factor in wound epithelization. Cells of the lower layer of mammalian epidermis, and the monocytes and fibroblasts from the subepidermal layer, might be subject to somewhat weaker fields whose polarity is the opposite of those along the surface of the outer living layer of the epidermis. (In cornea and oral epithelium, however, this might be where the significant lateral field would be.) Therefore migration of epidermal cells whose characteristic behavior would be to move toward the cathode in a field would be promoted, as would fibroblasts and possibly macrophages. It also is remotely possible that fields could, in causing an increase in capillary permeability, be responsible for augmenting the supply of white blood cells for debriding the wound and providing chemotactic factors to stimulate neoangiogenesis and the immigration of fibroblasts.

The migration of monocytes and macrophages in electrical fields deserves careful further study. Unlike the immigration of fibroblasts, where evidence for a chemotactic mechanism via factors produced by macrophages is quite strong (see pp. 177–178), immigration of monocytes (which become transformed into macrophages) appears to be independent of such an influence by the neutrophils that precede their arrival in the wound (Simpson and Ross, 1972). Therefore stimulation of monocyte and macrophage immigration by subepidermal electrical fields could be of considerable importance. Not only is it crucial to have these cells there to debride the wound, but also to serve as the means, via macrophage-derived factors, of promoting the immigration of fibroblasts and the ingrowth of capillaries into the wound.

The fact that fibroblast immigration is facilitated by macrophage-derived factors should not obscure the possibility that subepidermal fields could also have a role in promoting their migration into the wound. It will be recalled that when Leibovich and Ross severely depleted wounds of their macrophages, fibroblast immigration was delayed, to be sure, but it was not altogether absent (Leibovich and Ross, 1975). This implies that chemotaxis, while unquestionably a significant factor in promoting fibroblast immigration, is not the only means by which fibroblast migration is influenced.

Circumstantial Evidence That Fields Could Affect Wound Healing

Several observations are consistent with the hypothesis that epidermally generated lateral fields promote epithelization. In the healing of mammalian wounds, it has been reported (Christophers, 1972; Winstanley, 1975) that it is the *outer* layer of epidermal cells that is the first to migrate during wound healing. These cells would be exposed to the strongest field existing at the edge of a wound. On the other hand, it is the *inner* layer of amphibian epidermal cells (Radice, 1980a) that first migrates during wound healing. In amphibian skin, it will be recalled (see pp. 185–187), it is this layer that is closest to the lateral field under the epidermis.

Another bit of circumstantial evidence that is consistent with the hypothesis that lateral electrical fields promote epithelization has already been alluded to (see pp. 173–177 and pp. 185–187): dry wounds heal more slowly than wounds that are kept moist by occlusive dressings. Since drying a wound is an excellent way to shut off its currents and lateral fields (Barker et al., 1982), it is possible that this is at least one reason why dry wounds do not heal as rapidly as wounds that are not allowed to dry. Keeping the epidermis surrounding a wound moist, as well as the wound itself (as would be the case with an occlusive dressing) raises an interesting issue: If the stratum corneum is kept wet enough for it to be conductive, then its circuitry could be converted to one resembling the amphibian model, in which the significant potential drop is subepidermal (see previous section and Fig. 4). In such an instance, migration of cells during wound healing would be exclusively toward the negative pole of the field. If it turns out that mammalian epidermal cells do migrate towards the cathode in vitro, then this might be a quite significant reason why moist mammalian wounds heal more rapidly than dry ones do. Of course, as Eaglstein (1984a) has discussed, there are several other reasons why a moist wound would heal more rapidly than a dry one, one of which has been briefly discussed above (see pp. 173–177). An attempt to distinguish among these possibilities would be worthwhile.

A third hint suggesting that lateral electric fields might promote wound healing is found by interrelating the body of literature that describes the effect of diphenylhydantoin (DPH, an antiepileptic drug) on wound healing with the literature describing this drug's effect on sodium transport by the epidermis.[4] It has been known for some time that DPH can promote wound healing (reviewed by Houck et al., 1972; Pollack, 1982). Shapiro (1958), in a well-controlled double-blind experiment, found that administration of DPH accelerated both epithelization and fibrogenesis in the healing of human gingival wounds. Similar results were reported by Shafer et al. (1958) in a series of experiments with longitudinal wounds made in rat skin: 3 or 4 days after wounding, the tensile strength of wounds made in DPH-treated rats was significantly greater than in the controls. Kelln and Gorlin (1961), apparently unaware of these results, essentially repeated this experiment with virtually the same result. Simpson et al. (1965), on the basis of these previous findings, attempted to promote the healing of chronic leg ulcers (mean duration, 7.1 years) of hospital patients by DPH administration. They carried out a double-blind comparison of ulcers in patients receiving DPH and those receiving a placebo, with the overall result that, after 13 weeks, the control ulcers worsened significantly, while the DPH-treated lesions remained essentially the same. A result more in keeping with the earlier human and rat work was obtained by Kolbert (1968) in studies of the healing of corneal wounds in DPH-treated and control rabbits: at 1 week after wounding, the tensile strength of the DPH-treated wounds was significantly greater than that of the control wounds.

Much of the emphasis of these studies has been on the strength of wound

[4]I am indebted to K.R. Robinson for introducing me to this idea.

healing, suggesting an effect on fibroplasia, although Shapiro (1958) clearly saw an effect on epithelization, as well. The focus on fibroplasia is understandable; a well-known side effect of long-term DPH therapy is the production of a marked gingival hyperplasia in approximately 40–50% of patients receiving this medication, presumably via a stimulation of fibroblast proliferation (Kutt and Solomon, 1980). It is possible to stimulate fibroblast proliferation in tissue culture with DPH (e.g., Shafer, 1960), but this effect is not always seen by investigators (reviewed by Hassell and Gilbert, 1983). The variability of response of cultured fibroblasts to DPH and the inconstancy of the production of gingival hyperplasia in patients treated with DPH were considered by Hassel and Gilbert (1983). On the basis of a study of protein and collagen synthesis by fibroblasts derived from interdental papillae of 17 nonepileptic volunteers and cultured with and without DPH, they concluded that fibroblasts are genetically heterogeneous: they found that fibroblasts from some persons respond directly to DPH by increasing their rate of protein and/or collagen synthesis, those from others respond by decreasing these rates, while fibroblasts from others are predisposed not to respond directly to DPH at all.

This finding would suggest that DPH might have an indirect effect on fibroblasts and/or epidermal cells in the considerably more constant effect that it has on wound healing. A possibility germane to the thesis of this chapter is the fact that DPH, in therapeutic concentrations, markedly stimulates both the potential difference across amphibian skin and its short-circuit current (Carroll and Pratley, 1970; Watson and Woodbury, 1972; de Sousa and Grosso, 1973; Riddle et al., 1975; reviewed by Woodbury, 1982). This effect is sustained (de Sousa and Grosso, 1973) and appears to result from DPH's producing an increase in the Na^+ conductance of the outer membrane of outer living layer of the epidermis that is not unlike, but still distinct from, the effect that vasopressin and oxytocin have on amphibian skin (Watson and Woodbury, 1972; de Sousa and Grosso, 1973). It is likely that the effect of DPH is on existing Na^+ channels; when these channels are blocked by amiloride, the effect of DPH on the epidermal conductance is abolished (de Sousa and Grosso, 1973).

It is therefore possible that stimulation of wound healing by DPH could be mediated by its effect on the epidermal battery. This possibility is strengthened by the speculation of Kutt and Solomon (1980), who point out that the gingival hyperplasia produced by DPH is associated with poor oral hygiene. They propose that this gingival hyperplasia is a reparative response to the chronic inflammations extant in such gingiva.

A final hint in the literature that lateral wound fields might promote wound healing is provided by Scott et al. (1985), who found that the strength of healing in wounds that are not sutured (that is, left to close by secondary intention) ultimately is significantly greater than it is in wounds that are sutured. This increased strength of healing is probably attributable to the fact that by 6 days after wounding, the rate of collagen production by open wounds was found to be significantly greater than in the sutured wounds (Scott et al., 1985). No rationale is offered by Scott et al. to explain why these differences might have arisen, but one possibility comes to mind: in the open wounds, the completion of epithelization would have taken longer than in the sutured wounds, and so the lateral fields in the skin adjacent to the wound margins would have persisted longer in the open than in the closed wounds. To the extent that these fields could in one way or another have promoted fibroplasia, they should have been more effective in the open than in the closed wounds.

Healing Under Conditions of Reduced or Reversed Electrical Fields

More direct indication of the importance of lateral electrical fields in wound healing is provided by observing the effect of manipulating these fields. This manipulation has been accomplished by interfering with the function of the epidermal battery, by imposing fields via an external battery, and by a combination of these approaches, to perform a ''rescue'' experiment, using an exogenous current to replace the disabled endogenous one. In this section, experiments using the first and third approach will be considered; in the section that follows this, the second strategy will be discussed. The exploitation of the third approach has just begun, although potentially, it is the most powerful of the three.

L. R. Robinson (1985), studying wound healing in *N. viridescens* skin, has found that when wound currents are reduced by blocking Na^+ channels with 30 μM benzamil, wound healing is impaired. When healing of 200-μm wounds was studied under conditions in which wound currents were inward, rather than outward (30 μM benzamil to block the Na^+ channels and with $[Ca^{2+}]$ lowered to 0.5 mM), these wounds usually did not heal over the course of 1 hour, a time sufficient for healing of such wounds when normal outcurrents were produced by the skin. Larger (500-μm) wounds healed at about one-half the normal rate when epidermally driven currents were similarly reduced.

Stump and K. R. Robinson (1986) and, in somewhat more detail, Rajnicek et al. (1988) have conducted similar studies of wound healing in *Xenopus* tailbud and neurula embryos, respectively. As is the case for newts, wound currents of *Xenopus* embryos (driven by the embryos' epidermal TEP)

(Robinson and Stump, 1984) can also be attenuated by blocking their Na^+ channels (in these cases with amiloride or benzamil), as well as by replacing the Na^+ in the water with choline, or by eliminating Na^+ transport with ouabain. Rajnicek et al.'s (1988) observations establish a clear relationship between wound currents and wound healing: when these currents were diminished 70% by benzamil or amiloride, the percent of embryos healed at 7 hours diminished by a similar amount. When 100 μM ouabain was used to eliminate Na^+ transport, healing was abolished, as was the case when Na^+ was completely replaced by impermeant choline. With increasing concentrations of Na^+, the percentage of healing increased up to a plateau of 90% at 3 mM Na^+. Furthermore, Stump and Robinson (1986) report preliminary results of the important additional experiment of attempting to rescue healing in embryos whose wound currents were abolished by choline substitution for Na^+; they applied 100–200-nA currents in the normal (outward) direction, and found a decidedly better healing in 14 out of 16 of these embryos, compared with an equal number of matched controls, in which the current delivery electrode was grounded. (One control healed better than its paired counterpart, while in the other, the amount of healing in each was judged a tie.)

There are, then, a few early experiments that suggest that wound healing is promoted by epidermally generated fields. When these fields are reduced or reversed by various means, wound healing is impaired, and in one of these preparations, when fields are restored by current delivered from an external source, healing is at least partially restored.

Promotion of Wound Healing by Current Delivery From an External Source

In view of the fact that normal wound healing is impaired by reducing or reversing the epidermally generated lateral fields that exist in and/or below the skin at the edge of wounds, it is logical to inquire whether wound healing can be promoted by augmenting these fields. Actually, experiments testing this idea antedate by more than a decade the experiments that have only just begun to lay the theoretical foundation for such expectations (see Rowley et al., 1974b; Carley and Wainapel, 1985, for brief reviews). Perhaps in part because of this lack of theoretical underpinning, the experimental procedures with applied currents are as varied as the results they report, which range from excellent promotion of healing to a detrimental effect on this process.

To the best of this writer's knowledge, there are just three negative reports in the literature on the effect of applied currents on wound healing. In a briefly reported study, Carey and Lepley (1962) supplied 200–300 μA of current for 2, 3, and 5 days to two parallel, 4-cm full-thickness wounds made in rabbit back skin 2 cm to either side of the midline (one with the cathode, the other with the anode in the wound). The current density in the immediate vicinity of these electrodes (24-gauge stainless steel wires) would have been very high: ~ 320 $\mu A/cm^2$ to ~ 475 $\mu A/cm^2$. This procedure did not promote wound healing, either at the anode or the cathode. In fact, at the anode, a considerable amount of necrosis was observed (as might be expected from the electrode products that would be produced there). They noted a remarkable scarcity of inflammatory cells in wounds in which the cathode was implanted (which would argue that these cells should migrate to the anode in an imposed field) (Orida and Feldman, 1982). Wu et al. (1967) carried out a much more extensive study, but with healing not of the skin, but rather of the rectus abdominis muscle and underlying parietal peritoneum. They sutured bilateral laparotomy wounds made in rabbit musculoperitoneum with 3-0 stainless steel wire or 3-0 platinum wire and connected the wire from one wound to the positive, and the other to the negative, terminal of a battery. (The skin was then sutured with 2-0 silk.) Through these circuits 40, 80, or 400 μA of current was passed for 7 days. There were no statistical differences in the tensile strength developed during healing of the musculoperitoneal layer of control wounds receiving no current through the suture wires, or receiving current through either the anodes or the cathodes. Somewhat surprisingly, there was no significant necrosis observed. Regrettably, healing of the skin itself was not studied. The final negative study is with wound healing in horse skin (Steckel et al., 1984). In this experiment, full-thickness skin wounds 1.5 cm^2 or 2 cm^2 in area were produced, and currents of 10 or 20 μA supplied via stainless steel electrodes. In one series, the electrodes were sutures, with the anode threaded in the subcutaneous tissues ~ 5 mm outside the borders of the wound, and the cathode imbedded in the subcutaneous tissues in the center of the wound, which was not bandaged. With this preparation, supplying currents for 4 weeks (with the wound being cleansed about every 2 days), the wounds with electrodes in them, with or without current, healed considerably more slowly than control wounds with no electrodes, and in wounds with electrodes, there was no difference in healing between those with current and without. Infection was a problem in this series; all wounds with electrodes had purulent inflammation. In a second series, in which the anode and cathode were 18-gauge stainless steel needles implanted subcutaneously ~ 2 cm lateral to opposite edges of the wound, all wounds were reported to have healed at the same rate. However, the only assessment of this healing (except for the final biopsy histology at 28 days postwounding) was to photograph the wounds, and so wound contraction would probably have been the major aspect of wound healing studied. It is not likely that a clear view of even the progress of epithelization would have been possible.

These negative reports are not unduly disturbing. Carey and Lepley's study (1962) is not an extensive one, and there almost certainly was a problem with electrode products (which, from stainless steel, can include ions of manganese, chromium, nickel, and molybdenum [Weast, 1983–84, page F-112], all potentially toxic heavy metals [Beliles, 1979; Buck, 1979; Venugopal and Luckey, 1975; Nielsen, 1977]). Wu et al.'s study (1967) ignored the healing of skin, and, in studying the healing of the musculoperitoneum, electrode electrochemistry could not have been avoided, even with the platinum electrodes. In addition to wound infection, electrode products were also almost certainly a problem with Steckel et al.'s (1984) study, particularly in their first series. In their second series, the configuration of the electrodes was such that the field was imposed *across* the wound, so that processes that might be promoted at one side of the wound would be inhibited at the other side. Also, unfortunately, in this series the only parameter studied was wound contraction, a process not likely to be directly affected by electric fields (see pp. 191–195).

The reports in the literature of promotion of wound healing by supplying direct current from an external support are considerably more numerous than those in which findings were negative. These can be divided into three categories: those in which the cathode was placed in the wound for the duration of the experiment; those in which the anode was used throughout; and those in which first the cathode, and then the anode was placed in the wound (in some cases, with subsequent alternations). Most of these experiments have been carried out in humans; control experiments are in evidence in only a few. However, some of the experiments have been carried out in animals other than humans, and were well controlled.

Assimacopoulos (1968a) conducted experiments in which current (100 μA) was delivered to rectangular wounds via cathodes of stainless steel surgical gauze (2.9 $\times$ 2.9 cm). These electrodes were placed in the bed of full-thickness wounds of the same size made in the flank skin. This stainless steel gauze rectangle was placed in control wound beds, also, but no current was passed through it. The anode was a surgical steel wire fixed subcutaneously on the contralateral flank of the animal. Currents ($\sim$ 18 $\mu A/cm^2$) were delivered for 18 or 19 days. Assimacopoulos reported that wound healing was thereby accelerated by $\sim$ 25% (18 $\pm$ 0.5 [SEM] days for the treated, 25.5 $\pm$ 0.3 days for the control wounds to heal), and that the tensile strength of the wounds almost doubled (880 $\pm$ 177 g vs. 472 $\pm$ 74 g breaking load). The number of animals (four experimentals and four controls) in these experiments was regrettably small, and so the results must be interpreted with caution. However, they imply that epithelization was promoted by current delivery, since healing by wound contraction was virtually prevented by gluing the skin at the periphery of the wound to a plastic splint, and that fibrogenesis also was promoted by the treatment. As should be clear from earlier discussion (pp. 188–190 and 191–195), the polarity of the current delivery is such that promotion of fibroblast immigration into the wound would have been expected. However, the expected effect on epidermal cell migration is not so clear. For in vivo promotion of epidermal cell migration by fields in mammalian epidermis, the expectation is that these cells should migrate toward the anode. However, as discussed earlier (pp. 188–190), amphibian (K.R. Robinson, 1985) and teleost (Cooper and Schliwa, 1985) epidermal cells migrate to the cathode when a field is imposed on them in vitro. The actual preference of mammalian cells remains to be seen. Finally, regardless of the polarity, the imposition of such currents might reasonably be expected to increase capillary permeability to leukocytes, thereby enhancing the process of wound healing via the several means discussed above (pp. 190–191).

Assimacopoulos (1968b) subsequently reported three clinical cases in which the procedure worked out with the rabbits was used to promote healing of leg ulcers that had resulted from chronic venous insufficiency. The ulcers ranged from 4–96 cm^2, and had not yielded to standard procedures employed to promote healing for from 8 months to 5 years. In fact, the ulcers had been getting steadily worse during the previous unsuccessful treatment. With administration of slightly lower currents than those used in the rabbit (50–75 μA instead of 100 μA), and even lower current densities with the larger wounds (as low as 0.8 μA/cm^2 in the 96-cm^2 wound), these wounds healed in from 7 to 42 days. None of the ulcers was left untreated, so there were no controls in the strict sense. A biopsy of the healed region of one of the ulcers revealed quite dense connective tissue and a thin epidermis. This observation would be consistent with possibility that the treatment encouraged fibroblast immigration more effectively than immigration of epidermal cells.

Konikoff (1976) has essentially repeated Assimacopoulos's experiments, with similar results. Using ten rabbits with contralateral control wounds, he delivered 20 μA to full-thickness, 2-cm incisions (closed by sutures) via platinum gauze cathodes measuring 1.0 × 2.5 cm (yielding a current density of 8 μA/cm^2, slightly less than half the current density used by Assimacopoulos for his rabbit wounds). The anode was a 2-cm 28-gauge platinum wire implanted subcutaneously 4 cm from the cathode. The control wounds had exactly the same electrodes, but their batteries were inoperative. Five of the ten rabbits survived the 7 days of the experiment with intact stimulators and uninfected wounds, and only these were used for analysis of the tensile strength of the wound healing. The average load to rupture for the control wounds was 797 ± 36 g (SEM), while for the treated wounds, it was 1,244 ± 207 g, 1.6 times the control value.

Konikoff is cautious in interpreting his data, stating that although they indicate an effect of the current on wound healing, the results are not statistically significant. However, it would appear that they are: a one-tailed *t* test (Snedecor and Cochran, 1967) of the data reveals that the probability of being in error by rejecting the hypothesis that the treatment did not enhance the strength of the wound healing is 0.025, even correcting for the fact that the variances for experimentals and controls differ. Furthermore, the coefficient of correlation between the load to rupture and the mean current density actually delivered in each case (it ranged from 7.6 to 9.5 $\mu A/cm^2$) is a very respectable 0.97. It would seem, therefore, that although the number of cases is small, the effect seen in this experiment is real.

Alvarez et al. (1983b) have conducted a much larger scale study of the effect of applied electric current on wound healing in which the anode was placed in the wound, rather than the cathode. Shallow (0.3-mm-deep) rectangular wounds measuring 7 × 10 mm were made in pig skin, and the wounds divided into three groups: wounds with no treatment at all, not even an occlusive dressing; wounds with 50–300 μA DC delivered to a silver-impregnated nylon electrode positioned on the wound bed; and wounds with the electrode in the wound bed, but no current delivered to it. Electrodes were kept moist by saline-soaked gauze and wrapped with an Ace bandage. The cathode was a pregelled surgical grounding pad. All electrodes were replaced each day of the experiment. Currents were initially 300 μA (~ 430 $\mu A/cm^2$), but by 24 hours had dropped off to 50 μA (~ 70 $\mu A/cm^2$). Wound healing in ~ 30 wounds in each of the three conditions was assessed each day for 7 days, by estimating for each sample both the degree of epithelization and the capacity of the dermis for collagen synthesis. By day 3, epithelization of the placebo wounds had occurred significantly more rapidly than the untreated wounds, possibly because they were kept moist and the untreated wounds were not (see pp. 173–177 and pp. 195–198). However, epithelization of the wounds supplied with current was significantly more rapid than the placebos (and the untreated wounds) on days 2, 3, and 4. The time for complete epithelization of one-half of the untreated wounds was 4.6 days, and for the placebos, 4.1 days. However, 50% of the current-treated wounds had healed by 2.9 days, 29% faster than the placebos. Furthermore, between days 4 and 5, the collagen-synthesizing capacity of the dermis had become about twice that of the controls, when expressed on a per mg protein basis. However, when expressed on a per cell basis (via DNA assays; data not given), this difference was abolished. This finding suggests that immigration and/or proliferation of fibroblasts was promoted by the imposed fields.

The observed promotion of epithelization by placement of an anode in the wound would be predicted by the circuitry of the endogenous fields in guinea pig skin (Jaffe and Vanable, 1984) discussed above (see pp. 185–187 and pp. 191–195). However, the stimulation of fibroplasia observed would not be predicted. If mammalian fibroblasts behave the same way in an electric field as quail fibroblasts (Erickson and Nuccitelli, 1984), then their immigration should not have been promoted by the fields imposed in this experiment. The immigration of leukocytes and macrophages, however, might well be promoted by these fields, since under some conditions, it appears that they move toward the anode in an electric field (see pp. 188–190). It is also possible that the effect of increased capillary permeability described by Nannmark et al. (1985) could also have led to increased numbers of white blood cells in the wound. It is possible that the chemotactic effect of these cells on fibroblasts could override the presumably contrary effect of the imposed field.

The approach that apparently has been used most widely to stimulate wound healing in humans partially circumvents the difficulties in deciding which of the two polarities of the imposed fields should be used. In this approach, both polarities are used sequentially, with cathode in the wound bed first, and then the anode. The first account of this strategy was published by Wolcott et al. in 1969, and reports of its successful application in promoting the healing of recalcitrant ulcers have continued until as recently as 1985 (Carley and Wainapel, 1985).

The regimen for these studies is quite similar in all cases (Wolcott et al., 1969; Carley and Wainapel, 1985). The debrided and scrubbed wound is packed with several layers of sterile gauze, the cathode electrode (copper mesh) is sandwiched between these and several others placed on top of it, and the whole assembly is secured with waterproof tape. The anode, also sandwiched between sterile gauze, is taped to the skin 15 cm proximal to the ulcer. Both electrodes are then soaked in Ringer's solution and kept moist with additional Ringer's. The level of current delivered is determined empirically as the value between that which provokes a copious serious exudate from the ulcer (too low a level) and that which results in a bloody exudate (too high); as a practical matter, this is 200 μA below the setting that provokes the bleeding, and typically is ~ 400 μA. The electrodes typically are ~ 25 cm^2, providing a current density of 16 $\mu A/cm^2$, which is quite similar to those used by Assimacopoulos (1968a, b) and by Konikoff (1976).

They are ~ 25-fold lower than the initial current density used by Alvarez et al. (1983b), but only fourfold lower than the current density that ultimately prevailed in these experiments. Current is supplied in cycles of 2 hours on, 4 hours off and repeated three times a day for 3 days, unless the ulcer is infected. If infected, stimulation with the cathode in the wound bed is continued until 3 days after the ulcer is cleared of infection. At this point, the anode is placed in the wound, and the cathode 15 cm distal to the lesion, and the daily regimen continued, adjusting the current according to the character of the wound's exudate. In the event of a plateau of healing, the polarity is reversed until the next plateau occurs, at which time the polarity is again reversed. This alternation is then continued, with each switch occurring every 24 hours until healing is complete.

The rationale for using the cathode as the initial treatment, which has been investigated by Rowley (1972) and by Rowley et al. (1974a), has been that this procedure renders the wound free of infection. Rowley (1972) studied the effect of alternating and direct current that was passed through a three-compartment cylindrical chamber in which *E. coli* was growing. He found little effect with AC, but a maximum of ~ 40% decrease in the growth rate of *E. coli* over the course of a 2 1/2-hour passage of a 20-mA/cm^2 current. When the current density was at a more physiological level, 28 μA/cm^2, there was only a 4% reduction in the rate of bacterial growth. The effect was much less pronounced if the end compartments containing the platinum-10% iridium electrodes were partitioned off from the central compartment by Millipore filters, which would have greatly reduced the mixing between central and end compartments engendered by the vigorous aeration that was provided to the culture in the central compartment. This observation makes it likely that electrode products were responsible for the effect, although it is not clear what these products would have been. The pH of the central compartment was checked, and found to remain at 7 (although in chambers with filter barriers, the anode chamber became acidic, and the cathode alkaline). Similar studies by Spadaro et al. (1974) and by Berger et al. (1976) directly implicate electrochemical metal ion evolution in producing antibacterial effects, especially at low current ranges (0.4–4 μA, equivalent to 2–20 μA/cm^2), but at the anode rather than the cathode. They found that silver ion was the most effective of those they investigated, with copper and gold being much less effective. Clear effects with platinum electrodes were seen only at current densities 100–1,000 times greater than those showing clear effects with silver. At these high current levels, inhibition of bacterial growth was seen at both the anode and cathode, and was judged to result from toxic changes in the medium produced by electrolysis.

Conditions more closely matching those in effect during wound healing were used by Rowley et al. (1974a) with deliberately infected wounds. Rabbit skin was infected for 24 hours with gauze pads soaked in a suspension of 10^4 cells of *Pseudomonas aeruginosa,* and then treated with different combinations of antibiotic (polymyxin B sulfate) and current passage (~ 50 $\mu A/cm^2$, copper mesh cathode in the wound) for 72 hours. It is difficult to evaluate the outcome of these experiments. Variation in data from one run to the next was large, and so the statistical analysis somewhat complex, and, by necessity, nonparametric (Rowley et al., 1974a). Although statistical significance in the difference between the change in number of bacteria in wounds with and without current delivery exists when the ranks of the control group are swelled by adding in the cases in which no electrode (or antibiotic) was placed in the infected wound, the effect was not striking: 3 of 12 wounds in which an inactive electrode was placed had lower bacterial counts at 96 hours than at 24 hours, while 14 of 28 had lower counts at 96 hours after 72 hours of current passage. Antibiotic treatment was far more effective than current passage; the highest count per swab in these wounds was 4 (7 of 12 were 0), while the counts in the current-treated wounds ranged from a low of 1 (in 2 of 28 animals) to a high of 12,333. It is possible that the deliberate and severe infection of these wounds provided too demanding a test of the hypothesis that an active cathode in the wound effectively reduces its sepsis. It seems reasonable that it would do this; in addition to promoting an increase in vascular permeability (Nannmark et al., 1985), which would increase the general supply of phagocytic cells to debride the wound, it is possible (Dineur, 1892; Monguió, 1933) that the migration of activated macrophages would be promoted by having the cathode in the wound (see pp. 188–190).

However, it seems more likely that having the anode in the wound would have a beneficial effect on reducing bacterial contamination. In instances in which copper or silver electrodes are used, copper or silver ions are produced at the anode, and, as Spadaro et al. (1974) and Berger et al. (1976) have shown, these inhibit bacterial growth. In fact, Becker and Spadaro (1978) have used this strategy with excellent success in treating orthopedic infections. It is also likely that this mode of current delivery should promote the migration of nonactivated monocytes and macrophages to the wound (Fukushima et al., 1953; Orida and Feldman, 1982) (see pp. 188–190), in addition to producing a vascular permeability increase (Nannmark et al., 1985). In view of the uncertainty associated with it, it seems prudent to conclude that the actual evidence for the rationale proposed (Wolcott et al., 1969; Rowley et al., 1974a) for beginning the alternating cathode-anode procedure with the cathode in the wound requires further investigation.

In addition to the possible effect of the cathode on wound infection, it seems likely, in view of the behavior of fibroblasts in imposed fields (see pp. 188–190), that fibrogenesis should be promoted by having the cathode in the wound. Alternating it with the anode might well balance the promotion of fibroblast and activated macrophage migration into the wound that should be provided by having the cathode there with the epithelization that should be promoted by the anode's being in the wound (Alvarez et al., 1983b; Jaffe and Vanable, 1984). Capillary permeability should be increased regardless of the polarity of current delivery (Nannmark et al., 1985).

It is not clear just how many patients have been treated with this method, because in sequential publications from the same group of investigators it is not possible to determine in all instances whether the reports are for entirely new groups of patients or whether a cumulative number has been accrued. However, it seems safe to say that there are reports of promotion of healing of well over 300 ulcers with this approach (Wolcott et al., 1969; Wheeler et al., 1971; Rowley et al., 1974b; Page and Gault, 1975; Gault and Gatens, 1976; Carley and Wainapel, 1985). Of these, however, with the exception of Carley and Wainapel's study (1985), only a fraction of the total number of ulcers was left without treatment as controls. In Carley and Wainapel's study, 15 patients served as controls and 15 as the electrically treated. However, in no instance did any of these controls have an electrode in them (as did the wounds in the study of Alvarez et al. [1983b] discussed above); they were, rather, treated with standard therapeutic methods.

The results with the alternating cathode-anode procedure have been impressive. In the first study of Wolcott et al. (1969; Wheeler et al., 1971), 75 patients were treated; eight of them had bilateral ulcers that served as controls receiving standard treatment. In all eight cases with controls, healing of the electrically treated ulcer was more complete than its paired counterpart receiving standard treatment. The comparisons ranged from no healing in the controls vs. complete healing in the electrically treated ulcers (three cases) to 50% control vs. 70% experimental healing (one case). In all the electrically treated ulcers in this study, there was an 81.8% weighted mean volume decrease, at a mean rate of 13.4% healing per week, for a mean of 7.7 weeks of treatment. In 1974, Rowley et al. (1974b) briefly reported that the total number of electrically treated ulcers had increased to ~ 250, with 14 of them having paired controls. Their experience by that time was to see a nearly threefold increase in healing rate overall (presumably increasing over the rate seen before treatment), and a nearly fourfold increase in the instances with paired controls. Gault and Gatens (1976), summarizing their treatment of 106 ulcers (of which 6 had controls, and which may include cases reported earlier

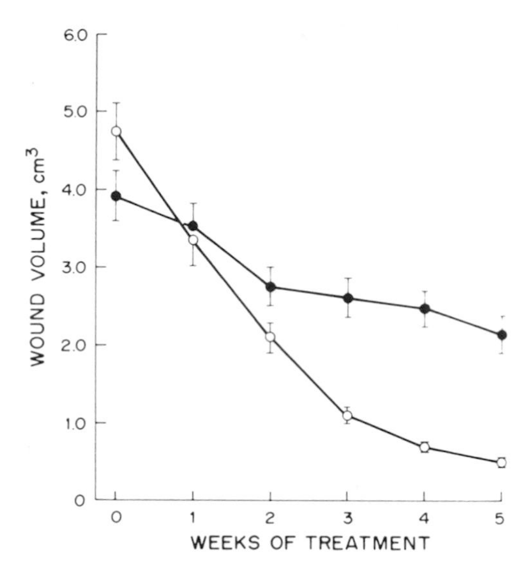

Fig. 5. Acceleration of wound healing with electrotherapy. Plotted from the data of Carley and Wainapel (1985), their Table 2. Error bars represent the standard errors of the mean, computed from the published standard deviations. ○, wounds with 30–110 $\mu A/cm^2$ current, first with the cathode, then with the anode in the wound; ●, wounds with conventional therapy. In addition to healing more slowly, the healed tissue of control wounds was more fragile than the electrically treated wounds.

in less detail [Page and Gault, 1975]), report a mean healing rate of 28.4% per week. Healing of their control ulcers occurred at an average rate of 14.8% per week, while the average healing rate of the paired electrically treated ulcers was 30.7% per week, approximately twice as fast. Carley and Wainapel (1985) compared the healing of ulcers in two paired groups of 15 patients each and found a clearly significant 1.5–2.5-fold increase in healing rate in the group that received the alternating anode-cathode treatment (Fig. 5). They further noted that, with the electrical treatment, the healed tissue was stronger than that of the controls.

It seems evident, then, that the healing of recalcitrant skin ulcers can be promoted by electrical fields, either by having the cathode in the wound for the whole period of treatment, by following a several-day treatment with the cathode in the wound with subsequent anodal treatment, or by continuing to alternate cathode and anode treatment. Finally, experiments with pigs show that both the epithelization and fibrogenesis aspects of wound healing can be promoted if just the anode is kept in the wound for the duration of the experiment. It should also be evident that there are reasonable rationales for expecting that both polarities of treatment should be effective in promoting one aspect or another of wound healing. This said, however, it remains true that there are many gaps in our understanding of the electrical aspects of healing wounds in the integument; filling in these gaps might considerably improve the efficacy of this approach.

FUTURE DIRECTIONS

Study of Simple Systems

Ultimately, from the point of view of learning how better to encourage wound healing in humans, it will be necessary to carry out studies of mammalian wound healing. However, particularly with regard to epithelization, there is much to be said for carrying out fundamental investigations of wound healing in simpler systems. In amphibians, for instance, in vivo optical conditions can be far superior to anything imaginable for mammals in vivo (Radice, 1980a). With adequate optics, epithelization can be studied in the same wound as it progresses, rather than by sampling different wounds at intervals. The speed of epithelization in amphibians, and the much simpler anesthesia (Vanable, 1985), also make them attractive subjects. A possible mammalian model that might come close to the amphibian would be the cornea, particularly if it could be successfully maintained in organ culture. It should be optically favorable, and in vivo, at least, the speed of epithelization is comparable to that in the amphibian. One caveat about using the mammalian cornea as a model for understanding the healing of mammalian skin wounds is that its circuitry may be more similar to amphibian than to mammalian skin (see pp. 191–195).

Further Study of Endogenous Fields During Normal Wound Healing

A good beginning has been made in studying the epidermally generated lateral fields in the vicinity of wounds made in mammalian (Barker et al., 1982) and in amphibian (McGinnis and Vanable, 1986b) skin, but much more must be done. In mammalian skin, it is important to determine whether in addition to the steep lateral field between the living and dead layers of the skin that has been measured (Barker et al., 1982), there is also a subepidermal lateral field corresponding to the one measured in amphibian distal limb stumps (McGinnis and Vanable, 1986b). Such a field, it will be recalled (pp. 191–195), could be relevant in promoting migration of leukocytes, macrophages, and fibroblasts, and, possibly, in promoting neoangiogenesis.

In amphibians, although lateral fields have been measured under the skin at the distal edge of an amputated limb stump, it is crucial to make such measurements at the edge of smaller wounds made in trunk, tail, and limb skin to complement studies of this more typical sort of wound healing that have already begun (Radice, 1980a,b; L.R. Robinson, 1985). It would also be of considerable interest to investigate the lateral fields in the vicinity of wounds made in mammalian cornea and in oral epithelia, to determine the extent to which, as speculated above, their circuitry is more similar to the amphibian than to mammalian skin.

Wound Healing With the Epidermal Battery Modulated

The potential developed by the epidermal battery, particularly that of amphibians, can be reduced by blocking Na^+ channels and by reducing the Ca^{2+} content of the water in which the skin is bathed (Eltinge et al., 1986). Studies of the effect of this procedure have just begun, and should be continued. It also is possible to increase the potential developed by the amphibian epidermal battery (Bentley, 1975; Borgens et al., 1977b), and this should be exploited, in conjunction with lateral field studies, in determining whether wound healing can be speeded up by this means. The effect of diphenylhydantoin (see pp. 195–198) should be studied from this point of view, both in amphibians and in mammals. The channels in mammalian skin have not yet been blocked as decisively as they have in amphibians (Barker et al., 1982), but it would be worth seeing whether the limited effect that Na^+ channel blockers have on the mammalian skin battery would retard wound healing. From the standpoint of Na^+ channel blocking, the corneal or oral mucosal epithelium could be ideal, since, as discussed above, they seem to resemble amphibian skin more closely than they do the rest of the mammalian integument.

It is crucial, also, to impose fields via an external source of current (as has been done by Stump and Robinson (1986) when the endogenous current has been modulated, to ensure that the effect of the channel blockers on wound healing is related to their effect on electrical fields, and not to some nonspecific effect.

Studies of the Effect of Imposed Fields on Cells in Culture

Examination of the effect of electrical fields imposed on cells in culture has been an active and fruitful area of study (K.R. Robinson, 1985). However, it should be apparent from the above discussion (see pp. 188–190 and 191–195) that there are several questions whose answers would provide valuable insight into the role of electrical fields in wound healing.

It is of extreme interest to determine whether mammalian epidermal cells migrate to the anode or to the cathode end of an imposed field. As discussed above (pp. 191–195), the prediction would be that at least those from the upper living layer of the epidermis should migrate towards the anode. This behavior would, to be sure, be contrary to that observed in the vast majority of in vitro studies (K.R. Robinson, 1985). However, the list of exceptions is growing, and so the possibility of migration toward the anode should not be dismissed out of hand: leukocytes, nonactivated macrophages (Dineur, 1892; Fukushima et al., 1953; Orida and Feldman, 1982), osteoclasts, (Ferrier et al., 1986) and perhaps a subpopulation of *Xenopus* epidermal cells (K.R. Robinson, 1985) respond to an imposed field by migrating preferentially to the anode. The behavior of leukocytes and macrophages in imposed electric fields also deserves further careful investigation, since there is uncertainty regarding the interpretation of experiments that so far have been conducted with these cells. Finally, it would also be worthwhile to study galvanotaxis in cultured adult mammalian fibroblasts, since the only fibroblast data that presently exist have been derived from studies with embryonic avian fibroblasts prepared from somite tissue (Nuccitelli and Erickson, 1983; Erickson and Nuccitelli, 1984).

Imposed Fields in Mammalian Wound Healing: Liquid Wires and Geometry

In two respects, it is remarkable that imposed fields have had any beneficial effect on mammalian wound healing at all. Without exception, metallic electrodes have been used to deliver the current to these wounds, and there is always a certain amount of chemistry associated with the transfer of charge between a metal electrode and a salt solution. At the very least, there would be changes in pH; at high current densities, evolution of oxygen and hydrogen occur, and, at the anode, there is often a production of the ions of whatever metals are in the electrode. Platinum is regarded as being "inert" in this regard, but certainly copper (used routinely in the "alternating polarity" technique (see pp. 199–209) and silver (used by Alvarez et al., [1983b]) yield their ions at the anode. Silver ions were, in fact, found in treated skin by Alvarez et al. (1983b). Normally, as a first approximation at least, one would consider copper and silver ions to be deleterious to normal cell activity. However, Berger et al. (1976) found that several types of cultured mammalian cells were relatively unaffected by 4 μg/ml silver ions, higher concentrations than those found by Alvarez et al. (1983b) in their studies with acceleration of wound healing in pig skin via silver anodes. It is conceivable, nonetheless, that the effect of having an anode in a wound could be improved if these ions were eliminated. The beneficial effects of silver ions in reducing wound contamination should not be ignored, but in wounds that are not grossly contaminated by pathogens, the normal microflora do not prevent normal wound healing (Mertz and Eaglstein, 1984; Eaglstein, 1984a).

Freedom from metallic ion contamination is made possible by supplying current to the wound via a ''liquid wire'' (Borgens et al., 1977a). This salt bridge connecting the battery to the wound can be made long enough to prevent the electrode products generated at the junction between the battery and the salt solution in the bridge from reaching the wound. This is simpler to do when the cathode is in the wound than when the anode is there, because in the latter instance, any positively charged metal ions produced at the anode would be electrophoresed from the site of their production to the wound. To reduce this possibility, it is desirable that the initial segment of the bridge be a relatively large-diameter conductor, so that the potential drop along it will be modest. To a limited extent, several of the studies of the effect of imposed currents on wound healing, particularly the ''alternating polarity'' studies, have used a salt bridge, by sandwiching the electrode in several layers of saline-soaked gauze. However, unless the gauze were to be frequently changed, it is likely that during anodal stimulation metal ions would have reached the wound bed and surrounding skin. It is therefore of considerable importance to rule out contamination by metal ions and other electrode products when attempting to promote wound healing by imposing fields. Until this is done, one cannot be confident that the effects seen are attributable to electric fields, and not to electrochemistry.

A final possible difficulty is in the geometry of the electrodes. When the whole wound bed is occupied by the electrode, the promotion of epithelization is probably not as effective as it would be with an electrode that occupies a relatively small fraction of the wound bed, placed in the center of a wound. With the whole wound area occupied by the electrode, the wound should be virtually isopotential. Under this circumstance, there would be less electrical impetus for epithelial cells at the edge of the wound to migrate into the wound than there would be if a smaller electrode were placed in the center of the wound, producing an electrical field between the edge and the center of the wound along which the epithelial cells could migrate.

There is, then, much to be done before it can be said that a solid theoretical and practical understanding of the role of electric fields in wound healing exists. Without such understanding, it is unlikely that the full potential of this approach to promoting wound healing will be achieved.

SUMMARY

1. Wound healing is a critical developmental process; it is important to understand how to promote this process, particularly in individuals with medical circumstances that are not favorable to wound healing. This chapter briefly reviews wound healing from the standpoint of asking whether electrical fields, both endogenous and applied, might be one factor useful in promoting the healing of wounds.

2. The epidermis of the skin acts as a battery, inwardly transporting sodium ions electrogenically, to generate a transepithelial potential (TEP) of ~70 mV. When a wound is made in the epidermis, this TEP provides the electromotive force to drive a steady current through the low-resistance path provided by the wound (as long as it is still hydrated), until epithelization covers the wound. In the vicinity of such wounds, steady lateral electrical fields of 100 mV/mm or more (in mammals) can be measured. In amphibians, these lateral fields are about 60 mV/mm.

3. Cells in culture respond to imposed electrical fields of ~ 10 mV/mm or more by moving preferentially toward one pole or another of the field, by growing toward one pole of the field (in the case of neuroblasts), or by orienting in the field. Among the cells whose migration is affected in vitro by imposed fields are several relevant to wound healing: epidermal cells, fibroblasts, macrophages, and leukocytes. In vivo, fields imposed across capillary beds have the effect of increasing the permeability of the capillaries to macromolecules and leukocytes. The strength of the fields causing this effect has not been measured, and it remains to be seen whether they are low enough to be relevant during normal wound healing.

4. The effect of the lateral fields measured in the vicinity of mammalian wounds could affect certain aspects of wound healing: as long as the wound is hydrated and not covered by epithelization (which restores a high resistance to the wound, and prevents current flow), the measured fields are sufficient to promote cell migration during the inflammatory phase, and during epithelization, fibroplasia, and neoangiogenesis. It is also possible that these fields could increase the permeability of local capillaries, helping to augment the supply of cells and macromolecules important to wound healing. Because wound contraction and collagen remodeling occur after epithelization is complete, it is not likely that these processes would be directly affected by wound fields.

5. There is circumstantial evidence that fields could affect wound healing: The first epidermal cells to migrate to cover the wound are those closest to the location of the lateral field. Hydrated wounds, in which electrical fields

would be sustained, heal considerably faster than do dry wounds. Wound healing is promoted by diphenylhydantoin, an antiepileptic drug that also augments the output of the epidermal battery. Finally, the strength of healing of wounds that are not sutured is significantly greater than that of those that are. This finding may be attributed to the fact that the completion of epithelization would be delayed in nonsutured wounds, thereby prolonging the time during which the wound fields could exert their effect(s).

6. The healing of amphibian wounds is delayed or prevented when the skin's battery is prevented from generating a TEP, either by Na^+ channel blocking, by reducing the Na^+ available for the epidermal battery, or by inhibiting the active transport of Na^+.

7. The healing of mammalian wounds can be promoted by the imposition of electrical fields. However, different polarities of field have been reported to be effective in these studies: the cathode in the wound throughout the treatment, the anode throughout the treatment, and finally, first the cathode and then the anode for, usually, the rest of the treatment. The fact that such disparate approaches have been reported to be effective has, perhaps, limited the acceptability of this mode of promoting wound healing.

8. It seems clear that before it can be said with confidence that electric fields are a significant factor in wound healing, a more solid understanding of the theoretical underpinnings of how they could affect the process is badly needed. This will require further studies with model systems using cell culture and whole animals, both amphibians (in which wound healing occurs quite rapidly and is readily observed) and mammals.

ACKNOWLEDGMENTS

I thank P.M. Mertz, G.J. Bourguignon, M.E. McGinnis, K.R. Robinson, and R.B. Borgens for their valuable comments and suggestions, and William Montagna, who many years ago planted the seed that grew into my realization of how important and interesting skin is.

REFERENCES

Alvarez OM and Biozes DG (1984) Cultured epidermal autografts. Clin. Dermatol. 2:54–67.

Alvarez OM, Mertz PM, and Eaglstein WH (1983a) The effect of occlusive dressings on collagen synthesis and re-epithelialization in superficial wounds. J. Surg. Res. 35: 142–148.

Alvarez OM, Mertz PM, Smerbeck RV, and Eaglstein WH (1983b) The healing of superficial skin wounds is stimulated by external electric current. Invest. Dermatol. 81:144–148.

Andersen L (1980) Ultrastructure of squamous epithelium during wound healing in palatal mucosa of guinea pigs. Scand. J. Dent. Res. 88:418–429.

Assimacopoulos D (1968a) Wound healing promotion by the use of negative electric current. Am. Surg. 35:423–431.

Assimacopoulos D (1968b). Low intensity negative electric current in the treatment of ulcers of the leg due to chronic venous insufficiency. Prelimary report of three cases. Am. J. Surg. 115:683–687.

Atnip KD, Mahan JT, and Donaldson DJ (1987) Role of carbohydrates in cell-substrate interactions during newt epidermal cell migration. J. Exp. Zool. 243:461–471.

Ausprunk DH, Falterman K, and Folkman J (1978) The sequence of events in the regression of corneal capillaries. Lab Invest. 38:284–294.

Baird A, Mormède P, and Böhlen P (1985) Immunoreactive fibroblast growth factor in cells of peritoneal exudate suggests its identity with macrophage-derived growth factor. Biochem. Biophys. Res. Commun. 126:358–364.

Banda MJ, Knighton DR, Hunt TK, and Werb Z (1982) Isolation of a nonmitogenic angiogenesis factor from wound fluid. Proc. Natl. Acad. Sci. U.S.A. 79:7773–7777.

Barker AT, Jaffe LF, and Vanable JW Jr. (1982) The glabrous epidermis of cavies contains a powerful battery. Am. J. Physiol. 242:R358–R366.

Becker RO (1960) The bioelectric field pattern in the salamander and its simulation by an electronic analog. IRE Trans. Med. Electronics, ME 7:202–207.

Becker RO and Spadaro JA (1978) Treatment of orthopaedic infections with electrically generated silver ions: A preliminary report: J. Bone Joint Surg. 60A:871–881.

Beliles RP (1979) The lesser metals. In Bentley PJ (ed.): *Toxicity of Heavy Metals in the Environment.* Marcel Dekker, New York, pp. 547–615.

Bentley PJ (1975) The electrical P.D. across the integument of some neotenous urodele amphibians. Comp. Biochem. Physiol. 50A:639–643.

Berger TJ, Spadaro JA, Chapin SE, and Becker RO (1976) Electrically generated silver ions: Quantitative effects on bacterial and mammalian cells. Antimicrob. Agents Chemother. 9:357–358.

Bitterman PB, Rennard SI, Hunninghake BW, and Crystal RG (1982) Human alveolar macrophage growth factor for fibroblasts: Regulation and partial characterization. J. Clin. Invest. 70:806–822.

Borgens RB, Vanable JW Jr., and Jaffe LF (1977a) Bioelectricity and regeneration. I. Initiation of frog limb regeneration by minute currents. J. Exp. Zool. 200:402–417.

Borgens RB, Vanable JW Jr., and Jaffe LF (1977b) Bioelectricity and regeneration: Large currents leave the stumps of regenerating newt limbs. Proc. Natl. Acad. Sci. U.S.A. 74:4528–4532.

Brown LF, Van de Water L, Harvey VS, and Dvorak HF (1988) Fibrinogen influx and accumulation of cross-linked fibrin in healing wounds and in tumor stroma. Am. J. Pathol. 130:455–465.

Buck WB (1979) Copper/molybdenum toxicity in animals. In Oehme FW (ed.): *Toxicity of Heavy Metals in the Environment, Part 1.* Marcel Dekker, New York, pp. 491–515.

Carey LC and Lepley D Jr. (1962) Effect of continuous direct current on healing wounds. Surg. Forum 13:33–35.

Carley PJ and Wainapel SF (1985) Electrotherapy for acceleration of wound healing: Low intensity direct current. Arch. Phys. Med. Rehabil. 66:443–446.
Carroll PT and Pratley JN (1970) The effects of diphenylhydantoin on sodium transport in frog skin. Comp. Gen. Pharmacol. 1:365–371.
Carter DM and Lin AN (1988) Wound healing and epidermolysis bullosa. Arch. Dermatol. 124:732–733.
Christophers E (1972) Kinetic aspects of wound healing. In Maibach HI and Rovee DT (eds.): *Epidermal Wound Healing*. Year Book Medical Publishers Inc., Chicago, pp. 53–69.
Chvapil M, Chvapil TA, and Owen JA (1986) Reaction of various skin wounds in the rat to collagen sponge dressing. J. Surg. Res. 41:410–418.
Clark RA, Stone RD, Leung DYK, Silver I, Hohn DC, and Hunt TK (1976) Role of macrophages in wound healing. Surg. Forum 27:16–18.
Cooper MS and Schliwa M (1985) Electrical and ionic controls of tissue cell locomotion in DC electric fields. J. Neurosci. Res. 13:223–244.
Croft CB and Tarin D (1970) Ultrastructural studies of wound healing in mouse skin. I. Epithelial behaviour. J. Anat. 106:63–77.
DeLustro F, Sherer GK, and LeRoy EC (1980) Human monocyte stimulation of fibroblast growth by soluble mediator(s): J. Reticuloendothel. Soc. 28:519–532.
Derby A (1978) Wound healing in tadpole tailfin pieces in vitro. J. Exp. Zool. 205:277–283.
de Sousa RC and Grosso A (1973) Effects of diphenylhydantoin on transport processes in frog skin (*Rana ridibunda*). Experientia 29:1097–1098.
Deuel TF, Senior RM, Chang D, Griffin DC, Henrickson RL, and Kaiser ET (1981) Platelet factor 4 is chemotoctic for neutrophils and monocytes. Proc. Natl. Acad. Sci. U.S.A. 78:4584–4587.
Deuel TF, Senior RM, Huang JS, and Griffin GL (1982) Chemotaxis of monocytes and neutrophils to platelet-derived growth factor. J. Clin. Invest. 69:1046–1049.
DiBona DR and Mills JW (1979) Distribution of Na^+-pump sites in transporting epithelia. Fed. Proc. 38:134–143.
Diegelmann RF, Cohen IK, and Kaplan AM (1982) Effect of macrophages on fibroblast DNA synthesis and proliferation. Proc. Soc. Exp. Biol. Med. 169:445–451.
Dinarello CA (1984) Interleukin-1. Rev. Infect. Dis. 6:51–95.
Dineur E (1892) Note sur la sensibilité des leucocytes à l'électricité. Bull. Séances Soc. Belge Microsc. (Bruxelles) 18:113–118.
Dohlman JG, Payan DG, and Goetzl EJ (1984) Generation of a unique fibroblast-activating factor by human monocytes. Immunology 52:577–584.
Donaldson DJ and Dunlap MK (1981) Epidermal cell migration during attempted closure of skin wounds in the adult newt: Observations based on cytocholasin treatment and scanning electron microscopy. J. Exp. Zool. 217:33–43.
Donaldson DJ, Mahan JT, Hasty DL, McCarthy JB, and Furcht LT (1985) Location of a fibronectin domain involved in newt epidermal cell migration. J. Cell Biol. 101:73–78.
Donaldson DJ, Smith GN Jr., and Kang AH (1982) Epidermal cell migration on collagen and collagen-derived peptides. J. Cell Sci. 57:15–23.
Eaglstein WH (1984a) Effect of occlusive dressings on wound healing. Clin. Dermatol. 2: 107–111.
Eaglstein WH (1984b) Current wound management: A symposium. Clin. Dermatol. 2: 134–142.
Eaglstein WH, Davis SC, Mehle AL, and Mertz PM (1988) Optimal use of an occlusive

dressing to enhance healing. Effect of delayed application and early removal on wound healing. Arch. Dermatol. 124:392–395.

Eckersley JRT and Dudley HAF (1988) Wounds and wound healing. Br. Med. Bull. 44: 423–436.

Eisinger M, Sadan S, Silver IA, and Flick RB (1988) Growth regulation of skin cells by epidermal cell-derived factors: Implications for wound healing. Proc. Natl. Acad. Sci. U.S.A. 85:1937–1941.

Eliason JA (1978) Leukocytes and experimental corneal vascularization. Invest. Ophthalmol. Vis. Sci. 17:1087–1095.

Eltinge EM, Cragoe EJ Jr., and Vanable JW Jr. (1986) Effects of amiloride analogues on adult *Notophthalmus viridescens* limb stump currents. Comp. Biochem. Physiol. 89A: 39–44.

Erickson CA and Nuccitelli R (1984) Embryonic fibroblast motility and orientation can be influenced by physiological electric fields. J. Cell Biol. 98:296–307.

Farquhar MG and Palade GE (1964) Functional organization of amphibian skin. Proc. Natl. Acad. Sci U.S.A. 51:569–577.

Farquhar MG and Palade GE (1965) Cell junctions in amphibian skin. J. Cell Biol. 26: 263–291.

Ferrier J, Ross SM, Kanehisa J, and Aubin JE (1986) Osteoclasts and osteoblasts migrate in opposite directions in response to a constant electrical field. J. Cell. Physiol. 129: 283–288.

Fishel RS, Barbul A, Beschorner WE, Wasserkrub HL, and Efron G (1987) Lymphocyte participation in wound healing. Morphologic assessment using monoclonal antibodies. Ann. Surg. 206:25–29.

Fisher LB and Maibach HI (1972) The effect of occlusive and semipermeable dressings on the cell kinetics of normal and wounded human epidermis. In Maibach HE and Rovee DT (eds.): *Epidermal Wound Healing*. Yearbook Medical Publishers, Chicago, pp. 113–122.

Folkman J and Klagsbrun M (1987) Angiogenic factors. Science 235:442–447.

Folkman J, Merler E, Abernathy C, and Williams G (1971) Isolation of a tumor factor responsible for angiogenesis. J. Exp. Med. 133:275–288.

Foulds LS and Barker AT (1983) Human skin battery potentials and their possible role in wound healing. Br. J. Dermatol. 109:515–522.

Friedenwald JS and Buschke W (1944) The influence of some experimental variables on the epithelial movements in the healing of corneal wounds. J. Cell. Comp. Physiol. 23: 95–107.

Fromer CH and Klintworth GK (1975) An evaluation of the role of leukocytes in the pathogenesis of experimentally induced corneal vascularization. II. Studies on the effect of leukocyte elimination on corneal vascularization. Am. J. Pathol. 81:531–544.

Fukushima K, Senda N, Inui H, Miura H, Tamai Y, and Murakami Y (1953) Studies on galvanotaxis of leukocytes. I. Galvanotaxis of human neutrophilic leukocytes and methods of its measurement. Med. J. Osaka Univ. 4:195–208.

Gabbiani G, Chaponnier C, and Hüttner I (1978) Cytoplasmic filaments and gap junctions in epithelial cells and myofibroblasts during wound healing. J. Cell Biol. 76:561–568.

Gabbiani G and Ryan GB (1974) Development of a contractile apparatus in epithelial cells during epidermal and liver regeneration. J. Submicrosc. Cytol. 6:143–157.

Garcia-Velasco J (1972) Keloids and hypertrophic scars. In Maibach HI and Rovee DT (eds.): *Epidermal Wound Healing*. Yearbook Medical Publishers, Chicago, pp. 281–289.

Garner A (1986) Ocular angiogenesis. Int. Rev. Exp. Pathol. 28:249–306.

Gault WR and Gatens PF Jr. (1976) Use of low density direct current in management of ischemic skin ulcers. Phys. Ther. 56:265–269.

Getman FH and Daniels F (1947) *Outlines of Physical Chemistry,* ed 7. John Wiley and Sons, Inc., New York.

Glenn KC and Ross R (1981) Human monocyte-derived growth factors for mesenchymal cells: Activation of secretion by endotoxin and Con A. Cell 25:603–615.

Goodson WH III, Radolf J, and Hunt TK (1980) Wound healing and diabetes. In Hunt TK (ed.): *Wound Healing and Wound Infection*. Appleton-Century-Crofts, New York, pp. 106–116.

Gospodarowicz D, Mescher AL, and Birdwell CR (1978) Control of cellular proliferation by the fibroblast and epidermal growth factors. Natl. Cancer Inst. Monogr. 48: 109–130.

Greenberg GB and Hunt TK (1978) The proliferative response *in vitro* of vascular endothelial and smooth muscle cells exposed to wound fluids and macrophages. J. Cell. Physiol. 97:353–360.

Grillo HC and Gross J (1967) Collagenolytic activity during mammalian wound repair. Dev. Biol. 15:300–317.

Grinnell F (1984) Fibronectin and wound healing. J. Cell. Biochem. 26:107–116.

Hassell TM and Gilbert GH (1983) Phenytoin sensitivity of fibroblasts as the basis for susceptibility to gingival enlargement. Am. J. Pathol. 112:218–223.

Hinman CD and Maibach H (1963) Effect of air exposure and occulsion on experimental human skin wounds. Nature 200:377–378.

Houck JC, Cheng RF, and Waters MD (1972) Diphenylhydantoin: Effects on connective tissue and wound repair. In Woodbury DM, Penry JK, and Schmidt RP (eds.): *Antiepileptic Drugs*. Raven Press, New York, pp. 267–281.

Huang JS, Huang SS, and Deuel TF (1983) Human platelet-derived growth factor: Radioimmunoassay and discovery of a specific plasma-binding protein. J. Cell Biol. 97: 383–388.

Illingworth CM and Barker AT (1980) Measurement of electrical currents emerging during the regeneration of amputated finger tips in children. Clin. Phys. Physiol. Meas. 1:87–89.

Jaffe LF and Vanable JW Jr. (1984) Electric fields and wound healing. Clin. Dermatol. 2:34–44.

Kelln EE and Gorlin RJ (1961) Healing qualities of an epilepsy drug. Dent. Prog. 1:126–129.

Kirschner LB (1983) Sodium chloride absorption across the body surface: Frog skins and other epithelia. Am. J. Physiol. 244:R429–R443.

Knighton DR, Hunt TK, Scheuenstuhl H, Halliday B, Werb Z, and Banda MJ (1983) Oxygen tension regulates the expression of angiogenesis factor by macrophages. Science 221: 1283–1285.

Kolbert GS (1968) Oral diphenylhydantoin in corneal wound healing in the rabbit. Am. J. Ophthalmol. 66:736–738.

Konikoff JJ (1976) Electrical promotion of soft tissue repairs. Ann. Biomed. Eng. 4:1–5.

Krawczyk WS (1971) A pattern of epidermal cell migration during wound healing. J. Cell Biol. 49:247–263.

Kulonen E and Niinikoski J (1968) Effect of hyperbaric oxygenation on wound healing and experimental granulomata. Acta Physiol. Scand. 73:383–384.

Kutt H and Solomon GE (1980) Antiepileptic drugs. Phenytoin: Relevant side effects. In Glaser GH, Penry JK, and Woodbury DM (eds.): *Antiepileptic Drugs: Mechanisms of Action*. Raven Press, New York, pp. 435–445.

Lambert WC, Cohen PJ, Klein KM, and Lambert MW (1984a) Cellular and molecular mechanisms in wound healing: Selected concepts. Clin. Dermatol. 2:17–23.

Lambert WC, Cohen PJ, and Lambert MW (1984b) Role of the epidermis and other epithelia in wound healing: Selected concepts. Clin. Dermatol. 2:24–33.

Lash JW (1955) Studies on wound closure in urodeles. J. Exp. Zool. 128:13–28.

Lassalle B (1979) Surface potentials and the control of amphibian limb regeneration. J. Embryol. Exp. Morphol. 53:213–223.

Leibovich SJ and Ross R (1975) The role of macrophages in wound repair: A study with hydrocortisone and antimacrophage serum. Am. J. Pathol. 78:71–100.

Leibovich SJ and Ross R (1976) A macrophage-dependent factor that stimulates the proliferation of fibroblasts *in vitro*. Am. J. Pathol. 84:501–513.

Leslie CC, Musson RA, and Henson PM (1984) Production of growth factor activity for fibroblasts by human monocyte-derived macrophages. J. Leuk. Biol. 36:143–159.

Mahan JT and Donaldson DJ (1986) Events in the movement of newt epidermal cells across implanted substrates, J. Exp. Zool. 237:35–44.

Martin BM, Gimbrone MA Jr., Unanue ER, and Cotran RS (1981) Stimulation of nonlymphoid mesenchymal cell proliferation by a macrophage-derived growth factor. J. Immunol. 126:1510–1515.

Martin TR, Altman LC, Albert RK, and Henderson WR (1984) Leukotriene B_4 production by the human alveolar macrophage: A potential mechanism for amplifying inflammation in the lung. Am. Rev. Respir. Dis. 129:106–111.

Martinet Y, Bitterman PB, Mornex J-F, Grotendorst GR, Martin GR, and Crystal RG (1986) Activated human monocytes express the c-*sis* proto-oncogene and release a mediator showing PDGF-like activity. Nature 319:158–160.

Martinez IR (1972) Fine structural studies of migrating epidermal cells following incision wounds. In Maibach HI and Rovee DT (eds.): *Epidermal Wound Healing*. Year Book Medical Publishers, Chicago, pp. 323–342.

Maumenee AE (1964) Repair in the cornea. In Montagna W and Billingham RE (eds.): *Advances in Biology of Skin, Vol. 5, Wound Healing*. Pergamon Press, Oxford, 208–215.

McGinnis ME and Vanable JW Jr. (1986a) Wound epithelium controls stump currents. Dev. Biol. 116:174–183.

McGinnis ME and Vanable JW Jr. (1986b) Electrical fields in *Notophthalmus viridescens* limb stumps. Dev. Biol. 116:184–193.

Mertz PM and Eaglstein WH (1984) The effect of a semi-occlusive dressing on the microbial population in superficial wounds. Arch. Surg. 119:287–288.

Miller TA (1980) The healing of partial-thickness skin injuries. In Hunt TK (ed.): *Wound Healing and Wound Infection*. Appleton-Century-Crofts, New York, pp. 81–96.

Mills JW, Ernst SA, and DiBona DR (1977) Localization of Na^+ pump sites in frog skin. J. Cell Biol. 73:88–110.

Monguió J (1933) Über die polare Wirkung des galvanischen Stromes auf Leukozyten. Z. Biol. 93:553–559.

Monroy A (1941) Ricerche sulle correnti elettriche dalla superficie del corpo di Tritoni adulti normali e durante la regenerazione degli arti e della coda. Pubbl. Stn. Zool. Napoli 18: 265–281.

Murray JC, Pollack SV, and Pinnell SR (1984) Keloids and hypertrophic scars. Clin. Dermatol. 2:121–133.

Mustoe TA, Pierce GF, Thomason A, Gramates P, Sporn MB, and Deuel TF (1987) Accelerated healing of incisional wounds in rats induced by transforming growth factor-β. Science 237:1333–1336.

Nannmark U, Buch F, and Albrektsson T (1985) Vascular reactions during electrical stimulation. Vital microscopy of the hamster cheek pouch and the rabbit tibia. Acta Orthop. Scand. 56:52–56.

Needham AE (1964) Biological considerations of wound healing. In Montagna W and Billingham RF (eds.): *Advances in Biology of Skin, Vol. 5, Wound Healing*. Pergamon Press, Oxford, pp. 1–29.

Nielsen FH (1977) Nickel toxicity. In Goyer RA and Mehlman MA (eds.): *Advances in Modern Toxicology, Vol. 2, Toxicology of Trace Elements*. Hemisphere Publishing, Washington, DC, pp. 129–146.

Niinikoski J (1980) The effect of blood and oxygen supply on the biochemistry of repair. In Hunt TK (ed.): *Wound Healing and Wound Infection—Theory and Surgical Practice*. Appleton-Century Crofts, New York, pp. 56–70.

Nuccitelli R and Erickson CA (1983) Embryonic cell motility can be guided by physiological electrical fields. Exp. Cell Res. 147:195–201.

Odland G and Ross R (1968) Human wound repair. I. Epidermal regeneration. J. Cell Biol. 39:135–151.

Orida N and Feldman JD (1982) Directional protrusive pseudopodial activity and motility in macrophages induced by extracellular electric fields. Cell Motil. 2:243–255.

Page CF and Gault WR (1975) Managing ischemic skin ulcers. Fam. Physician 11:108–114.

Peacock EE (1984) *Wound Repair, ed 3*. W.B. Saunders, Philadelphia.

Pierce GF, Mustoe TA, Senior RM, Reed J, Griffin GL, Thomason A, and Deuel TF (1988) In vivo incisional wound healing augmented by platelet-derived growth factor and recombinant c-*cis* gene homodimeric proteins. J. Exp. Med. 167:974–987.

Pollack SV (1982) Systemic medications and wound healing. Int. J. Dermatol. 21:489–496.

Pollack SV (1984) The wound healing process. Clin. Dermatol., 2:8–16.

Polverini PJ, Cotran RS, Gimbrone MA Jr, and Unanue ER (1977) Activated macrophages induce vascular proliferation. Nature, 269:804–806.

Polverini PJ and Leibovich SJ (1984) Induction of neovascularization *in vivo* and endothelial proliferation *in vitro* by tumor-associated macrophages. Lab Invest. 51: 635–642.

Radice GP (1980a) The spreading of epithelial cells during wound closure in *Xenopus* larvae. Dev. Biol. 76:26–46.

Radice GP (1980b) Locomotion and cell-substratum contacts of *Xenopus* epidermal cells *in vitro* and *in situ*. J. Cell Sci. 44:201–223.

Rajnicek AM, Stump RF, and Robinson KR (1988) An endogenous sodium current may mediate wound healing in *Xenopus* neurulae. Dev. Biol. 128:290–299.

Reibel J, Dabelsteen E, Birkedal-Hansen H, Ellegaard B, and Mackenzie I (1978) Demonstration of actin in oral epithelial cells. Scand. J. Dent. Res. 86:470–477.

Rick R, Dörge A, von Arnin E, and Thuran K (1978) Electron microprobe analysis of frog skin epithelium: Evidence for a syncytial sodium transport compartment. J. Membr. Biol., 39:313–331.

Riddle TG, Mandel LJ, and Goldner MM (1975) Dilantin-calcium interaction and active Na transport in frog skin. Eur. J. Pharmacol. 33:189–192.

Robinson DH and Mills JW (1987) Ouabain binding in tadpole ventral skin. II. Localization of Na pump sites. Am. J. Physiol. 253:R410–R417.

Robinson KR (1985) The responses of cells to electrical fields: A review. J. Cell Biol. 101: 2023–2027.

Robinson KR and Stump RF (1984) Self-generated electrical currents through *Xenopus* neurulae. J. Physiol (Lond.) 352:339–352.

Robinson LR (1985) The effects of electrical fields on wound healing in *Notophthalmus viridescens*. M.S. Thesis, Purdue University.

Rose SM and Rose FC (1974) Electrical studies on normally regenerating, on X-rayed, and on denervated stumps of *Triturus*. Growth 38:363–380.

Ross R (1968) The fibroblast and wound repair. Biol. Rev. 43:51–96.

Ross R (1980) Inflammation, cell proliferation, and connective tissue formation in wound repair. In Hunt TK (ed.): *Wound Healing and Wound Infection—Theory and Surgical Practice*. Appleton-Century-Crofts, New York, pp. 1–8.

Ross R (1986) The pathogenesis of atherosclerosis. An update. N. Engl. J. Med. 314: 488–500.

Ross R and Benditt EP (1961) Wound healing and collagen formation. I. Sequential changes in components of guinea pig skin wounds observed in the electron microscope. J. Biophys. Biochem. Cytol. 11:677–700.

Ross R and Odland G (1968) Human wound repair. II. Inflammatory cells, epithelial-mesenchymal interrelations, and fibrogenesis. J. Cell Biol. 39:152–168.

Ross R, Raines EW, and Bowen-Pope DF (1986) The biology of platelet-derived growth factor. Cell 46:155–169.

Rowley BA, (1972) Electrical current effects on *E. coli* growth rates. Proc. Soc. Exp. Biol. Med. 139:929–934.

Rowley BA, McKenna JM, and Chase GR (1974a) The influence of electrical current on an infecting microorganism in wounds. Ann. N.Y. Acad. Sci. 238:543–550.

Rowley BA, McKenna JM, and Wolcott LE (1974b) The use of low level electrical current for enhancement of tissue healing. Biomed. Sci. Instrum. 10:111–114.

Sariban-Sohraby S and Benos DJ (1986) The amiloride-sensitive sodium channel. Am. J. Physiol. 250:C175–C190.

Sather MR, Weber CE Jr., and George J (1977) Pressure sores and the spinal cord injury patient. Drug Intell. Clin. Pharm. 11:154–169.

Sawyer PN, Suckling EE, and Wesolowski SA (1960) Effect of small electric currents on intravascular thrombosis in the visualized rat mesentery. Am. J. Physiol. 198:1006–1010.

Sawyer PN and Wesolowski SA (1964) Electrical hemostasis. Ann. N.Y. Acad. Sci. 115: 455–469.

Schweigerer L, Neufeld G, Friedman J, Abraham JA, Fiddes JC, and Gospodarowicz D (1987) Capillary endothelial cells express basic fibroblast growth factor, a mitogen that promotes their own growth. Nature 325:257–259.

Schoefl GI and Majno G (1964) Regeneration of blood vessels in wound healing. In Montagna

W and Billingham RE (eds.): *Advances in Biology of Skin, Vol. 5, Wound Healing*. Pergamon Press, London, pp. 173–193.

Sciubba JJ, Waterhouse JP, and Meyer J (1978) A fine structural comparison of the healing of incisional wounds of mucosa and skin. J. Oral Pathol. 7:214–227.

Schultz GS, White M, Mitchell R, Grown G, Lynch J, Twardzik DR, and Todaro GJ (1987) Epithelial wound healing enhanced by transforming growth factor-α and vaccinia growth factor. Science 235:350–352.

Scott PG, Chambers M, Johnson BW, and Williams HT (1985) Experimental wound healing: Increased breaking strength and collagen synthetic activity in abdominal fascial wounds healing with secondary closure of the skin. Br. J. Surg. 72:777–779.

Shafer WG (1960) Effect of dilantin sodium on growth of human fibroblast-like cell cultures. Proc. Soc. Exp. Biol. Med. 104:198–201.

Shafer WG, Beatty RE, and Davis WB (1958) Effect of dilantin sodium on tensile strength of healing wounds. Proc. Soc. Exp. Biol. Med. 98:348–350.

Shapiro M (1958) Acceleration of gingival wound healing in non-epileptic patients receiving diphenylhydantoin sodium (Dilantin, Epanutin). Exp. Med. Surg. 16:41–53.

Shimokado K, Raines EW, Madtes DK, Barrett TB, Benditt EP, and Ross R (1985) A significant part of macrophage-derived growth factor consists of at least two forms of PDGF. Cell 43:277–286.

Sholley MM and Cotran RS (1978) Endothelial proliferation in inflammation. II. Autoradiographic studies in X-irradiated leucopenic rats after thermal injury to the skin. Am. J. Pathol. 91:229–242.

Sholley MM, Gimbrone MA Jr., and Cotran RS (1978) The effects of leukocyte depletion on corneal neovascularization. Lab. Invest. 38:32–40.

Silver IA (1972) Oxygen tension and epithelization. In Maibach HI and Rovee DT (eds.): *Epidermal Wound Healing*. Year Book Medical Publishers, Chicago, pp. 292–305.

Silver IA (1980) The physiology of wound healing. In Hunt TK (ed.): *Wound Healing and Wound Infection—Theory and Surgical Practice*. Appleton-Century-Crofts, New York, pp. 11–28.

Simpson DM and Ross R (1972) The neutrophilic leucocyte in wound repair: A study with antineutrophil serum. J. Clin. Invest. 51:2009–2023.

Simpson GM, Kunz E, and Slafta J (1965) Use of sodium diphenylhydantoin in treatment of leg ulcers. N.Y. State J. Med. 65:886–888.

Sminia T and Dijkstra CD (1986) The origin of osteoclasts: An immunohistochemical study of macrophages and osteoclasts in embryonic bone. Calcif. Tissue Int. 39:263–266.

Snedecor GW and Cochran WG (1967) *Statistical Methods*, ed. 6. The Iowa State University Press, Ames, Iowa.

Spadaro JA, Berger TJ, Barranco SD, Chapin SE, and Becker RO (1974) Antibacterial effects of silver electrodes with weak direct current. Antimicrob. Agents Chemother. 6:637–642.

Steckel RR, Page EH, Geddes LA, and Van Vleet JF (1984) Electrical stimulation on skin wound healing in the horse: Preliminary studies. Am. J. Vet. Res. 45:800–803.

Stump RF and Robinson KR (1986) Ionic current in *Xenopus* embryos during neurulation and wound healing. In Nuccitelli R (ed.): *Ionic Currents in Development*. Alan R. Liss, Inc., New York, pp. 223–230.

Thakral KK, Goodson WH III, and Hunt TK (1979) Stimulation of wound blood vessel growth by wound macrophages. J. Surg. Res. 26:430–436.

Thiyagarajan C and Silver JR (1984) Aetiology of pressure sores in patients with spinal cord injury. Br. Med. J. 289:1487–1490.

Tsutsumi O, Tsutsumi A, and Oka T (1988) Epidermal growth factor-like, corneal wound healing substance in mouse tears. J. Clin. Invest. 81:1067–1071.

Turck CW, Dohlman JG, and Goetzl EJ (1987) Immunological mediators of wound healing and fibrosis. J. Cell Physiol. [Suppl.] 5:89–93.

Tuszynski GP, Rothman V, Murphy A, Siegler K, Smith L, Smith S, Karczewski J, and Knudsen KA (1987) Thrombospondin promotes cell-substratum adhesion. Science 236:1570–1573.

Ubels JL, Edelhauser HF, and Austin KH (1982) A comparison of healing of corneal epithelial wounds stained with fluorescein or Richardson's stain. Invest. Ophthalmol. Vis. Sci. 23:127–131.

Vanable JW Jr. (1985) Benzocaine: An excellent amphibian anesthetic. Axolotl Newsletter 14:19–21.

Van Driessche W and Zeiske W (1985) Ionic channels in epithelial cell membranes. Physiol. Rev. 65:833–903.

Venugopal B and Luckey TP (1975) Toxicology of non-radioactive heavy metals and their salts. In Coulston F and Korte F (eds.): *Environmental Quality and Safety, Suppl., Vol. I: Heavy Metal Toxicity, Safety and Hormology*. Georg Thieme, Stuttgart, pp. 4–73.

Viziam CB, Matoltsy AG, and Mescon H (1964) Epithelialization of small wounds. J. Invest. Dermatol. 43:499–507.

Wall RT, Harker LA, Quadracci LJ, and Striker GE (1978) Factors influencing endothelial cell proliferation *in vitro*. J. Cell Physiol. 96:203–214.

Watson EL and Woodbury DM (1972) Effects of diphenylhydantoin on active sodium transport in frog skin. J. Pharmacol. Exp. Ther. 180:767–776.

Weast KC (ed.) (1983–84) *CRC Handbook of Chemistry and Physics*, ed. 64. CRC Press, Boca Raton, FL.

Wharton W, Gillespie GY, Russell SW, and Pledger WJ (1982) Mitogenic activity elaborated by macrophage-like cell lines acts as competence factor(s) for BALB/c 3T3 cells. J. Cell. Physiol. 110:93–100.

Wheeler PC, Wolcott LE, Morris JL, and Spangler MR (1971) Neural considerations in the healing of ulcerated tissue by clinical electrotherapeutic application of weak direct current: Findings and theory. In Reynolds DV and Sjoberg AE (eds.): *Neuroelectric Research*. CC Thomas, Springfield, IL, pp. 83–99.

Winstanley EW (1975) The epithelial reaction in the healing of excised cutaneous wounds in the dog. J. Comp. Pathol. 85:61–75.

Winter GD (1964) Movement of epidermal cells over the wound surface. In Montagna W and Billingham RE (eds.): *Advances in Biology of Skin*. Pergamon Press, Oxford, pp. 113–127.

Winter GD (1972) Epidermal regeneration studies in the domestic pig. In Maibach HT and Rovee DT (eds.): *Epidermal Wound Healing*. Year Book Medical Publishers, Chicago, pp. 71–112.

Wolcott LE, Wheeler PC, Hardwicke HM, and Rowley BA (1969) Accelerated healing of skin ulcers by electrotherapy: Preliminary clinical results. South. Med. J. 62:795–801.

Wolf JE Jr. and Harrison RG (1973) Demonstration and characterization of an epidermal angiogenic factor. J. Invest. Dermatol 61:130–141.

Woodbury DM (1982) Phenytoin: Mechanisms of action. In Woodbury DM, Penry JK, and Pippenger CE (eds.): *Antiepileptic Drugs*. Raven Press, New York, pp. 269–281.

Wu DT, Go N, Dennis C, Enquist L, and Sawyer PN (1967) Effects of electrical currents and interfacial potentials on wound healing. J. Surg. Res. 7:122–128.

Zitelli JA (1984) Secondary intention healing: An alternative to surgical repair. Clin. Dermatol. 2:92–106.

Electric Fields in Vertebrate Repair, pages 225–284

CHAPTER 6

The Nature and Effects of Electricity in Bone

Michael E. McGinnis

Center for Paralysis Research, Department of Anatomy, School of Veterinary Medicine, Purdue University, West Lafayette, Indiana 47907

INTRODUCTION

The bones of vertebrates perform several vital functions, including hemopoiesis, mineral homeostasis, and, of course, the skeletal functions of support, protection, and muscle attachment. To maintain the ability to perform these functions, especially the skeletal ones, bone possesses several repair or regenerative processes that respond to the traumas and insults of normal life. This ability to respond to trauma is by no means unique to bone. It is necessary for survival at cell, tissue, organ, and organism levels and is common to most forms of life. Recently it has been suggested that trauma or damage necessarily disrupts the normal electrical pattern of the cell, tissue, or organism (Borgens, 1982). This altered electrical profile is perceived by some to serve either as a signal for or a causative agent in the repair or regenerative response. This mechanism has been proposed for nerve regeneration (Borgens, et al., 1980), limb regeneration (Borgens, et al., 1977), epithelial wound healing (Barker, et al., 1982; Jaffe and Vanable, 1984), and bone fracture healing (Friedenberg and Brighton, 1966). A closely allied concept has been proposed for certain phenomena in normal development (Jaffe and Nuccitelli, 1977; Nuccitelli, 1983), in which an altered electrical environment leads to a developmental response, as in axial development in fucoid eggs (Nuccitelli and Jaffe, 1974), pollen tube growth (Weisenseel, et al., 1975), remodeling in bone (Bassett and Becker, 1962; Yasuda, 1953), and limb development in *Xenopus* (Robinson, 1983) and *Ambystoma* (Borgens et al., 1983). The hypothesis that electrophysiological changes are a common denominator of normal development and healing has applications to the study of bone physiology and repair.

This chapter is intended as a critical overview (*not* an exhaustive review) of the role that applied and endogenous electric fields may play in bone physiology, especially in the repair process. Considering the vast literature and many reviews on the subject, this chapter will attempt to provide an understanding and synopsis of basic principles rather than to develop a complete analysis of the literature. Relatively greater attention has been given to the "classical" experiments that have formed the foundation of the present state of understanding, rather than to recent works that do not significantly alter this understanding. Besides the difficulties presented by the unwieldy size of this literature, another problem arises from the nebulous affiliation of scientific, medical, and commercial concerns. This issue is addressed by Connolly: ". . . an orthopedic-industrial union, like any marriage, brings mixed blessings. The scientific processes of observation, reflection, verification and generalization suddenly become condensed . . . ." (Connolly, 1981b). This "condensation" of the scientific process in the study of the bioelectricity of bone has, almost of necessity, resulted in a literature marred by poorly controlled studies, incomplete reporting of experimental details, a profusion of nearly incomparable model systems, and a pathologic reliance on statistical significance without the attending biological significance necessary to give a finding validity. Also, electrical measurements and their interpretation in biological systems are endeavors whose complexities have often been overlooked by investigators unfamiliar with conventional electrophysiological techniques. However, when the fog of scientific imprecision is cleared away, a number of basic, well-accepted observations about the bioelectric nature of repair and regeneration do emerge.

Bone Processes

Before considering the bioelectric phenomena in bone, it is useful to review some of the major processes in bone development and response to trauma. Frost (1980) lists and describes several such processes: growth, modeling, remodeling, and repair. The first process, growth, is the accumulation of an amount of tissue that is adequate for normal function. Largely under endocrine control (Frost, 1980), growth produces the basic tissue mass on which the other processes of modeling, remodeling, and repair act.

Modeling. Modeling is defined by Frost as "the local influences that alter the growth pattern and organization of a tissue or organ and thereby produce macroarchitectural features" (Frost, 1980). The crude shape as well as the fine details of the biomechanical structure of bone are determined by the modeling process. Modeling results from the differential rates of bone deposition by osteoblasts and bone resorption by osteoclasts in a local area.

The terms "external remodeling" and "anatomical remodeling" also refer to this process (LaCroix, 1971) and seem to characterize better its dynamic, continuous nature. Modeling is characteristic of active growing bones and is almost nonexistent after skeletal maturity is reached.

Modeling is the mechanism by which Wolff's Law is realized. Wolff's Law, in its simplest form, states that function dictates structure. As presented by Wolff in 1892, it reads: "Every change in the . . . function of a bone . . . is followed by certain definite changes in . . . internal architecture and external conformation in accordance with mathematical laws" (Treharne, 1981). Modeling is the means by which the bone adapts to the various stresses placed on it throughout periods of growth or repair in the life of the animal. The control mechanism for modeling has not been established. Some "pressure transducer" must be present in bone to detect local stresses and initiate a local response of bone cells to counteract the applied load in a typical negative feedback fashion. Various transducer mechanisms have been proposed and are reviewed by Treharne (1981). The proposal most appropriate to the present discussion suggests that stress-generated potentials (SGPs) are produced in bone in response to loads and that bone cells are able to detect and respond appropriately to these electrical signals. Although this electrically mediated system is widely accepted, other systems are possible. Especially attractive is a microfracture control system, as discussed by Treharne (1981).

Remodeling. A third major process in bone physiology is the microscopic turnover of bone that is termed remodeling by Frost, or Haversian remodeling, internal remodeling, or histological remodeling by others (LaCroix, 1971). All of these terms refer to the local resorption of bone and subsequent deposition of new bone in small packets. Individually, each of these microscopic remodeling events has no effect on the bone structure as a whole, although cumulatively they ensure the long-term integrity of the whole bone. Bone remodeling is necessary because, in addition to the acute trauma of fracture, bone experiences chronic "microtrauma" as a result of the loads and stresses experienced during normal activity. Over time, dynamic loading leads to mechanical fatigue, resulting in microcracks and other forms of microdamage (Frost, 1973). The integrity of bone is compromised by accumulated microdamage but is restored through the process of remodeling.

Remodeling is accomplished by a functional entity known as the basic multicellular unit (BMU). The BMU consists of a group of osteoclasts that removes a packet of old bone, and a group of osteoblasts that follows and deposits new bone. The sequence of activation, resorption, and formation results in a new bone packet with a fresh complement of osteocytes derived from entrapped osteoblasts. This new bone packet (or bone structural unit) serves as the basic building block of Haversian bone. Just how the BMU is activated and directed to an area of microdamage remains a critical question.

Fracture repair. In spite of the modeling process that allows a bone to adapt its architecture to the normal loads placed on it and the remodeling process that allows for the incremental repair of areas suffering mechanical fatigue, gross fracture or damage to a bone occasionally occurs. At this point a complex repair process begins. The sequence of events leading to functional recovery varies with the type of bone and the type of fracture. For this general discussion an overview of the repair process in a simple, closed, traumatic fracture of a mammalian long bone will be given. Most authors divide the process into three phases, which usually overlap to some extent in both time and location (Cruess, 1984; McKibbin, 1978; Simmons, 1980). Although the terms and divisions vary from author to author, the three phases include an initial aseptic inflammatory phase, a secondary reparative phase in which an osteogenic repair tissue develops, and finally, a remodeling phase that restores the original state of the bone.

Trauma immediately severs the local vasculature of the bone and the surrounding soft tissues (Cruess, 1984; Sevitt, 1981). The osteocytes on either side of the lesion die from lack of circulatory support and thus form a zone of necrotic bone at the fracture site. McKibbin emphasizes the fact that the broken ends of the bones do not actively participate in bone repair, but "play, at best, only a passive role in what is an essentially bridging process between the more distant regions of living bone" (McKibbin, 1978). Following the initial hemorrhage, an inflammatory phase begins. This response has aptly been called an "aseptic traumatic inflammation" (Sevitt, 1981). A fibrin-rich proteinaceous exudate, polymorphonuclear leukocytes, macrophages, lymphocytes, and other cell types as well as plasma components accumulate near the fracture as a result of vasodilation and extravasation. Although the role of the inflammatory response is not clear, it appears to be of greater consequence than simply providing a mileau for removing the necrotic tissue resulting from the initial trauma.

The second phase is characterized by the organization of the hematoma into a granulation tissue, or callus, capable of bridging the fracture gap. Within hours of the injury there is increased cell division in the periosteum, both adjacent to the fracture and far away. This proliferation returns to background levels at sites distant from the injury after several days, but remains elevated near the fracture for several weeks (Tonna and Cronkite, 1961). This periosteal proliferative activity, along with similar activity in the endosteum, contributes to a pool of osteogenic cells. These cells combine with an accumulating mass of connective tissue cells to produce a granulation tissue with osteogenic potential (Sevitt, 1981). (Sevitt points out that while the concept of an osteogenic "granulation tissue" may be meaningful to pathologists and clinicians, the term osteogenic "blastema" may also be useful, especially for biologists). Concurrent with the accumulation of this cellular mass, a new vascular network of small capillaries invades the area and contributes to the forming callus. As differentiation and matrix production occurs, both cartilage and bone are produced. It appears that cartilage is preferentially produced in areas of low oxygen tension, but that as vascular ingrowth occurs, the cartilage is converted to bone by endochondrial ossification. Stability is achieved by this fibrocartilage callus that eventually ossifies.

The final phase of the repair process involves modeling and remodeling of the callus to restore functional strength to the bone. These usually continue long after functional bony union has occurred. Of the approximately 2,000,000 fractures occurring in the United States each year, the natural repair process, assisted by competent orthopedic practice, results in functional union in 95% of the cases (Hall, 1983). The remaining 5%, or 100,000 fractures, are termed nonunions or delayed unions. Frost distinguishes between technical failures and biological failures (Frost, 1980). Technical failures are those instances in which the competency of the repair process has been compromised by some aspect of fracture management. In these cases a callus normally forms, but union is prevented. The solution is usually to correct the technical problem and allow the natural process to proceed, with or without the stimulatory effect of bone grafts, drilling, or other procedures.

Biological failures, representing only 10% or less (Frost, 1980) of all nonunions, are those caused by a deficiency in the biological repair process. These nonunions can result because the callus is inhibited by a lack of either cell division, capillary invasion, or proper cellular differentiation. In other cases the callus forms naturally; however, subsequent modeling and/or remodeling is disrupted and leads to a defect.

Bone Bioelectricity

The modern study of bioelectric phenomena in bone was begun in 1953 by Yasuda, a Japanese orthopedic surgeon (Yasuda, 1953). His early investigations established the basis for most subsequent work. Yasuda initially observed that bone, in response to applied stress, was capable of producing electricity, which he termed "piezoelectricity of bone." Knowing of other work that demonstrated stress-induced osteogenesis, Yasuda theorized that stress leads to osteogenesis by way of electricity. The missing piece of evidence in this theory, that electricity alone could lead to osteogenesis, was tested by inserting two needle-shaped electrodes into the medullary canal of a rabbit femur. One microampere of current was passed for 3 weeks, and then histological sections were analyzed. New bone was found around both electrodes, although substantially more was present at the negative electrode, or cathode. Now the equation was apparently complete:

$$
\begin{array}{c}
\text{Force} \rightarrow \text{Osteogenesis} \\
\text{Force} \rightarrow \text{Electricity} \\
\text{Electricity} \rightarrow \text{Osteogenesis} \quad \therefore \\
\hline
\text{Force} \rightarrow \text{Electricity} \rightarrow \text{Osteogenesis (Treharne, 1981).}
\end{array}
$$

A similar thought process was followed in the first comparable American publication by Bassett and Becker in 1962, and in a companion paper in 1964 (Bassett et al., 1964), even though these studies were initiated without knowledge of Yasuda's earlier work. It will be useful to enumerate the various findings from these seminal works.

1. Stress applied to bone produces measurable electric potentials.
2. These stress-generated potentials occur in living, dead, dry, or wet bone.
3. The SGPs differ significantly from signals expected from classic piezoelectric elements.
4. The area of bone under compression tends to become electronegative and that under tension electropositive.
5. Small DC currents of 1–3 μA result in a local osteogenic response with an apparently greater response at the cathode than at the anode.

The subsequent 25 years of research has expanded on this base and has clarified many of the details. An extraordinary number of reports have been generated by investigators studying the bioelectricity of bone. Several reviews have been presented (Bassett, 1971; Cochran, 1972; El Messiery, 1981; Eriksson, 1976b; Hall, 1983; Herbst, 1978; Lavine and Grodzinsky, 1987; Pollack, 1984; Singh and Saha, 1984; Spadaro, 1977; Watson, 1981), along with reports of symposia and conferences (Brighton, 1977a, 1984a; Brighton et al., 1979a; Burny et al., 1978; Connolly, 1981a; Liboff and Rinaldi, 1974).

Most of this research conveniently falls into one of three categories. The first group of studies consists of attempts to measure and characterize the electrical output of bones and bone tissue. Studies in the second category investigate the effects of experimentally applied electricity on bone growth or other measures of bone physiology. The final group applies electrical therapy to clinical situations. The literature also includes implications for general orthopedics.

CHARACTERIZATION OF THE ELECTRICAL SIGNAL

Two major types of electrical signals have been detected in bone. The first and most thoroughly studied is the SGP, which is a property of bone as a tissue and is independent of bone viability. The other type of signal measured from bone is a steady direct current (DC) voltage called the bioelectric potential or BEP, which is dependent on bone metabolism. The two will be discussed separately.

Stress-Generated Potentials

Stress-generated potentials were apparently first demonstrated in skeletal tissue in 1912 by Gayda (from Cerguiglini et al., 1967) and later by Yasuda (1953). Yasuda termed the effect piezoelectricity or ''pressure''electricity. This phenomenon was confirmed by Fukada and Yasuda in 1957, by Bassett and Becker in 1962, and by Shamos et al. in 1963. If electrodes are placed on bone, either dry or wet, these investigators found that a transient voltage can be measured upon loading or unloading the bone. Early explanations attributed this effect to the piezoelectric properties of the collagen component of bone tissue.

Piezoelectricity. Classically, piezoelectricity refers to the charge separation that results from deformation of a crystal lattice that lacks a center of symmetry (Fig. 1). This definition has been extended to noncrystalline materials displaying similar properties, although the term ''piezoelectric texture'' (Cochran et al., 1968) may be more appropriate in describing them. Bone, as well as other biological tissues, has been shown to have these piezoelectric properties. Bassett reviews much of this literature (Bassett, 1968). In addition, pyroelectric (Lang, 1966), electret (Athenstaedt, 1970), ferroelectric (El Messiery et al., 1979), and solid-state semiconductor (Becker et al., 1964) properties have been claimed for bone tissue. These phenomena, either singly or in combination, are valid candidates as the source of SGPs in dry bones only. In physiologically wet bone, the shunting effects of mobile ions would minimize the potentials produced from these sources. A piezoelectric effect does occur in dry bone, and the study of this and related phenomena does contribute to our understanding of bone from

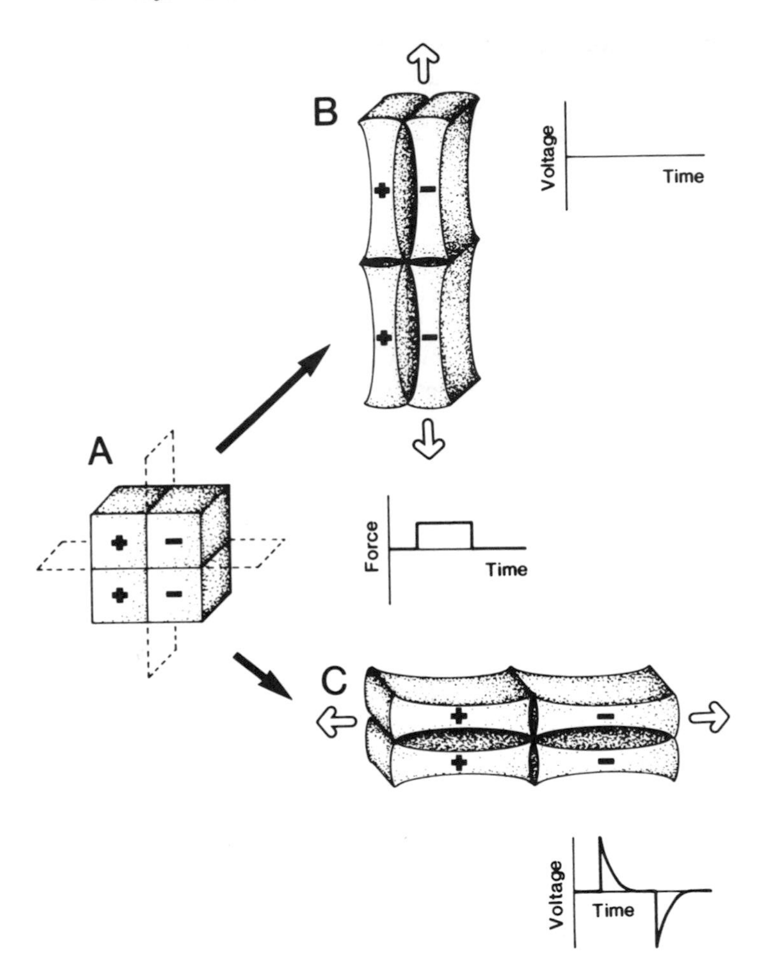

Fig. 1. Piezoelectricity, or "pressure electricity," is the production of a voltage by the application of force to a material. Although the term originally referred only to inorganic crystals without a center of symmetry, it now is used in a more general sense for any substance that produces a voltage in response to applied force. **A:** For illustrative purposes, a substance with a plane of symmetry in the horizontal plane, but not in the vertical plane, is shown being stretched in two directions. **B:** The material is stretched perpendicular to the plane of symmetry, so no net charge separation occurs, and no voltage is detected. **C:** The same material is stretched parallel to the plane of symmetry. This results in a separation of charges that can be detected as a voltage difference between the two ends of the material. Upon release of the stretching force, the material returns to its original state. This also produces a voltage spike, but in the opposite direction.The true piezoelectric response is a square pulse voltage. The spikes are produced by the shunting effect of electrical leakage paths, either through the measuring device or the material itself (Williams and Breger, 1975). Since there is a leakage path in most bone studies, the typical spikes that are reported were chosen for illustration purposes instead of the more classic square pulse response.

biophysical and material science viewpoints. Contrary to the theories developed by the early investigators, however, these insights only indirectly apply to the biological processes of growth and repair. Instead, the weight of available evidence argues that streaming potentials are probably the dominant source of SGPs in physiologically moist bone and almost definitely in living bone.

Streaming potentials. The streaming potential is an electrokinetic phenomenon that results from the flow of a liquid containing electrolytes past a charged surface. The interaction of two different properties of the fluid-solid interface accounts for the streaming potential. The first is that ions in solution having a charge opposite in polarity to the fixed surface charge are attracted to this surface and form a layer called the electrical double layer. This bound layer of ions leaves the bulk liquid depleted to some extent of ions of that charge and enriched with ions of the other charge, resulting in a nonuniform potential profile that can be measured perpendicular to the surface (Fig. 2A). The second contributing property of the fluid-solid interface is that the fluid adjacent to the surface tends to remain stationary when fluid is forced past the solid. The boundary between this stationary layer and the bulk of the fluid that is in motion is called the hydrodynamic slip plane. Since, in the case of the electrical double layer, there is a different ionic composition in the fluid on either side of this slip plane, their relative motion produces a charge separation that establishes a potential difference that is parallel to the direction of fluid flow (Fig. 2B). The magnitude and polarity of this streaming potential depends on the electrical potential of the double layer at the slip plane with reference to the bulk fluid. This value is called the zeta potential and varies with the surface charge and specific adsorption properties of the solid, and the ionic composition of the fluid. The magnitude of the streaming potential also depends on the velocity, viscosity, and conductivity of the fluid as well as the geometry of the solid phase.

It seems that Cerguiglini et al. (1967) were the first to specifically suggest that the SGPs in bone were the result, at least in part, of electrokinetic rather than piezoelectric phenomena. They compared the effects of bending to the effects of fluid perfusion in plant stems and in bones of toad, frog, rat, rabbit, and chicken. Qualitatively, they discovered that perfusion could mimic the SGPs produced by bending. They therefore concluded that the fluid flow resulting from bending the bones could produce streaming potentials and thus were an important component of the in vivo electrical response. In 1968, Anderson and Eriksson demonstrated that the SGPs they measured in tendon were dependent on the pH of the fluid in the tendon. As the pH was lowered, the magnitude of the SGP decreased, became zero, and finally reversed in

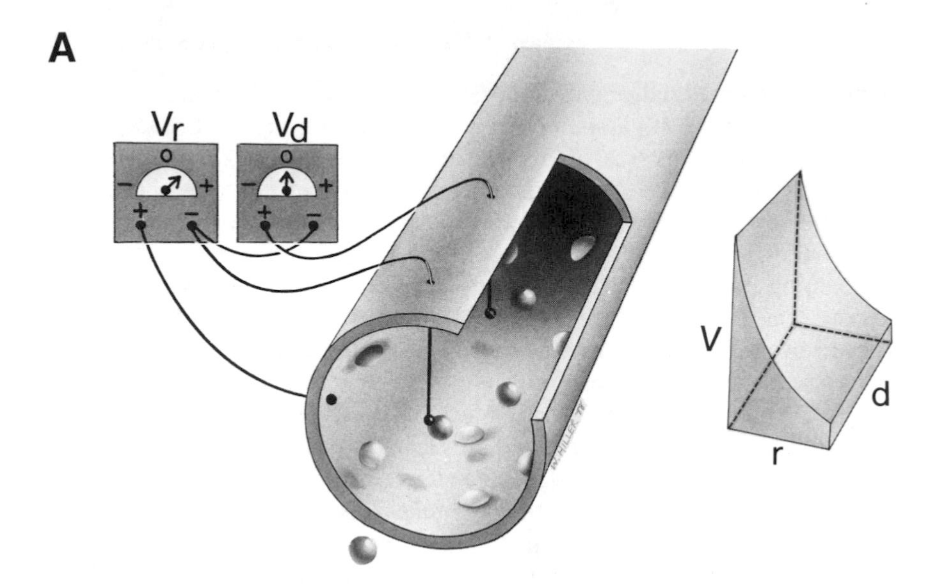

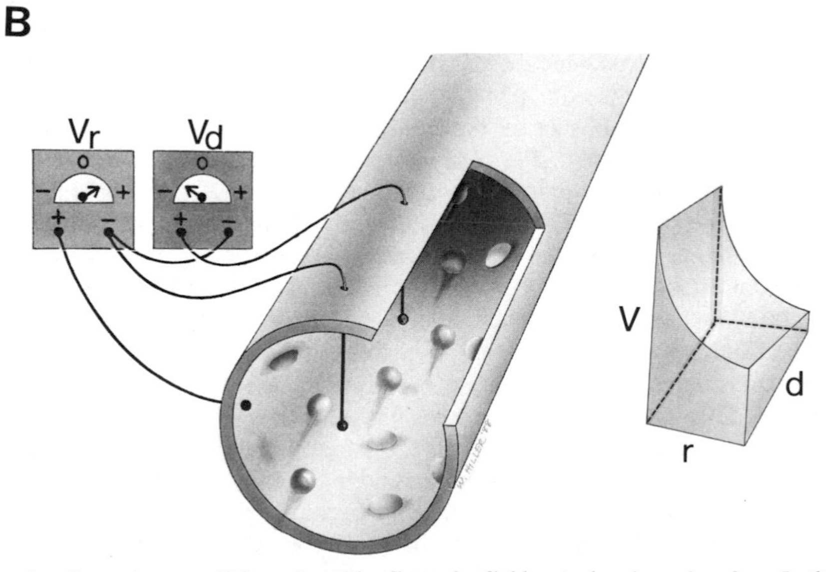

Fig. 2. Streaming potentials produced by flow of a fluid passed a charged surface. In this highly diagrammatic illustration, a tube with a positively charged surface has three point electrodes inserted through the walls to measure the potential difference radially, V_r, and longitudinally, V_d. **A:** When there is no fluid flow, there is a potential registered between the wall and the center of the tube (V_r), as a result of the negative (light-colored) ions being preferentially bound to the charged surfaces, which leave the bulk fluid enriched with positive (dark-colored) ions. However, there is no potential along the length of the tube, V_d. **B:** When fluid flows, the bound ions remain in place, while the mobile ions are swept away. This flow of a charged fluid constitutes an electric current that produces a potential difference along the length of the tube, now measured as V_d.

polarity and increased in magnitude when the pH was very low. The pH at which the SGP was zero was the isoelectric point for collagen. Since the zeta potential of collagen decreases with pH, becomes zero at the isoelectric point, and increases with opposite polarity upon further reduction, it was concluded that the SGP was dependent on the zeta potential of the collagen. This conclusion was consistent with the hypothesis that the SGPs are the result of streaming potentials. In 1970, they extended this work to bone (reviewed by Eriksson, 1976b). Gross and Williams in 1982 directly tested the streaming potential theory by perfusing electrolytes of various ionic concentrations and viscosities through discs of bones. Pienkowski and Pollack (1983) also tested the streaming potential theory by systematically varying the conductivity, viscosity, and NaCl concentrations of slabs of bone subjected to bending. The careful work of these two groups along with others has firmly established that streaming potentials are the predominant electrogenerative mechanism of stress-generated potentials in physiologically moist bone. Work continues to be done to further characterize this phenomenon.

Stress-generated potentials in vivo. From the perspective of one interested in the biology of SGPs, and not their biophysics, it is apparent that a shortcoming of most of the studies on SGPs is that the potentials measured by the investigator are most likely not those detected by the cell. Most studies involve placing large contact electrodes on the surface of polished slabs of bone and measuring the potential between the areas under the electrodes. Considering the heterogeneity and complex microarchitecture of bone, the potentials measured under large electrodes are a summation of local potentials of various magnitudes, phases, and possibly orientations. This fact has been confirmed for cortical bone by Starkebaum et al. (1979) and others from Pollack's group (Pienkowski and Pollack, 1983) using a microelectrode (5 μm in diameter) to measure the voltage at many points on the surface of bone samples. They found local SGPs radially concentric about the Haversian canals. The magnitude of the potentials is similar to the macropotentials, as typically measured. However, since these potentials form a gradient over very small distances, the electric fields are higher locally than would be predicted from the macroscopic measurements. The fields they measure are between 100 and 300 mV/cm, the same order as those found necessary to affect cells in vitro (Robinson, 1985). These data complement the anatomical model for streaming potentials based on osteonal structure (Pollack et al., 1984). This model involves fluid flow through the Haversian canal and canaliculi in response to bending. The measurements with the microelectrode technique give a reasonable indication of the type of electrical signal present at the osteonal level in dead bone. As Pollack points out, however, ". . .

these observations have yet to be verified in living bone. The presence of cells and cell processes in the lacunar-canalicular systems may alter the fluid flow characteristics in this region as well as the zeta potential. The presence of charged macromolecules and complex ion concentrations in the porous fraction of bone may also alter these properties. Therefore, the extrapolation of these results from machined 'dead ' bone to living bone must, at this time, be made with caution'' (Pollack, 1984). Another point to be remembered when considering the anatomical models for streaming potentials is that not all bones have Haversian systems. Frog and rat bones, which do not have Haversian systems, still produce SGPs (Cerguiglini et al., 1967).

Although the analytical techniques of Pollack's and Williams's groups have not been applied to living bone, some qualitative attempts to measure SGPs in vivo have been made. One study, by Lanyon and Hartman (1977), involved placing both a silver/silver chloride recording electrode and a three-element rosette strain gauge on the cranial aspect of the radii of sheep. Then, while the sheep walked on a treadmill, both surface strain and charge were recorded simultaneously. They found that during slow locomotion the SGPs were small, variable, and not related directly to the strain; with a moderate gait the signals were larger but still variable and seemed sensitive to minor mechanical changes; during fast locomotion or trotting, the SGPs gave a faithful reflection of strain as shown by the strain gauge output. They also noticed during these and companion in vitro studies that the SGPs depend both on the magnitude of the strain as well as the strain rate. Another interesting aspect of their work is that the results obtained in vivo were quite comparable to in vitro measurements on dead wet bone. Others have had problems distinguishing between SGPs measured from bone and stray electrical signals derived from muscle and nerve action potentials. Cochran et al. (1968) isolated the bone from surrounding soft tissues with a sheet of silicone rubber and then mechanically produced ''stepping'' movements in an anesthetized cat, since stimulation of the muscles produced interfering potentials. They also reported that SGPs measured in vivo were similar to those measured in vitro.

In summary, bones do produce electrical potentials in response to stress both in vivo and in vitro. It also seems to be well established that streaming potentials are the predominant source of the SGPs under physiological conditions. Furthermore, the spatial distribution or pattern of these potentials would suggest that they are capable of serving as the postulated pressure transducer inherent in Wolff's Law. The one missing piece of evidence is whether SGPs play any physiological role. As Brighton (1977b) states, ''Whether or not stress generated potentials or endogenous electricity induces osteogenesis has not been proven: the classical experiment showing cause and effect between stress generated potentials and osteogenesis has yet to be performed.''

Bioelectric Potentials

Besides the stress-generated potentials measured in bone, there is evidence of a second type of electrical activity in bone that, unlike the SGP, is dependent on bone metabolism and viability. These steady DC voltages have been studied in a wide variety of organisms and tissues, including both plants and animals, since the turn of the century (Rosene, 1947) and were generally thought to play a role in growth and development (Lund, 1947). Recent investigations using the vibrating probe have again demonstrated the presence of steady electric currents and voltages in development and repair processes (Borgens, 1982; Jaffe, 1979; Nuccitelli, 1983). In most cases, the potentials in multicellular organisms are due to potentials developed across epithelia by active processes. These transepithelial potentials (TEPs) are ubiquitous in the animal kingdom and are present with any epithelial compartmentalization. The separation of two areas usually occurs in response to the need for different local environments to be maintained. This compartmentalization is the work of epithelia and/or membranes and invariably involves the production of electrical potentials across the epithelia. The potential derived from an epithelium can be detected when the voltage between two electrodes contacting the tissue in each compartment is measured. The term ''bioelectric potential'' (BEP) is often used to designate the voltage along a surface of a compartment. Such a voltage is not itself a TEP, but is ultimately derived from a TEP. Although it has a decidedly imprecise ring to it, the term bioelectric potential is so well engrained in the bone literature that it will be used in this discussion.

The bone membrane concept. Studies by Neuman's group have established that bone maintains a bone fluid compartment separate from that of extracellular fluid (Geisler and Neuman, 1969). The bone fluid appears to have a very high potassium concentration, low calcium concentration, and a general hypertonicity, in comparison with extracellular fluid (Talmage, 1970). Because of the large electrochemical gradient maintained by potassium, Canas et al. (1969) have concluded that ''the extracellular fluid compartment of bone is functionally separated from the circulation by an active cellular barrier or membrane.'' For many years, investigators have proposed that such a membrane or epithelium effectively separates bone from the extracellular fluid of the rest of the body (Talmage, 1969). This ''bone membrane,'' as it has come to be called, theoretically plays a major role in general mineral metabolism, ion fluxes in the bone, and especially calcium homeostasis. Some morphological evidence for the existence of such a cell layer has been presented (Norimatsu et al., 1979).

If an electrochemical gradient is maintained across a bone membrane, it would be expected to produce a measurable electrical potential. An attempt has been made by Trumbore et al. (1980) to measure this postulated bone membrane potential by an indirect method. Using embryonic chick calvaria (skull caps) in vitro, they measured the distribution of charged and uncharged radioactive tracers in both the bone fluid compartment and the bathing medium. By comparing these distributions in normal calvaria with those in calvaria whose metabolic activity has been disrupted with potassium cyanide, they were able to determine the change in distribution of the extracellular tracers due to cell metabolism. Although the distribution of the uncharged tracer was not affected by cell metabolism, the distribution of the negatively charged tracer was. Using thermodynamic equations, they calculated that a potential of 4 mV across the bone membrane could account for the difference, with the bone fluid compartment being 4 mV positive with respect to the bathing medium.

Recent work by the same group demonstrated that parathyroid hormone, which increases calcium release from bone, increases the calculated bone membrane potential by 0.6 mV (Peterson et al., 1985). Since their model assumes that calcium passively follows the potential developed by actively pumped potassium, the increase in membrane potential would lead to the release of bone calcium. Although this work is indirect, it does suggest that bone is capable of producing metabolic potentials.

Digby (1966) also reported the measurement of bioelectric potentials in bone, although his work is equivocal because of a lack of documentation and apparent technical flaws. He used glass capillary electrodes filled with saturated KCl to measure potentials at different points on a rat tibia in vivo. The details are sketchy, but it seems that a potential of 10–15 mV was present across the growing end of the tibia, the marrow cavity being negative to the synovial cavity. "Similar potentials, but rather lower" were found across the periosteum. The marrow cavity was essentially isopotential with the outside surface of the bone under the periosteum. Digby found that the potential across the end of the tibia was decreased by 1 mV by compression of the femoral artery and vein of that leg. If the femoral artery and vein of the opposite leg were compressed, the potential rose by 0.2 mV. Since either a decrease or increase in blood flow to the tibia resulted in a decrease or increase in the potential across the bone, he concluded that the blood flow is the source of the potential. He hypothesized that streaming potentials due to blood flow through asymmetric capillary beds create a net charge in bone that is stored by the semiconducting properties of bone as a diffusion potential. After blood flow was stopped, the potential across the head of the tibia fell from 10–15 mV to 5–10 mV in 1–2 hours and to 0.1–1.5 mV after several days. This finding he attributes to slowly decaying diffusion potentials.

His work is compromised because the effect of injury potentials from the surrounding tissues damaged during surgery is not known. Drill holes must have been used to gain access to the marrow cavity, undoubtedly causing further damage. Also, to attain the extraordinarily low noise levels depicted in his figures, Digby must have used open capillary tubes (i.e., not pulled to a fine point) that would have allowed rapid diffusion of the saturated potassium chloride solution into the tissue, possibly leading to an unphysiological situation. The high external potassium concentration could cause a depolarization of the resting cell membrane potential. Apparently, potassium chloride did diffuse from the electrodes, because Digby detected negative potentials when one of the electrodes was pressed against muscle tissue. He attributed this effect to streaming potentials that may have been generated in the muscle due to the deforming force of the electrode. However, a more likely explanation is that negative potentials were generated by the diffusion of the potassium chloride past the charged surface of the muscle. A similar phenomenon has been documented in amphibian skin (McGinnis and Vanable, 1986; Nunes and Lacaz Vieira, 1975).

Bone Surface Potentials

Measurements. Other, more direct techniques have been widely used to detect the electrical potential profile distributed along the length of the surface of a bone rather than between the surface and the interior. These BEPs are measured by recording the potential difference between two electrodes placed on the bone surface. Usually one electrode is left stationary as a reference, and the other is moved incrementally along the surface to map out the potential profile. The electrodes consist of calomel (Friedenberg and Brighton, 1966; López-Durán Stern and Yageya, 1980; Rubinacci et al., 1984) or Ag/AgCl (Friedenberg et al., 1973; Friedenberg and Smith, 1969; Lokietek et al., 1974) electrodes connected via salt bridges to the bone. The salt bridges should be Ringer's agar or a Ringer's solution-soaked wick, since these large-diameter bridges are contacting living tissue. However, the salt bridges that have been used are reported to contain "isotonic saline" (Friedenberg and Brighton, 1966) (which is assumed to be 0.9% NaCl), "saline" (Friedenberg et al., 1973: Friedenberg and Smith, 1969; Lokietek et al., 1974) (which again, although with some reservation, is assumed to be

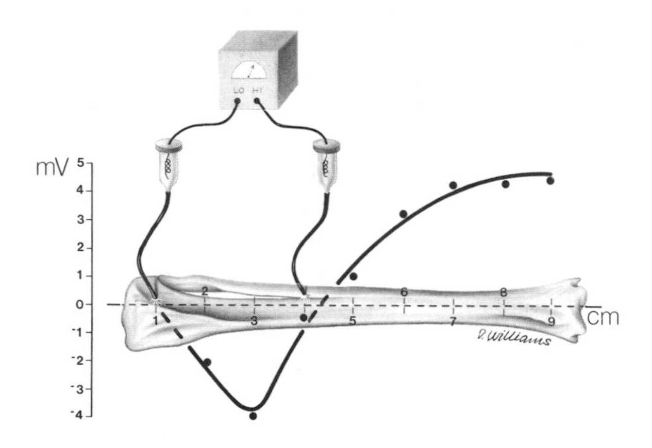

Fig. 3. Illustration of the technique and results of measuring the bioelectric potentials, BEPs, of a rabbit tibia. Typically, the tibia is exposed through an anterior skin incision, and salt bridge electrodes are placed in contact with the bone either supra- or subperiosteally. The salt bridges electrically connect a reservoir containing Ag/AgCl electrodes to the bone. The Ag/AgCl is in turn connected to a high-impedance voltmeter or electrometer. The end of the salt bridges may have a tuft of cotton or silk for making contact with the bone. The proximal electrode (left side) remains stationary and serves as the reference. The other electrode is moved incrementally along the shaft of the bone, and the voltage difference between the two electrodes is recorded.

0.9% NaCl), and "saline" (Rubinacci et al., 1984) that the authors identify as being 3 M KCl! The latter is simply not an acceptable means of making contact with living tissue, unless of course microelectrodes are being used. A high-input impedance ($>10^9$ ohms) electrometer or voltmeter measures the differential voltage between the two electrodes.

The typical BEP pattern that is described for the rabbit tibia is shown in Figure 3. The proximal metaphysis is negative with respect to the rest of the bone, whereas the distal half of the bone is electropositive. Both the proximal epiphysis and the midshaft are intermediate in magnitude. There is much variability in both the magnitude of the potentials measured and the shape of the potential profile. Rubinacci et al. (1984) attempted to establish a standard curve for the BEP profile of the adult rabbit tibia. They used six animals, with the reference electrode placed in three different points on each tibia. The shapes of the curves they obtained were fairly consistent, even when different reference points were used. They then compared their curve with those reported by others (Friedenberg and Brighton, 1966; Friedenberg et al., 1973; Friedenberg and Smith, 1969) and noted a great similarity, which added validity to their hypothesized standard curve. A critical evaluation of this comparison, however, reveals some discrepancies. In 1966, Friedenberg and Brighton showed a profile similar to that of Rubinacci et al. (1984). However, this profile was of immature rabbits weighing less than 1.8 kilograms. For adult rabbits weighing over 3.0 kilograms, similar to the 3.5-kilogram rabbits used by Rubinacci et al. (1984), a different BEP profile was obtained, being negative both at the proximal and distal metaphyses with respect to the diaphysis. Oddly enough, the adult rabbits in the 1969 Friedenberg and Smith study have a profile similar to the adults in the Rubinacci et al. study (1984) and the immature rabbits of the earlier study (Friedenberg and Brighton, 1966).

Rubinacci et al. (1984) also favorably compare their BEP curve with that found by Friedenberg et al. in 1973. Unfortunately, two distinct curve patterns were found by Friedenberg et al., only one of which was similar to that in Rubinacci et al. This curve could be obtained only if 7 cm of the 9-cm-long tibia was surgically exposed. If the full 9 cm was exposed, both the shape and magnitude of the curve was altered significantly. The investigators state that possibly the articular structures are negatively charged and that the monitoring of these voltages accounts for the change in the potential profile. This explanation is not at all reasonable, because the ''monitoring'' of the negativity of articular structures should have no effect on the electrical polarity of a point 1 cm distal to these structures. In any case, the fact remains that the BEP profile measured was dependent on the surgical technique used. This observation is quite disturbing if a case is to be made that the BEP profile represents the natural physiology of bone. A further critical reading of these papers reveals that the magnitude of the potentials varied with time (Friedenberg and Brighton, 1966), depth of anesthesia (Friedenberg and Brighton, 1966), surgical manipulation (Friedenberg and Brighton, 1966; Friedenberg et al., 1973; Lokietek et al., 1974), and exact placement of the electrodes (Friedenberg and Smith, 1969), and could even be reversed in polarity if anesthesia was too deep (Friedenberg and Brighton, 1966).

Several experiments have attempted to determine the BEP's character and origins. Friedenberg et al. (1973) claims that denervation and interruption of the blood supply does not affect the BEPs within 30 minutes. However, the magnitudes are reduced 30 minutes after administration of cytotoxic poisons

(2,4 dinitrophenol or iodoacetate), 6 hours after death of the animal, or 4 weeks after local application of ultrasound. They interpret the results to indicate that the BEPs depend on a viable bone cell population. A critical reading of the paper reveals that although the conclusions seem reasonable, the data as presented are not convincing. No indication of the amount of variability is given (e.g., standard error bars), although it appears to be extreme.

An explanation both for the variability and the source of the BEPs was offered by Lokietek et al. (1974). They suggested that the potentials measured on bone were the result of injury potentials from the nearby cut skin and muscle. They showed that as more soft tissue was damaged, the bone BEP increased. If the bone was insulated from the surrounding tissue by a sleeve of silicone, or removed from the animal completely, the potential profile was largely eliminated. Friedenberg and Brighton (1966) had also removed the bone from the body and claimed there was no change in the BEP profile 10 minutes after removal. Inspection of their Figure 3 however, shows a quite striking difference between the "in vivo" and "ten minutes after death" curves. Weigert and Werhahn (1977), however, comment that they were able to verify the existence of a BEP profile similar to that found by Friedenberg even when the bone was isolated from the soft tissue by a polyethylene foil.

Another inconsistency in these studies involves the question of effects of hemodynamics on the BEPs of bone. Lokietek et al. (1974) raised the intramedullary pressure by clamping the inferior vena cava of rabbits and found that the maximum potential difference on the bone surface decreased by a few millivolts. After clamping the abdominal aorta to lower the intermedullary blood pressure, a 1-mV increase in the BEP was measured. In the Friedenberg et al. (1973) study, however, ligation of the femoral artery caused a decrease, which they considered insignificant, in the BEP. Digby (1966), although measuring "across the head of the tibia" in rats, found a decrease in potential when the femoral artery and vein were compressed. Thus it has been found that decreasing the tibial intermedullary pressure can increase the BEP, decrease the BEP, or not affect the BEP.

The origin of these surface potentials measured along bone remains obscure. They may be as much a result of surgical or measuring technique as of the natural physiology of bone. The most, and perhaps only, consistent finding is that the proximal metaphysis is negative with respect to the midshaft (Friedenberg and Brighton, 1966; Friedenberg et al., 1973; Friedenberg and Smith, 1969; López-Durán Stern and Yageya, 1980; Rubinacci et al., 1984). The significance of this observation is not at all apparent. It must be remembered that these potentials are measured along a surface and not across an electrogenerative structure. They could well be the result of differences in local ionic concentrations, local surface charges, local vascularity, local injury, or local ionic transport mechanisms. The point is that the measurement of an electrical potential between two points may simply be a reflection of different local conditions under the two electrodes and is not necessarily indicative of the existence of an active biological mechanism that performs useful work either by means of energy or information transfer. In other words, the BEP of bone could simply be an artifact. However, it may well be a useful artifact and reveal much information about bone physiology. In a similar way, many histological techniques may be considered useful artifacts.

On the other hand, there may be a bioelectric system in bone that performs a useful and necessary role in the development, maintenance, growth, or repair of bones. The most convincing evidence so far is from a study by Rubinacci and Tessari (1983) of the correlation between BEPs and bone formation rates. They demonstrated that the spatial pattern of new bone formation, as revealed by tetacycline labeling, corresponded quite closely with the BEP profile, areas of relative electronegativity having a higher percentage of labeled bone. The critical question is whether the process of bone formation led to the measurement of local electronegativity or whether a local negativity led to bone formation. Rubinacci and Tessari (1983) mentioned experiments by Eriksson on the surface charge of bone particles that may be relevant to this question; they are discussed below.

Surface charge effects. Eriksson (1976a) found that demineralized bone specimens that had bone morphogenetic activity (i.e. were competent to induce new bone growth) carried a higher-than-normal negative surface charge, whereas bone samples lacking bone morphogenetic ability were consistently less negatively charged. In addition, chemical treatments that removed the bone morphogenetic ability also drastically lowered the magnitude of the negative surface charge. Consistent with the hypothesis that negative surface charges are a component of the bone morphogenetic ability, Eriksson also found that Proplast, an artificial porous composite material that can induce bone ingrowth, has a highly negative surface charge.

Subsequently, Eriksson and Jones (1977) demonstrated that demineralized bone with bone morphogenetic ability spontaneously recalcified when placed in a metastable calcium phosphate solution, while samples lacking this morphogenetic ability did not recalcify. Passing 10 μA of constant current through a bone sample with bone morphogenetic ability enhanced the recalcification at the cathode, or negative electrode, and eliminated it at the anode, or positive electrode. In a sample lacking bone morphogenetic ability, the current had no effect. Noda and Sato (1985a) reported that 10 μA of DC current induced significant calcification at cathodes inserted in the diaphyses of chick bones in organ culture, compared with sham controls. This calcification occurred in embryonic avian cartilage that normally does not calcify directly. They concluded that the calcification was an ''electromechanical phenomenon,'' and did not result from ''promotion of the physiologic process.''

Besides effects on chemical processes, the surface charge of a substrate can also have an influence on cell adhesion, motility, and spreading (Ericksson, 1976b). Norton et al. (1979) placed bone cell cultures in a capacitive field that they assumed acted predominantly by altering the surface charge of the plastic walls of the culture tubes. The rate of cell adherence to the walls, the strength of the adherence, and the level of DNA synthesis were all found to be increased in the stimulated tubes, compared with sham controls. It was also determined that all of these effects occurred primarily on the cathodal or negative side of the tubes.

It is also possible that the osteogenic ability of implanted electrets is the result of their surface charge characteristics. Yasuda (1977) and others (Inoue et al., 1977, 1979) have found that Teflon electrets (Teflon strips that have been electrically polarized) can induce callus formation just by being placed adjacent to bone. Nonpolarized Teflon has no effect. Negligible current flows from one side of the electret to the other (2 pA/cm^2) (Inoue et al., 1979), so the observed cell stimulation is likely to be the result of the surface charge on the electret. If this is true, it would be expected that one side of the electret would have more osteogenic ability than the other. Although this notion has not been directly tested, inspection of Table 1 of Inoue et al. (1979) illustrates that the horns of callus seem to grow on one or the other side of the electret and not both.

It is possible then, to speculate that the difference in rates of bone formation observed by Rubinacci and Tessari (1983) could be the result of local differences in fixed surface charge. Could the pattern of bioelectric potentials be similarly explained? Fixed surface charges are known to complicate biological potential measurements. Fixed charge in the dead cell layer of amphibian epithelium, the stratum corneum, coupled with KCl flowing from the electrode tip produces negative potentials (McGinnis and Vanable, 1986; Nunes and Lacaz Vieira, 1975) that in the past have been mistaken for intracellular potentials. Similarly, Nelson et al. (1978) found artifactual potentials produced by the combination of fixed charge and KCl leakage from the electrode. Finally, Tasaki and Singer (1968) discuss the Pallman or suspension effect that occurs in the presence of colloidal polyelectrolytes. The fixed charges of the polyelectrolyte (or charged surface) interact with the mobile charges in the fluid and produce a potential difference between the two phases. "Even small variations in the nature of local concentrations of polypeptides . . . can produce significant changes in the measured potential differences" (Tasaki and Singer, 1968).

It may follow, then, that BEPs represent local difference in surface charge that reflect other aspects of the bone's overall physiology. Although this concept is not consistent with the observations that the BEPs are directly dependent on metabolic processes (Friedenberg, et al., 1973; Lokietek et al., 1974), the hypothesis of an active cellular electrogenic process is not well supported either. It is evident that laboratory studies have yet to establish even the existance, much less the role, of such a system.

Effect of fractures. Before leaving the subject of the measurement of surface BEPs in bone, something must be said about the effects of a fracture on the BEP profile of bone. Friedenberg and Brighton originally observed in 1966 that a peak of electronegativity occurred at the fracture site, along with a general increase in the electronegativity of the entire bone, when referred to the proximal epiphysis. This notion, that the fracture site is strongly electronegative, has persisted in the literature (Borgens, 1984; Eriksson, 1976b; Friedenberg et al., 1973; Pollack, 1984; Watson, 1981). However, these peaks in electronegativity were initially measured on the skin over tibial fractures in both rabbits and humans, and not on the bones themselves (Friedenberg and Brighton, 1966). In a subsequent study, Friedenberg and Smith (1969) found no correlation between potentials measured on the skin with those measured on the bone itself. This observation alone removes much of the value of the original study. To make matters worse, Friedenberg and Smith (1969) then measured the BEPs on the surface of the tibia both before and after fracture. After fracture, they found a general increase in negativity along the whole bone. This increase was greatest just distal to the fracture site. However, when the information in their Figure 2 (increase in electronegativity along the bone) is combined with that in their Figure 1 (BEP profile along the bone), no peak in electronegativity is demonstrated (Fig. 4). The fracture was negative with respect to the distal portion of the bone, but positive with respect to the proximal portion.

López-Durán Stern and Yageya (1980) also found that fracture causes a shift in the overall potential of the bone with respect to the proximal epiphysis, rather than a peak at the fracture. Their study on rat tibias found the entire distal two-thirds of the bone to be positive. Fracture through this positive area caused an "immediate" increase in this positive potential. When next measured, 1 day after fracture, potentials were reduced all along the bone. Furthermore,

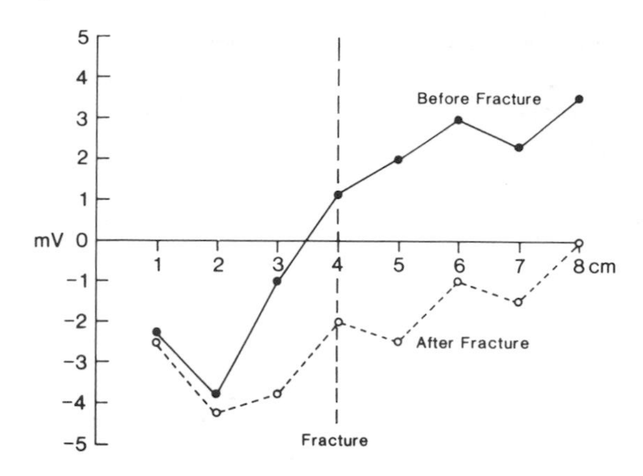

Fig. 4. The bioelectric potential (BEP) profile of rabbit tibiae before and after fracture. The data are taken from Figures 1 and 2 of Friedenberg and Smith (1969). The measurement technique was similar to that shown in Figure 3 (this chapter) with the proximal reference placed at the 0 cm point. For the "before fracture" data, an average was taken of the right and left tibia data presented by Friedenberg and Smith. The fracture was made at either 3.5 cm or 4.5 cm distal to the proximal epiphysis. However, since the data for the two fracture sites were combined in the original figure, 4.0 cm was chosen as the fracture site for this illustration.

Lokietek et al. (1974) measured the change in BEPs due to an osteotomy. Although this study involved only two animals, they found the BEP profile was altered by the osteotomy. However, this response could then be eliminated in part by using a silicone sheet to electrically insulate the site from the surrounding soft tissues. In conclusion, fracture of a bone does affect the potential profile along its surface. It is not true, however, that a distinct locus of negativity is found at the fracture site. Whether the fracture is positive or negative depends on the point of reference.

Further problems with interpretation of BEP data are engendered by the fact that a variety of unconventional connections have been used between the electrodes and the measuring device. The most straightforward and traditional method would have been to connect the negative (or low) input of the voltmeter, electrometer, etc. to the reference electrode, which remains stationary. The positive or high input would then be connected to the movable recording electrode. Rarely has this arrangement been used (Friedenberg et al., 1973; Lokietek et al., 1974), and when it was, it was not explicitly stated. Apparently, the positive input was used as a stationary recording electrode twice (Friedenberg and Brighton, 1966; Friedenberg and Smith, 1969) and as a stationary reference once (López-Durán Stern, 1980);

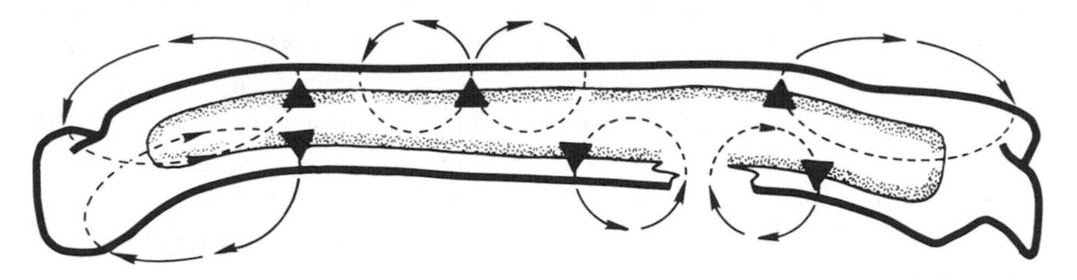

Fig. 5. Current pattern around and in a mouse metatarsal as inferred from Borgens's measurements made with the vibrating probe (Borgens, 1984). The bone was isolated from the mouse and placed in a dish of Ringer's solution at 37°C. Steady currents were measured external to the bone, and the internal current patterns were inferred.

Lokietek et al. (1974) clearly establish a proper sign convention by using a battery and noting the direction of the oscilloscope deflection. It would have been reassuring had other investigators taken the same care in reporting their sign conventions when nonstandard references are used.

Bone electric currents. Borgens (1984) has studied the bioelectric activity of bone isolated in short-term organ culture to avoid the problems of soft tissue injury potentials. He selected the weanling mouse metatarsal because it has no muscle attachments and can be easily dissected free from the body without damage to the bone membranes. The electrical fields around these small (4–7-mm-long) bones were mapped using an ultrasensitive vibrating probe (Jaffe and Nuccitelli, 1974), and current densities and directions were calculated. His study showed the presence of both stress-generated currents, which he considered to be the result of streaming potentials, and steady currents in undamaged and damaged bones. The stress-generated currents could be produced in either dead or living bone, whereas the steady currents were detected only in living bone. These steady currents, with densities of 0.5–12 $\mu A/cm^2$, entered the articular surfaces, while smaller currents both entered and left along the shaft of the bone (Fig. 5).

If the bone was fractured, large currents (100 $\mu A/cm^2$) entered the damaged region but decayed by 30 minutes to a plateau of 5 $\mu A/cm^2$ that was stable for hours. These steady fracture currents were not produced by damage just to the periosteum. The notch or fracture had to extend through the cortical bone before currents could be detected. Borgens also found that an isolated diaphysis or shaft could maintain its currents even with the cartilagenous ends removed. These observations allowed him to conclude that the source of the currents was not the periosteum or the growing ends of the bones, but was probably located in either the endosteum or lining cell layer of the marrow cavity. He also found these steady currents to be temperature-dependent, decreasing dramatically as the bath temperature was lowered from 37°C to 20°C. If the ionic composition of the bathing medium was altered, striking change in the currents resulted. Replacing the Na^+ in the medium with an impermeant ion such as choline or Tris increased the current density, while replacing Cl^- with gluconate reversed the direction of the current. Removal of Mg^{2+} or Ca^{2+} also caused increases in current density at the fracture.

Borgens's data support the hypothesis that bone maintains a voltage across itself, the marrow negative with respect to the outside, by metabolically dependent ion transport. Digby (1966) also reports a negative interior in bone, although his measurements indicate that the potential is developed across the periosteum. Borgens (1984), Friedenberg and Brighton (1966), and Rubinacci et al. (1984) all find the periosteum to be relatively inert as far as electrical properties are concerned. Trumbore et al. (1980) also found a transcortical potential, although the bone compartment was positive to the bathing medium.

Problems with measurements of BEPs. There have been other attempts to measure bioelectric potentials in bone that have suffered more than the usual degree of technical problems. A few of these will be reviewed to point out some of the serious errors that can be made in such measurements. For example, one group used six stainless steel electrodes chronically implanted in rabbit tibia to measure the BEP profile of the bone (Behari and Singh, 1981). They state that "the choice of the electrode material has been made so as to reduce the polarization effect. Also since both the electrodes are made of identical materials, the polarization due to the galvanic cell is identical in the vicinity of both the electrodes . . . . Care was taken to record observations within minimal time after putting electrodes in circuit" (Behari and Singh, 1981). The choice of stainless steel as an electrode material to reduce polarization was ill advised. Stainless steel is *not* an acceptable electrode material for measuring DC potentials in fluids. Even though the half-cell potential at both electrodes should be of the same magnitude, and theoretically should cancel each other out, this does not actually occur; in practice large offset potentials result. Their comment that the readings were taken immediately after placing the electrodes in the measuring circuit is an indication that they were experiencing polarization effects, i.e., a rapidly changing potential as current is drawn through the measuring system. Also, their measurement of 80–120 mV between two electrodes spaced 1 cm apart in the marrow cavity is far higher than is reasonable to expect and is very similar to the 100 mV measured by Friedenberg and Brighton (1966) when they implanted two stainless steel electrodes 1 cm apart in rabbit tibia. Fortunately, Friedenberg and Brighton recognized these large potentials as artifactual and switched to calomel electrodes.

In a similar fashion, the electrical properties of the electrode-fluid interface dominate measurements of tissue impedance or resistance when the same electrodes are used both for delivering and measuring the current and voltage. For instance, in the above study (Behari and Singh, 1981) an attempt was made to measure the impedance between two of the electrodes by applying 1.6 volts alternating current (AC) to the electrodes and measuring the resulting current through these electrodes. This method is totally unacceptable for measuring biological impedances. Such measurements will be completely dominated by the impedance of the electrode-electrolyte interface. Furthermore, these measurements were made in the range of 0.5–100 kHz, which does not yield useful information when bioelectric or DC signals are being studied. Proper impedance measurements require the use of four electrodes, two to pass current and two to measure voltage (Geddes and Baker, 1975). The use of AC in this technique allows for the measurement of the impedance of the tissue, from which the DC resistance can be calculated by extrapolation from impedances at very low frequencies (< 10 Hz). Finally, to assign a resistance value to any tissue or structure, the geometry of current flow must be rigorously determined or limited. This was not possible, or even attempted, with the preparation studied by these investigators. All of the data presented in this paper were dominated by, if not completely attributable to, phenomena at the electrode-fluid interface and were not measurements of the biological system. Similar measurements would probably have been obtained by sticking six stainless steel needles in any ionic solution.

Other serious errors are illustrated in another publication (Weinberg et al., 1982). In this study the authors drilled holes in the femur of anesthetized rabbits and measured the intramedullary potential between the holes. They initially tried using polarizable electrodes, platinum and lead, but quickly found artifactual potentials. They measured voltages of 30–400 mV with platinum and 7–131 mV of both polarities with lead. These measurements are similar to the 100-mV potentials reported with stainless steel electrodes in the previously discussed report. To their credit, the authors finally abandoned the platinum and lead for Ag/AgCl electrodes. However, in subsequent measurements of resistance and current, the authors apparently used the ohmmeter and ammeter functions of a multimeter to obtain their data. Ohmmeters cannot be used for measuring resistances in biological systems, or, for that matter, any system containing fluid. An ohmmeter determines resistance by passing a small known current through the load and measuring the resulting voltage drop. Since the resistance of the electrode-fluid interface depends on the current passing through it, ohmmeter readings reflect the dynamic changes occurring at the electrode surface. Furthermore, the potentials generated by the "battery" that is produced between the two metal electrodes in solution interfere with the ohmmeter measurements. Therefore, an ohmmeter can measure neither the electrode resistance nor the tissue resistance. Again, to measure a biological resistance, a tetrapolar electrode arrangement must be used (Geddes and Baker, 1975).

An ammeter only gives valid measurements of current when all of the current in a branch of a circuit is made to flow through it by placing it in series with that branch. If, however, two electrodes are placed in a biological system and connected by an ammeter, only a portion of the biological current will flow through the measuring system. The magnitude of this biological current that passes through the ammeter will depend on the relative resistances of the two parallel paths, the one normally available through the tissue and the new one through the ammeter. Therefore, only a portion of the biological current will be measured. The resistance of the path through the ammeter will be determined by the electrode resistances. Furthermore, creating a parallel current path will lower the overall resistance of the system, with possible biological consequences.

Current measurements in biological systems are complex and difficult to make. One method is to measure the voltage at two points and, knowing the resistance of the material between them, calculate the current. This technique is the basis of current measurements using the vibrating probe (Jaffe and Nuccitelli, 1974). The real difficulty is determining the effective resistance of the tissue or fluid in the current path of choice. Another way to measure a biologic current is to use a bucking current from an artificial source to drive the voltage difference between two electrodes in the current path to zero. The amount of current in the bucking circuit is then taken as a measure of the current in the biological circuit. A problem with this technique is that the geometries of the two currents need to be the same, which means the biologic sources and sinks must be known and mimicked by the bucking circuit. This can rarely be achieved except under the most artificial of conditions.

With these and other pitfalls in mind, several suggestions can be made concerning the measurements of DC biological potentials and currents:

1. Reversible, or nonpolarizable, electrodes must be used for DC potential measurements. Ag/AgCl electrodes are the most convenient and should

be used in most cases. Calomel electrodes are an acceptable alternative but are usually more difficult to work with. Stainless steel and platinum simply are not adequate.

2. High-impedance voltmeters, electrometers, voltage followers, etc. should be used. The minimum input impedance should be 10^8 ohms, preferably higher.

3. Ohmmeters must not be used for measuring biological resistances. They are, however, valuable for trouble-shooting equipment and checking the continuity of circuit elements.

4. Biological resistance and/or impedance measurements should only be made with a tetrapolar measuring system (Geddes and Baker, 1975) using a defined and/or limited biological geometry. Care should be used in assigning resistivity values to specific structures or components of a heterogeneous tissue sample.

5. Ammeters cannot be used for making quantitative measurements of biological currents, except in the rare case in which the geometry allows for complete shunting of the current through the meter. Even in this case there may be problems because of the artificial nature of the geometrical arrangement and impedance matching between the normal biologic pathway and the substituted ammeter pathway. Ammeters can be used for measuring experimentally applied currents.

6. When possible, the Ag/AgCl electrode should contact the preparation indirectly, through a salt bridge or "liquid wire." This prevents contamination of the electrode by proteins, oils, etc. and also mechanically stabilizes the electrode-fluid interface, which is sensitive to vibrations and movements. Also, since Ag/AgCl acts as a chloride ion-specific electrode and environments with different Cl^- concentrations yield different electrode potentials, a salt bridge allows a known stable ionic composition to be in contact with the Ag/AgCl. The salt bridge should be a biologically compatible fluid such as Ringer's solution. The use of high concentrations of KCl is completely unwarranted when using large-bore electrodes, which leak an unacceptable amount of KCl into the tissues being measured. Saturated or 3–4 M KCl is specifically used to overcome the problems of tip potentials and high resistance of small-diameter (<1 μm) glass microelectrodes used for intracellular recordings. Since extracellular fluid typically has a low potassium concentration, around 3 mM, the use of 3 M KCl in large electrodes could well have detrimental effects, including depolarization of cell membranes and collapse of cells as a result of the extreme hypertonicity.

7. All voltage measurements should be taken with awareness of the injury potentials from soft tissues damaged during surgery.

8. An understanding and appreciation for the volume conduction properties of biologic tissue is necessary for the interpretation of any current paths or field patterns.

ELECTRICALLY INDUCED OSTEOGENESIS

One of the major elements in the argument that electricity is involved in the response of bones to trauma is that artificially applied electric currents or fields do affect bone. There is extensive accumulated evidence that applied electricity induces an osteogenic response in bone. The term "osteogenic" is used in its broadest sense and simply refers to a demonstrable accumulation of bone tissue. Two general observations about the literature are immediately apparent. First, almost any application of electricity, in any form, seems to be capable of eliciting an osteogenic response, although some optimization of the stimulus parameters has been achieved. Secondly, none of the effects are repeatably striking. Most require statistical techniques to demonstrate the significance of an effect because of large levels of variability.

The many attempts at experimental electrogenesis have fallen into two major areas. The first involves the direct application of a DC current to tissues. This technique is by far the more common and has yielded the most consistent results. All other techniques fall into the second category, which includes attempts to produce a more "compatible" stimulating system than invasive DC applications. In some of these cases, "compatibility" means producing a signal that is thought to be more similar to the natural stress-generated potentials. Other studies used pulsed DC or AC not because it seemed more biologically appropriate, but because of electrode considerations. By using pulsed or alternating currents, some of the problems of electrode polarization and electrode products are minimized. Finally, some consider the ultimate in compatibility is to use a completely noninvasive method, such as capacitive or inductive coupling, to apply electricity to the system. The plates and coils used in these techniques to produce electric fields in the tissue are kept external to the body, and require no surgical insertion.

DC Stimulation

A discussion of some of the early investigations of electrically induced osteogenesis by direct currents will introduce the general technique and help define the area. Although earlier work in Japan by Yasuda (1953) showed that small electric currents could elicit an osteogenic response in vivo, the more influential work was done by Basset et al. in 1964. Their stimulation unit consisted of two bare iridium-platinum electrodes connected in series with a resistor and a 1.4-V mercury battery. The battery and resistor were encapsulated in epoxy and silicone, leaving the two electrodes protruding. These electrodes were inserted in two drill holes in the femurs of dogs. Currents of approximately 0.7, 2, or 3 μA were passed between the two electrodes in one femur. In the contralateral femur, an inactive stimulator was implanted. Two and three weeks later new bone had formed around the inactive electrodes as well as around the anodes and cathodes of the active stimulators. The only difference was that at the cathode delivering 2 or 3 μA, there was substantially more bone than at any of the other electrodes. At 2 weeks, this new mass was quite cellular, but at 3 weeks it was maturing fibrous bone. Furthermore, between the electrodes of the 2- or 3-μA packs, increased cortical remodeling was stimulated.

Several conclusions can be drawn from this one study:

1. Osteogenesis can be elicited by exogenous DC currents delivered through metal electrodes.
2. Some bone forms at the inactive electrodes.
3. Bone preferentially forms near the cathode, although some does form at the anode.
4. Bone is altered in between the electrodes as well as in the area immediately surrounding the electrodes.
5. A minimum current level is necessary for demonstrable effects.

Verifying or repudiating these findings has been the basis of much subsequent work.

This study, however, also introduces several problems that recur in later work. First of all, the work demonstrates that the electrode-electrolyte interface can dominate the current delivery system. The battery packs were designed to produce 100, 10, and 1 μA of current. Within 15 minutes of placement in the body, the current fell to 3, 2, and 0.7 μA, respectively, due to polarization of the electrodes. Of related interest is the problem of monitoring the current actually being delivered to the tissues. No details about measuring techniques were given by Basset et al. (1964), although the current apparently was measured both before and after the 3-week stimulation period.

The most serious problem Basset et al. faced was documentation of their results. They chose to simply state the qualitative results and show a photograph of one bone sample to support their observation. Adequate quantification of the amount and quality of induced osteogenesis is a persistent problem in this literature. In a seminal work such as that of Basset et al., the lack of quantification could possibly be excused. In subsequent work, however, documentation of the sometimes subtle effects of electrical stimulation is of paramount importance. The next paper in this series demonstrates this point in particular.

O'Connor et al. (1969) attempted an exact replication of the Basset et al. (1964) work. Their qualitative results were similar, although quantitatively the results differed. They ranked the cathode, anode, and two control electrodes from each dog as to the relative amount of new bone present. The scoring was done by "naive observers." In agreement with Basset et al., they found the most new bone at the cathode. However, instead of the uniform result implied by Basset et al., they found the cathode had the highest rank in only 7 of 12 animals. In two animals, the cathode actually had the least bone. Although the cathode effect did prove to be statistically significant by a binomial expansion, it was also shown to be a variable effect. This extreme variability has proved to be the rule. The grossly obvious response to DC stimulation originally described by Basset et al. (1964) is the exception.

A convenient context in which to discuss subsequent work is as it supports or contradicts the basic findings about electrical osteogenesis that were listed and discussed above.

Osteogenesis elicited by exogenous currents. The observation that osteogenesis can be elicited by exogenous electric fields is supported by numerous and varied experiments. Several reviews from the 1970s (Cochran, 1972; Herbst, 1978; Spadaro, 1977) and a recent listing by Brighton et al. (1985) include studies that have demonstrated electrically induced osteogenesis. In 1977, Spadaro listed only 6 of approximately 75 experimental papers that described negative or equivocal results. One of these studies (Hambury et al., 1971) is especially interesting, because it was an almost complete replication of the Basset et al. (1964) study, using rabbits instead of dogs. No differences in osteogenesis were found between the cathode, anode, or inactive electrodes, even using a sensitive bone scan of ^{85}Sr uptake. A possible explanation of this discrepancy is that the amount of new bone resulting from trauma was much larger than that resulting from electricity, thereby obscuring any direct electrical effect.

Osteogenesis at the inactive electrodes. This observation leads to the second point, namely that bone forms at the inactive as well as the active electrodes. It has long been recognized that either trauma or the presence of a foreign body (e.g., an electrode) can lead to local bone formation. Most studies have attempted to circumvent this problem by comparing the effects of an active electrode with a sham electrode. A serious consideration, however, is that the amount and degree of trauma is not easy to quantify or standardize. One of the major advantages of using electricity is that the investigator has considerable control over the stimulus parameters. Current, voltage, frequency, duty cycle, etc. are under experimental control. If several stimulators are used in an experiment, they can be assumed to match the specifications of the protocol, or they can be measured. The trauma of anesthesia, surgery, drilling of holes, and placement of electrodes, however, would be expected to involve many variables beyond the investigator's control. Therefore it is possible that much of the variability, both within and between studies, is the result of implant-induced osteogenesis.

Three recent papers address these issues. Brighton and Hunt (1986) and Esterhai et al. (1985) placed active (20 μA) and inactive electrodes in the medullary canal of rabbit tibia. They followed the changes that occurred in the marrow near the electrodes. The trauma of inserting the electrode itself led to the formation of a mass of polymorphic cells (Brighton and Hunt, 1986), followed by a replacement of the polymorphic cells with osteoblastic new bone formation. This new growth occurred similarly at both the active and inactive electrodes through the ninth day. From then on, the amount of new bone at the inactive electrode declined while that at the active electrode increased (Brighton and Hunt, 1986; Esterhai et al., 1985). If the current was not applied until 21 days after the electrodes were inserted, new bone would begin forming 2 to 3 days later, demonstrating that current by itself is capable of leading to osteogenesis even long after the original trauma. Interestingly, "there were very few discernible morphological differences . . . between osteogenesis induced by trauma and osteogenesis induced by electricity" (Brighton and Hunt, 1986).

Another study, by Spadaro et al. (1986), discusses the effect of mechanical movement of electrodes implanted in the medullary canal of rabbits. It was found that simple mechanical motion could elicit and maintain an osteogenic response for up to 12 weeks in the vicinity of the electrodes. Similar electrodes that were not allowed to move elicited very little osteogenesis. Motion can be transmitted to implanted electrodes by the wire leads that connect them to distant power supplies that are strapped to the body or implanted. Body movements and muscular contraction are coupled to the electrodes by these connecting wires. Spadaro et al. (1986) found that the use of very flexible wires prevented the transmission of motion to the electrodes and eliminated much of the mechanically induced osteogenesis. They also found that the bone formation ceased after fixation of the implant by tissue growth.

Several guidelines for future work can be drawn from these studies. To reduce electrode motion, flexible leads and adequate fixation of the electrode at the implant site should be used. Also, to maximize the apparent electrically induced osteogenic effect, the onset of current application should be delayed until about 3 weeks after the electrodes have been implanted. This delay should allow for a decrease in the trauma-induced osteogenesis to low levels and ensure adequate mechanical fixation of the electrode. Finally, the sham electrodes should always have the same mechanical fixation as the active electrodes, or the experimentor risks inducing different degrees of mechanically induced osteogenesis. This fixation should include leads that are extended to and connected to a dummy power source, similar in size and weight to the active one, to ensure that similar stresses are applied to both electrodes.

Osteogenesis at the active electrodes. Nearly all studies have reported that electrically induced osteogenesis occurs in bone at the cathode. Since the cathode is the site of obvious electrogenesis in most studies, the anode has been relatively ignored. Conflicting evidence exists as to whether the anode also stimulates osteogenesis, causes necrosis, or has little effect. At the typical current levels of 20 μA or less, Friedenberg's group has consistently found necrosis at the anode (Friedenberg et al., 1970, 1971b; Friedenberg and Kohanim, 1968), although there is usually no gross resorption. At 50 μA or more, there can be bone necrosis (Buch et al., 1986; Friedenberg et al., 1974) and/or resorption at both the anodes and cathodes. Some have attributed this tissue death to toxic electrode byproducts liberated at these high current levels and/or a direct effect of high fields. On the other hand, others have reported electrically induced osteogenesis at the anode (Janssen et al., 1978; Petersson et al., 1982; Yasuda, 1953), although the response at the anode is usually less than that at the cathode.

An argument has been advanced by Brighton and Friedenberg (1974) that the enhanced osteogenesis at the cathode, or cathodal effect, is a result of electrochemical changes in the microenvironment near the electrode. Specifically, they postulate a decrease in the oxygen tension and an increase in hydroxyl concentration near the electrode. A chemical reaction must occur to transfer the current carried by electrons in the wire electrodes to ions in solution. According to Brighton and Friedenberg, hydrolysis occurs at the cathode with one of the following reactions.

$$2H_2O + O_2 + 4\,e^- \rightarrow 4OH^-$$

or

$$O_2 + 2H^+ + 4\ e^- \rightarrow 2OH^-$$

In either case, oxygen is consumed and the pH increased. Brighton and Friedenberg (1974) consider both of these shifts to favor osteogenesis. These reactions, however, only occur within a certain voltage range for each type of metal. For stainless steel, for example, if the electrode voltage approaches 1 volt or higher, the electrochemistry shifts from one of the previous oxygen-consuming reactions to one of the following hydrogen-producing reactions

$$2H_2O + 2\ e^- \rightarrow H_2 + 2OH^-$$

or

$$2H^+ + 2\ e^- \rightarrow H_2$$

This may explain why higher currents and therefore higher electrode voltages lack osteogenic properties. Recently an effort was made to evaluate the osteogenic ability of several metals at both low and high current densities, but because of technical difficulties, the results are equivocal (Spadaro, 1982).

As mentioned, Brighton and Friedenberg attribute cathodal osteogenesis to reduced oxygen tension and increased pH of the local environment (Brighton and Friedenberg, 1974). The measured O_2 consumption is about 2.5×10^{-11} mole/sec at a stainless steel cathode delivering 10 μA. The normal concentration of free oxygen in body fluids, however, is about 1.4×10^{-7} mole/ml (Spadaro and Becker, 1979). This means that, given no circulation, after 1 minute of current there would be a 1% change in oxygen in the cubic centimeter of tissue around the electrode. Clearly, however, circulation is present in tissues; therefore any small changes in P_{O_2} may be overridden by the normal oxygen delivery system. Spadaro and Becker (1979) argue that it is not this small change in oxygen but rather the relatively large change in concentration of reaction intermediates such as peroxide (H_2O_2) that affects cells. In either case, products and/or intermediates of the reduction of molecular oxygen at the metal electrode surface can be considered possible mediators of cathodal osteogenesis.

It is possible that there are several means by which DC electric current stimulates osteogenesis. The cathodal effect could be the result of the mechanical trauma of electrode insertion, the liberation of noxious electrode products, irritant effects of high local fields, or direct electrical or chemical effects on bone cells. The first three possibilities could all be considered forms of local insult to the tissue. The liberation of electrode products or cell damage because of high fields could act as chronic irritants and elicit osteogenesis similar to the mechanically induced osteogenesis reported by Spadaro et al. (1986). On the other hand, the cathodal effect may be the result of nontraumatic osteogenesis, i.e., a direct stimulatory effect on bone cells by competent electrical or chemical (PO_2, pH, H_2O_2) signals.

Bone Alteration Between Electrodes

Evidence. In addition to osteogenic effects immediately adjacent to the electrodes, researchers have found a significant osteogenic response far removed from the actual electrode sites. It is interesting that the original Basset et al. paper (1964) notes that besides the callus at the electrodes, there was enhanced remodeling occurring in the cortex between the electrodes. Also, early reports by Yasuda (1974, 1953) showed effects both at the electrodes and in between. Lavine et al. (1969) reported preliminary evidence that DC current could enhance the healing of a bone defect placed in between two electrodes. A later report substantiated this finding (Lavine et al., 1971). Their experiments were the first designed specifically to distinguish between true current effects and other possible electrode effects.

Weigert and Werhahn (1977) also provided evidence that a bone defect, in this case an experimental osteotomy, heals faster when electrodes are placed on either side of the defect. They found an increase in endosteal and periosteal callus, vascularization, amount of mature lamellar bone in the fracture gap, and uptake of fluorochromes in the stimulated as opposed to control bones. Similarly, the amount of new bone found in a bone growth chamber implanted in rabbit tibia was increased by current from electrodes placed 5 mm on either side of the chamber (Buch et al., 1984). No indication was found of a bias of bone growth in the chamber toward the cathode. Connolly et al. (1977), in describing their similar experience, state, "the reaction to electrical stimulation appeared to be a diffuse, irritative periosteal response, more a response to the electrical field than an accumulation of bone at either pole."

Friedenberg et al. (1971b), on the other hand, found no effect of 10-μA direct current on a fibular osteotomy unless the cathode was located directly in the defect. If electrodes were placed on either side of the osteotomy, osteogenesis was not stimulated at the lesion.

While these contradictory results are difficult to reconcile, the weight of evidence supports the notion that true electrical, as opposed to electrode-mediated, osteogenesis does occur. Little data on the geometry of electrical callus formation is available. Indeed, most investigators have had enough trouble just documenting the existence of electrically stimulated osteogenesis without the added difficulty of mapping out the three-dimensional pattern. The induction of osteogenesis away from the electrodes could be the result of direct electrical stimulation or possibly of more global changes arising secondarily to electrode effects. However, it appears from the literature that osteogenesis is limited to loci near the electrodes or to areas, especially defects, between two closely spaced electrodes (approximately 1–2 cm apart). Therefore a general systemic stimulation arising secondary to direct electrode effects is unlikely. The most likely mechanism for the induction of osteogenesis away from the electrodes is a direct electrical effect on the tissue between the electrodes. Although current from implanted electrodes will spread in the tissue via volume conduction, areas within a few centimeters of the electrode will experience higher fields because of current convergence. This may be why even these supposedly "true" current effects are usually limited to areas near the electrodes.

Vascular changes. Although most attention is directed toward the bone cells, as the receivers of the electrical signals, there is some very interesting evidence that possibly another component of bone, namely the vascular system, is a target for the electrical signal. Weigert and Werhahn (1977) observed that the electrically enhanced healing of an experimental osteotomy involved more extensive revascularization. Similarly, Zichner and Happel (1979) found the first reaction to an electrical stimulus was a dilation of blood vessels with a subsequent increased vascularity in the electrically induced callus. More recently, a Swedish group (Buch et al., 1985; Nannmark et al., 1985), using an optical bone chamber, was able to observe directly the vascular network in new living bone and its response to applied electricity. The optical chamber was a titanium cylinder (with two quartz glass rods glued inside) that was threaded through the tibia of a rabbit until a transverse hole between the glass rods was at the level of the cortical bone. After 6 weeks the implant had stabilized, and new bone had infiltrated the space between the pieces of glass. This new bone was repeatedly observed with a microscope by transillumination.

It was found that passing 5 or 20 μA DC between electrodes placed 5 mm on either side of the chamber caused more bone to form in the chamber than formed in control chambers (Buch et al., 1985). There were no immediate or obvious changes in the blood vessels, capillaries, or blood flow. However, over time it was found that the vascular density increased in those chambers being stimulated. Another delayed effect of the current was that the average diameter of small vessels (15–25-μm-diameter) doubled. A higher current, 50 μA, which caused bone resorption, also caused a decrease in the vascular density.

In another study (Nannmark et al., 1985) using these same techniques, they injected a fluorescent macromolecule (FITC-dextran; 70,000 daltons) into the blood stream of the rabbits. In control animals, the observed fluorescence was always restricted to the blood vessels. However, within 1 hour of applying 5 μA of DC, the fluorescent molecule began leaving the vascular system and became diffuse within the tissues. This macromolecular leakage occurred at the capillary level and did not show any bias toward either electrode.

In parallel studies using the more accessible vascular network of the hamster cheek pouch, it was found that white blood cells appeared in the tissue concurrently with the macromolecular leakage identified by fluorescence. The leakage was induced by either DC or AC current and was inhibited by indomethacin, a nonsteroidal antiinflammatory drug. Indomethacin inhibits prostaglandin synthesis and has been shown to inhibit fracture repair and bone remodeling. Since both bone repair (Ro et al., 1976; Sudmann and Bang, 1979) and electrically induced macromolecular leakage in the vascular system (Nannmark et al., 1985) are both blocked by indomethacin, it is plausible that a similar mechanism is involved in both systems. Ro et al. (1976) state that "the fracture trauma is followed by an aseptic inflammation which initiates the process of fracture repair. The effect of indomethecin (inhibition of fracture repair) may, consequently, be caused by an interference with the inflammatory process." A pertinent hypothesis may be that electrical stimulation can elicit an "aseptic inflammatory" response on a local level. An important point is that since the inflammatory type of reaction did not occur until 1 hour after the current was applied, it is possible that the vascular effect is secondary to some other effect.

In conclusion, it can be said that small DC electric currents can elicit an osteogenic response from bone tissue. There is a direct effect on bone cells that stimulates them to begin laying down new bone when they are in the presence of a sufficiently large field. This direct electrical effect is probably modified by electrochemical and mechanical effects at the electrodes. At the cathode, changes in oxygen tension, pH, and H_2O_2 concentration may dominate and increase the magnitude of the direct electrical response.

Effective current levels. There seems to be a general rule that low currents, less than 1 μA, have no effect on bone, and high currents, greater than 30–50 μA, cause bone resorption. These guidelines specifically refer to stainless steel cathodes as used by Brighton, Friedenberg, and coworkers (Friedenberg, et al., 1970; Friedenberg et al., 1974). Most other workers have accepted these standards and conform to them.

However, it appears that there may not be a current threshold below which osteogenesis is not elicited. Spadaro (1982) obtained osteogenesis with 0.5 μA DC when Pt, Co-Cr, or Ag was used as the cathode, but not when stainless steel was used. Baranowski et al. (1983) however, found a slight osteogenic response to 0.075 μA at a stainless steel cathode. In earlier studies that failed to demonstrate electrically induced osteogenesis at low current levels, the small effect due to electrical stimulation may have been swamped by the larger effects from the trauma of electrode insertion. Unfortunately, though, complete confidence cannot be placed in the study by Baranowski et al. (1983), because instead of comparing the osteogenic response at the 0.075-μA cathode with the response at a sham electrode in the contralateral tibia, they compared it with the response at sham electrodes in a study reported 9 years earlier. Although the use of "historical controls" is valid in many circumstances, this was not one of them: the contralateral tibia was available (in fact, it was used as an "active control," adding nothing to the study) and, as they noted, "even age- and weight-matched rabbits show a considerable range in osteogenic response to uniform stimulus."

Electrical Stimulation Other Than by Constant DC

All modern attempts to stimulate bone growth by electricity are predicated, at least historically, on the observation that bone produces endogenous electrical signals. On the basis of the assumption that bone tissue is capable of responding to these electrical signals, many investigators have tested the osteogenic competency of electrical signals that are more analogous to the naturally occurring stress-generated potentials than is DC, such as AC and pulsed DC. Other stimulation modalities have arisen from efforts to avoid the use of implanted electrodes, both for practical and theoretical reasons. Both capacitive and inductive stimulation techniques have been developed to noninvasively generate an electric field in tissue. Many of these different stimulation modalities, along with galvanic DC, have been imposed on bone cell or organ cultures. These in vitro experiments will be briefly discussed as a group at the end of this section.

Pulsed DC and AC. Besides constant direct current, several investigators have used implanted electrodes to deliver either AC or pulsed DC signals to bone. In general, AC or pulsed DC can stimulate osteogenesis, although the typical cathodal effect seen with constant DC now occurs to some extent at the anode also when the DC is pulsed. Herbst (1978) reviews much of this literature.

Levy (1971) found that a DC pulse with a frequency of 0.5 Hz is substantially more effective in inducing new bone formation than constant DC. From calculations of the transfer function of stress-generated potentials, he determined that 0.7 Hz is the frequency at which the transfer function is maximized. He assumed that if SGPs do affect bone, then bone should respond optimally to a signal around 0.7 Hz; hence the 0.5 Hz used in this test. Weigert and Werhahn (1977), however, used a SGP-simulating current pulse at 0.3 Hz and found the effects to be essentially the same as the effects of a constant DC. Neither report was quantified, so comparison is difficult. In a similar vein, Kenner et al. (1975) used a 5-Hz square pulse signal to modify an experimentally induced disuse osteoporosis. The pulsed voltage caused a dramatic modification of the osteoporosis, even though it was only applied for 1 hour each day. Finally, Klapper and Stallard (1974) report that a 1-Hz AC current causes more new bone formation than 1-Hz pulsed DC. They state that there was no difference between anodal and cathodal effects when the pulsed DC is used.

Taken as a whole, the literature appears to indicate that pulsed DC or low-frequency AC has osteogenic properties similar to steady DC. The difference lies in the effect close to the electrodes. With the time variant signals the electrode effects are less pronounced, perhaps because there is time between pulses for electrode products to diffuse away, and their accumulation at the electrode is avoided. Furthermore, with pulses or AC, electrode polarization is less pronounced. This lowers the electrode access resistance and hence the electrode voltage for a given current. Perhaps more of the ''true'' current effects are demonstrable when electrode effects are reduced or avoided. It is also quite possible that cells are more receptive to a time variant signal than a constant one, thus enhancing the electric effects. Another view is that cells respond to the ''steady'' field portion of a pulsed signal. Only to the extent that the pulsed signal or asymmetric AC signal approaches DC is it a component signal for osteogenesis. In a convincing series of experiments, Brighton et al. (1981) found a linear relationship between the percent of the medullary canal filled with bone and the frequency of a 1-msec-wide pulse. The authors concluded that the total amount of current delivered to the tissue was the pertinent factor.

Stimulation with noninvasive electrodes. All of the current applications discussed so far involve the insertion of an electrode into the tissue to deliver the current. There have been several techniques developed, however, that allow an electrical signal to be coupled to the tissues without requiring an implanted electrode. The most popular of these methods are capacitive and inductive coupling.

Capacitive coupling utilizes two insulated metal plates placed externally on either side of the bone defect. By applying a voltage between the plates, an electric field is produced in the intervening space. This field can be static or dynamic, depending on the signal applied to the plates. Bassett (1971) reported that electrostatic fields (1,000 V/cm) enhanced DNA synthesis and collagen production in cultured fibroblasts. When the field was pulsed at 1 Hz, the reported effect was even more striking. On the basis of these observations, Bassett and Pawluk (1975) attempted to modify fracture repair in rabbit fibulae with an electrostatic field of 100 V/cm. They reported the observation that with the lateral plate negative and the medial plate positive, there was an increase in the quality of "healing" after 21 days. Oddly, however, reversing the polarity of the field resulted in a decrease in bone strength, with more cartilage being found in the osteotomy site, as opposed to the fibrous bone that was found with the other polarity. Little subsequent work was done with electrostatic fields because the large voltages necessary would probably not be permissible in a clinical situation. Also, since most of the research is based, at least superficially, on the concept of simulating the natural electrical phenomenon present in bone, electrostatic fields are not as interesting as dynamic fields or the steady fields established by current flow. Low-frequency capacitively coupled fields have been thought to fulfill the requirements of being both biologically appropriate and noninvasive. To ensure that sufficient current was produced in the tissue, however, high voltages had to be used in many experiments (McElhaney et al., 1968). Various studies have demonstrated significant effects of capacitive fields on cultured bones or cells. Many of these reports are listed by Brighton et al. (1985).

Besides the low-frequency, high-voltage capacitive fields, there has been some experimentation using high-frequency, low-voltage capacitive fields. An advantage of high frequency is that it permits the current to enter the tissue with a lower driving voltage than is needed at low frequencies. For example, Brighton et al. (1983) used a 60-KHz symmetrical sine wave to stimulate growth plate elongation at a voltage of 5–10 V peak-to-peak and fracture healing with a peak-to-peak voltage of only 220 mV (Brighton et al., 1985). In both cases, apparent dose-response curves were obtained using different voltages. A critical examination of the data presented, however, reveals that the reported significance is tenuous. Just the presence of the capacitor plates in the sham controls caused a significant increase in growth plate growth. The increase in growth of the bone with active electrodes was determined to be greater than the sham-induced growth. However, the details of the statistical process are not given; for example, it is unclear which data were paired. In the other study, significance was demonstrated only at a specific voltage and a specific frequency. The disturbing point here is that the number of cases is doubled for these points, because the same combination of frequency and voltage was used twice. Therefore, the question is raised whether statistical significance would have been achieved if the two samples had not been combined. In any case, it should be noted that a 60-KHz sinusoidal signal does not simulate any known biological signal, and, unless a remarkable rectification process is involved, it is not likely to simulate the competent portions of other artificial osteogenic wave forms.

The other popular technique for the noninvasive application of electricity to bone is the use of inductive coupling or electromagnetic stimulation. An electric field can be induced in a conductor by placing it within a time-varying magnetic field. Conversely, a magnetic field is produced by current flowing through coils of wire. Therefore, a time-varying current flowing through coils of wire outside of the body can produce a time-varying magnetic field that penetrates the tissues and produces an electric field and current flow in the conductive tissue of the body.

The first use of pulsed electromagnetic fields (PEMF) on bone tissue was reported in 1974 by Basset et al. They stimulated fibular osteotomies in dogs for 28 days with two different PEMF signals. The first had a 1-msec pulse at 1 Hz that produced a field of 2 mV/cm in the bone, and the second had a 150-μsec pulse at 65 Hz that produced a 20-mV/cm field. The first signal proved ineffective in increasing the rate of healing, even though it more closely resembled the SGPs produced by ambulation. The second signal was reported to have had a statistically significant stimulatory effect. Whether the difference was in the duration of the pulse, the frequency, or the magnitude of the induced field was not determined. Again, there are concerns about data analysis; for example, the claim that P values of less than 0.07 or 0.10 are to be considered significant is questionable.

Another study (Dehass et al., 1979) used the healing of an osteotomy of the radius of a rabbit as the assay for analyzing the competence of different frequencies of PEMF. In this case, frequencies of 0.1 and 1 Hz were found effective, and 5 Hz was ineffective. The slight acceleration seen in the stimulated bones occurred at 2 weeks after surgery and was gone by 3 weeks. These investigators also chose to consider $P < 0.10$ as the level for significance.

These studies indicate that PEMFs may have a slight influence on the rate of healing of fresh bone defects. It has also been shown that PEMF can alter the course of experimentally induced disuse osteoporosis. Basset et al. (1979) showed that PEMF could prevent and possibly reverse the loss of mechanical strength in the cancellus bone of rat tibia in a disuse model. Cruess et al. (1983), using the same model, analyzed biochemical aspects of the PEMF-induced amelioration of the osteoporosis and found significant differences in several areas.

In his recent review, Bassett (1984a) states: "It is now quite clear that the actions of PEMFs and DC electrodes may be quite different . . . . In fact, more and more substance is being added all the time to the concept that changes in pulse characteristics impart high levels of specificity to biologic effects of these time-varying fields." It is amply apparent that imposed electric currents or fields can have a pronounced osteogenic effect on bone. However, the high information content and specificity implied by Basset seems extreme. The term "osteogenesis," for this discussion, has been used in the broadest sense possible, including nearly any positive effect, such as growth, stability, indicator uptake, healing, loading capacity, radiolucency, etc. Nearly all reasonable applications of electricity to bone have, in someone's hand, elicited an osteogenic response. This indicates that electricity may not be acting by specific pathways but rather is acting through general mechanisms established for growth and repair. In this light it is understandable that there are few, if any, reports of growth retardation or inhibition outside of those using high levels of DC, which are almost certainly attributable to local toxic effects. If highly specific cellular mechanisms exist that are coupled to specific electrical signals, then it seems that inappropriate electrical signals should, under at least some of the myriad testing protocols and parameters, produce an inappropriate response. Kenner et al. (1975) report excessive growth of immature bone and discuss other cases of inappropriate growth, but not electrically inhibited growth. It may be useful to consider that applied electrical signals serve as irritants that activate or enhance bone processes that normally react to the more natural irritations of damage, infection, anoxia, microdamage, etc. That natural electrical signals such as SGPs and biopotentials would act by a similar process is a reasonable, although at this point untested, hypothesis.

Electrical effects on bone cells and tissue in vitro. Many investigators have begun to use in vitro cell cultures or organ cultures to study the effects of electricity on bone tissue. In vitro work allows the researcher experimental control over many variables and makes the system both accessible and isolated from other tissues. Besides the typical in vitro culture of isolated cells, it is possible to maintain embryonic bones in organ culture for long periods of time. They continue to grow and develop, remaining responsive to outside influences, such as hormones (Stern and Raisz, 1979) and even mechanical stress (Basset, 1968).

Treharne et al. (1979) used an organ culture system to compare the effects of constant DC, pulsed DC, and SGP-shaped pulses. Pointed stainless steel electrodes inserted into embryonic rat tibiae served as cathodes, while the stainless steel rafts that supported the bones functioned as anodes. After passing currents for 8 days, the investigators noted a statistically significant increase in the thicknesses of the bone cortex near the active electrodes, compared with sham controls. While this increase in cortical thickness was stimulated by all three current modalities, the constant DC was by far the most effective.

Using similar techniques, Noda and Sato (1985b) performed a study in which chick tibiae were cultured not in vitro, but on the chorioallantoic membrane of a host chicken embryo. Two platinum electrodes were inserted into the isolated bone, and the bone was then lowered onto the highly vascularized chorioallantoic membrane of the host egg through a window cut through the shell. They found that after passing 10 μA of constant DC for 10 days the tissue experienced enhanced remodeling of the original bony trabeculae and an accumulation of osteoblasts near both the anode and cathode. The bone near the cathode and between the electrodes demonstrated a thickened proliferative periosteum. The anode was surrounded by a 300–500-μm acellular necrotic zone. Surrounding this zone was an area of bone that was significantly enriched in numbers of both osteoblasts and osteoclasts, with the density of osteoclasts triple that seen elsewhere. The only response generated from 1 μA of constant DC was a 100-μm necrotic zone around the anode. No effect was seen at either the inactive sham electrodes or at the electrodes passing AC.

Organ culture of embryonic bones has also been used to test the effects of capacitively and inductively coupled electrical fields (Gerber et al., 1978; Watson et al., 1975). These studies are equivocal, however, because of the unquantified effects of heat produced by the fields. Gerber et al. (1978) demonstrated the artifactual heat effect by applying electromagnetic fields to embryonic rat femur rudiments in vitro. Femurs placed in the magnetic field for 1 hour five times per day experienced an increase in wet weight 20% higher ($P < 0.01$) than that of controls kept in the same 37°C incubator. When precautions were taken to prevent temperature increase in the stimulated culture dishes, however, the effect completely disappeared.

The effect of static and dynamic capacitive fields on fractional increases in the length of embryonic chick tibiae was investigated by Watson et al. (1975). No increase in growth was demonstrated in bones in the static field

(1,000 V/cm) when compared with paired sham controls. The dynamic field, 1,000 V/cm pulsed at 1 Hz, apparently did cause a slight increase in the length of the bones. While only three of five batches of bone showed statistically significant increases, the total of all five groups was highly significant. Because the temperature was not monitored, whether the slight acceleration in bone growth was a direct effect of the fields or of small temperature increases in the stimulated dishes is unclear.

Just as organ culture studies are a useful complement to in vivo studies, so are cell cultures a powerful complement to organ cultures in studies of mechanisms of electrically mediated effects. Both the behavior of individual cells and the biochemical consequences of electrical fields can be examined. Ferrier et al. (1986) have recently demonstrated a striking difference between the responses of osteoclasts and osteoblasts to an imposed electric field in vitro. Cells from two osteoblast-like cell lines were found to orient perpendicular to the field and migrate toward the cathode in fields of 1,000 mV/mm and 100 mV/mm. Osteoclasts, however, migrated toward the anode when subjected to the same fields. On the whole, osteoclasts were much more responsive to the fields than the osteoblasts, having a shorter lag time before responding (2–10 minutes vs. 5–45 minutes), faster migration (138 μm/hr vs. 5–32 μm/hr), and quicker response to field reversal (3–10 minutes vs. 30–120 minutes) in the 1,000-mV/mm fields. All of the effects were also seen at 100 mV/mm, although the lag times were longer and the migration slower. A multinucleated osteoclast moving toward the anode at 47 μm/hr in a 100-mV/mm field is shown in Figure 6. Of importance is the differential response to electric currents shown for osteoclasts and osteoblasts. Whether similar galvanotaxis is functional in vivo is not known, although the observation by Noda and Sato (1985b) of osteoclast accumulation at the anode is intriguing.

Several research groups have used bone cell cultures to determine the biochemical means by which physical strain and electrical signals alter cellular function. Rodan et al. (1975) have shown that mechanical perturbation of bone cells causes changes in ion fluxes and cyclic nucleotide levels. Somjen et al. (1980) then demonstrated that mechanical stretching of the culture substrate enhanced prostaglandin synthesis. Recently Yeh and Rodan (1984) confirmed this finding. They seeded osteoblast-like cells onto collagen ribbons that were subsequently stretched 5–10% of their length several times. In response to stretching, the cells produced 3.5 times the amount of prostaglandin E (PGE) as control cells. In a similar study, Binderman et al. (1984) found that a slight tensile force increased both cAMP levels and DNA synthesis. From these observations, the hypothesis was developed that PGE_2

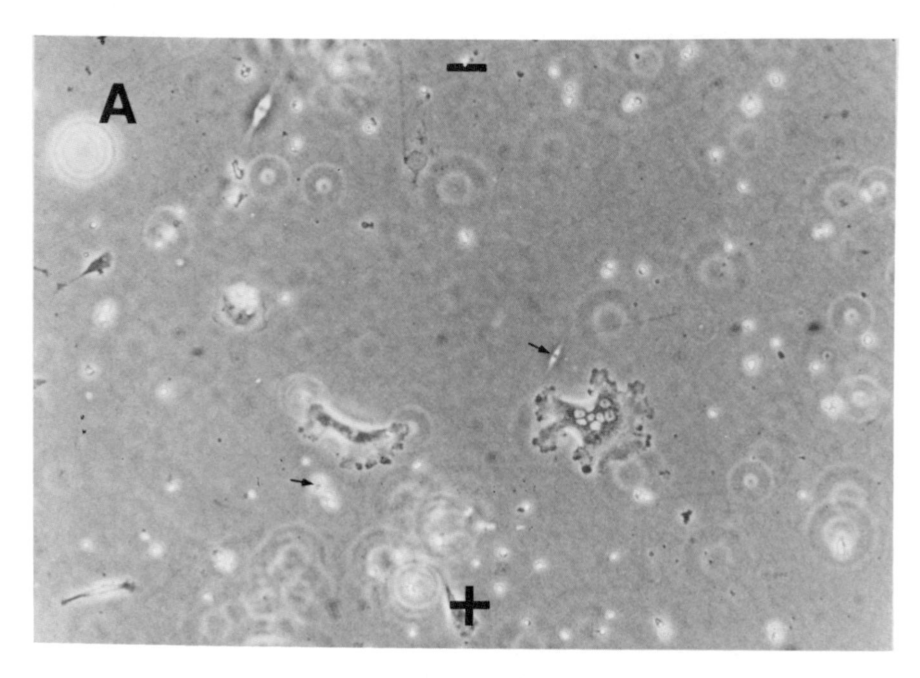

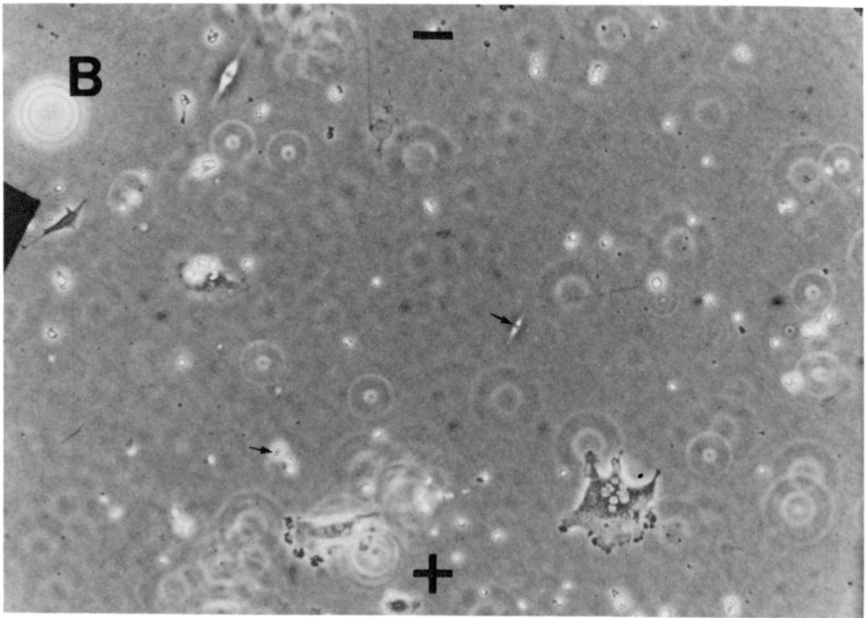

Fig. 6. Osteoclasts migrating in an electric field. **A:** Two osteoclasts have been in a 100-mV/mm field for 136 minutes. **B:** The same two cells 2 hours later. Debris on the substrate can be used for registration of the micrographs. The osteoclasts are migrating toward the anode at a rate of 47 μm/hr. (Micrographs provided by the courtesy of J. Ferrier.)

acts as a "first messenger" in the transduction of physical strain to cell activity, whereas ion fluxes, specifically Ca^{2+}, and cAMP act as "second messengers." PGE_2 can mediate changes in cAMP and also lead to enhanced DNA synthesis either directly or by way of cAMP. Some evidence indicates that responses may be specific for different cell types (Binderman et al., 1984).

It appears that electrical signals may act by altering cAMP levels and ion fluxes through perturbation of the cell membrane, instead of via PGE_2 production, as seen for physical strain. Rodan et al. (1978) showed that oscillating capacitive fields (1,000 V/cm, 5 Hz) could increase the incorporation of tritiated thymidine into the DNA of embryonic chick chondrocytes. The effects of the electric field could be blocked by verapamil, a calcium antagonist, and tetrodotoxin, a sodium channel inhibitor, indicating that the electrical field effect may be mediated through altered fluxes of these ions. Luben et al. (1982) used electromagnetic fields to block both the increase in cAMP production and the inhibition of collagen synthesis by osteoblast-like cells that is induced by parathyroid hormone (PTH). They concluded that both effects are mediated at the plasma membrane, either by direct influences on the hormone receptor or by alterations in ionic fluxes. Binderman et al. (1984) arrived at a similar conclusion. (Interestingly, Binderman et al. [1984] found that indomethacin, a PGE_2 inhibitor, could inhibit the effects of physical strain on DNA synthesis and cAMP production, but does not alter effects induced by electrical fields. This observation allowed them to conclude that the electrical effect is not mediated by PGE_2. Nannmark et al. [1985] demonstrated, however, that indomethacin can inhibit the electrically induced macromolecular leakage of capillaries in bone. Therefore it appears that an electrical stimulus may induce some cell types to produce PGE_2, which then may be able to act directly on bone cells.)

The main conclusions that can be drawn from these and similar studies are that both electrical and mechanical stimuli can affect bone cells directly; the second messenger for both stimuli is probably cAMP; the first messenger for physical strain is PGE_2 (in osteoblasts) and for electrical stimuli is altered ion fluxes; electrical signals can interfere with or alter hormonal effects on cells; and osteoclasts and osteoblasts have distinctly different responses to an applied electrical field. It is important to note that osteoblasts are able to respond directly to mechanical force *without* electricity as an intermediary. However, electrical and mechanical signals apparently act along related biochemical pathways. The distinction between a purely mechanical mechanism and an electrically mediated one as the effector of Wolff's Law will be extremely difficult to determine.

CLINICAL APPLICATIONS OF ELECTRICITY TO BONE

The final topic to be considered is that of the use of electrical therapy as an orthopedic tool (see Lavine and Grodzinsky, 1987 for a well-written, concise review). The history of the clinical use of electricity in stimulating bone healing has been traced by Peltier (1981) to the early 1800s. Apparently, in 1812, a Mr. Birch applied "shocks of electric fluid" to an unhealed fracture of 13-months duration. After 8 weeks of treatment, the patient walked out of the hospital. Peltier also includes an account from Lente in 1850. The additional details in this case describe the application of the electrical stimulus through acupuncture, needles being inserted transcutaneously to the periosteum. Lente states, "Electricity, by the ordinary galvanic apparatus, is easy of application, not very painful, and in no way dangerous; it is therefore one which, I think, should always precede other means" (Peltier, 1981). After 135 years, the picture presented by the clinical literature is little different. Most of the reports are anecdotal accounts of delayed unions or nonunions that went on to consolidation after electricity was applied. The assumption has been made, either explicitly or implicitly, that the electric current or field is directly responsible for the salvage of a long-standing injury. Therefore, since electricity is "easy of application, not very painful, and in no way dangerous," it has become a very popular means of treating delayed unions, and congenital or acquired pseudoarthroses.

Because of ethical considerations, much of medical "experimentation" becomes empiricism, as opposed to the more controlled experimentation possible under laboratory conditions. By its nature, empiricism allows only superficial correlations among observations. This necessarily places much of the clinical literature outside the realm of traditional scientific inquiry.

In more recent times, because electrical therapy has been experimental, review boards initially allowed it to be used only in extreme cases of nonunion when many other therapies had failed and, in many cases, when amputation was imminent. The salvage of these tragic cases by simple electrical stimulation gave impetus to further clinical trials.

In 1979 the Food and Drug Administration approved the clinical use of electrical treatment for nonunions. In 1984, it was estimated that 30% of all nonunions in the United States were being treated with some form of electrical therapy (Brighton, 1984b). With 100,000 nonunions in the U.S. each year (Hall, 1983), perhaps as many as 30,000 Americans are being treated with electrical therapy annually. This figure does not include the substantial number of cases from European and Japanese medical practices, or any veterinary applications.

There are three main forms of electrical therapy in wide medical use. These are termed invasive, semiinvasive, and noninvasive (Brighton, 1984b). The invasive and semiinvasive modalities both use electrodes inserted directly into the defect site and connected to a power source, usually delivering 10–20 μA DC. Surgical implantation of both the electrodes and the power source is used in the invasive technique for a totally internal system (Paterson, 1984). The semiinvasive technique on the other hand, uses an external power source and anode, and percutaneous electrodes (cathodes) inserted into the defect site without surgery (Brighton, 1984c). Electrodes are not used in the noninvasive mode, the electric currents being induced in the tissue by capacitive plates (Brighton and Pollack, 1985) or PEMF coils placed outside the body (Bassett, 1984a). Many variations on these techniques have been used, all seemingly with success. The typical success rates quoted in the literature are between 70% and 90% (Brighton et al., 1979b).

Friedenberg et al. (1971a) are credited with the first modern attempt to heal a nonunion in a human by applying DC current through implanted electrodes. In 1979, this group (Brighton et al., 1979b) reported on their accumulated experience of treating nonunions with constant direct current. Of the 160 nonunions, 134 healed, giving a success rate of 84%. Later, in 1984, they reported a success rate of 73% on 429 nonunions (Brighton, 1984c). In both cases, failures due to inadequate electricity and/or a large gap or synovial pseudoarthrosis were removed from the statistical pool (although successes in this group were not). The investigators clearly imply that a sham treatment without electricity would have yielded a much lower percentage of healing. In fact, some may assume that none of these nonunions would have healed without the electrical therapy. However, Brighton does mention that in cases in which a patient has both a tibial and a fibular nonunion, if the tibia is successfully treated with electricity and heals, "the fibular nonunion heals spontaneously without any further treatment" (Brighton, 1984c). This observation is made in spite of the fact that this group concludes that the electrical effect is local to the cathode.

Likewise, Becker et al. (1977) note that in two cases in which more than one nonunion was present, both the electrically treated and the remote nonunion healed (Becker et al., 1977). In fact, one of the cases, the untreated nonunion healed faster than the treated site. In an attempt to explain this finding, Becker et al., suggest that the optimum "energy input" was achieved at the remote site, while too much "energy" was put into the treated site. This conclusion is based on a faulty understanding of the current distribution. Most likely, all electrical effects were confined to the vicinity of the cathode, and the remote site received very little, if any, stimulation. It is interesting to note, however, that the low current levels (200–900 nA) used by Becker et al. (1977) in this study are two orders of magnitude below the 80 μA commonly used by Brighton's group (four cathodes at 20 μA each). If the reasonable assumption is made that these low-current treatments are essentially the same as no-current sham controls, it may be that Becker et al. have inadvertently supplied the only sham controls for the DC electrical treatment of nonunions in humans. Although Baranowski et al. (1983) and Spadaro (1982) report osteogenesis in response to low currents (< 1 μA), the effect was minimal and restricted to the area immediately adjacent to the cathode. It is unlikely that such a small local effect could significantly alter nonunions in the larger bones of humans. Becker et al. (1977) report a success rate of 77%, and in a rewrite of much of the same data a few years later, they report a 72% success rate with these low currents (Becker and Spadaro, 1979). These results, combined with the observation that even nonunions remote from the site of current application can heal, leads one to question how much healing would occur in the absence of electric currents, but in the presence of the competent standard orthopedics practiced by these clinicians.

Furthermore, nonunions with large fracture gaps, infections, synovial pseudoarthroses, or other complications are frequently excluded from clinical trials or, if included, represent a good portion of the failures. The "uncomplicated" nonunions (probably technical, and not biological, nonunions) which respond most favorably to electrical therapy, are also the ones most likely to respond to simple, sound orthopedic management.

The completely invasive system (Paterson, 1984) is also credited with high success rates for uniting nonunions of long bones (89%). However, even though Paterson warns that "it was essential that these new devices be used responsibly lest the same fate occur as befell electrical stimulation in the nineteenth century," (Paterson, 1984) and says that "a controlled clinical trial commenced in Australia in 1976 using the implanted bone growth stimulator," it appears that no sham controls were ever done. The term "controlled clinical trial" apparently refers to the strict adherence to a set of protocols for patient selection and treatment, and not to the use of a sham or placebo group (Paterson et al., 1980). Although this is commendable, it is not adequate for establishing that electrical therapy can enhance the clinical outcome for a treated human nonunion. Fortunately, in this case a convincing laboratory study in dogs has been presented that bolsters confidence in their claims (Paterson et al., 1977).

Because of the widespread use of the noninvasive PEMF system, there has been a call by some to reevaluate the therapeutic advantages of this treatment

(Barker, 1980; Editorial, 1981). In fact, some suggest that there has never been an adequate initial evaluation of the PEMF technique (Editorial, 1980). It has proved very difficult to develop animal models of nonunions, so little experimental evidence exists to corroborate the clinical claims of PEMF-induced healing of long-standing nonunions. It is reported that approximately 80% of PEMF treated nonunions eventually heal. The clear implication is that none of these would have healed otherwise.

In the only double-blind controlled study yet performed, Barker et al. (1984) found no enhancement of the rates, or degree or percentage of consolidation of nonunions due to PEMF. The experiment was designed so neither the patients, staff, nor physicians knew whether an active or inactive stimulator was used. Of the nine patients treated with PEMF, five of these were judged to have reached clinical union by 24 weeks. However, in the sham control group, five out of seven healed similarly. It seems that the simple conservative management regime that is an adjunct to PEMF therapy was as effective alone as with the PEMF. Some may criticize Barker et al. (1984) for prematurely ending the study, suggesting that a greater number of patients should have been included. Barker, however, makes a convincing argument that if the success rate with PEMF is 80% and the success rate without PEMF is negligible, then large numbers of patients are not necessary to demonstrate a clear statistical difference between the groups. Even this preliminary study clearly indicates that the effectiveness of PEMF therapy can not be wholly attributed to electrical or magnetic effects.

The critical importance of the Barker et al. (1984) paper cannot be overemphasized. Simple, conservative orthopedic management resulted in union of long-standing nonunions *without* the addition of PEMF, or any electrical therapy in the control group. In fact, one patient in the sham control group had had an ununited tibia for 11 years with sepsis present at the time he entered the trial. Twenty-four weeks later the tibia was united after placebo treatment with PEMF. It appears that the non-weight-bearing regime advocated by Bassett (1984a) as an adjunct to PEMF therapy may alone be responsible for the union.

The negative side of the clinical picture must be emphasized to counteract the overzealous arguments put forward by those in favor of electrical therapy (Bassett, 1984a,b). In a chapter in a medical text on fracture management, Bassett (1984b) lauds PEMF as the treatment of choice for nonunions. He also predicts the coming of "a nirvana" in which the need for a chapter on treatment of ununited fractures "would cease to exist," presumably because of elimination of all nonunions by early treatment with PEMF.

Bassett acknowledges that "the relative contributions of PEMFs, on the one hand, and good-casting and non-weight-bearing, on the other, in the successful management of a patient cannot be identified with precision.'' He then dismisses this concern because it will not matter to the patient, as long as his/her problem is dispensed with ''in short, safe, and comfortable order'' (Bassett, 1984a). He further emphasizes the absolute, unequivocal necessity of rigid casting or fixation and non-weight-bearing to the success of PEMF therapy (Bassett, 1984a,b). It is interesting to note that he then denigrates another system of electrical therapy by saying ''internal fixation is always a part of the method, so it is difficult, if not impossible, to differentiate between effects of the surgical procedure and those of the electrical currents'' (Bassett, 1984b).

A final pair of quotes from Bassett (1984a) will conclude this argument. In response to the call for a sham-controlled study, he replies that ''the effectiveness of PEMFs in large numbers of long-standing (more than two years), multiply operated, and infected nonunions attests amply to their biologic action without requiring morally difficult to justify double-blind methods in such desperate individuals . . . [A] few inexperienced individuals still demand double-blind studies of nonunion before they will accept the overwhelming evidence that such studies are no longer needed.'' Simply put, the study by Barker et al. (1984) indicates otherwise. Such studies are still needed to determine whether conservative management or PEMF is the ''real'' therapy. Lavine and Grodzinsky (1987) state that ''randomized trials have been performed successfully for the study of many problems that are more serious than the nonunion of fractures'' and that such trials ''will ultimately be essential in order to establish its [i.e., electrical stimulation] proper clinical role in the treatment of orthopaedic problems.''

In conclusion, the effectiveness of electrical therapy in consolidating and healing nonunions is still open to question. Some laboratory evidence, along with accumulated case histories of recalcitrant nonunions that went on to union following treatment, argue favorably for the clinical use of electrical therapy. One must critically examine this evidence, however, remembering that appropriate nonunion models are difficult to produce in animals, that the laboratory studies are in many cases flawed and unconvincing, and that adequate sham controls in clinical trials have not been used. Anecdotal accounts of nonunion repair are not adequate substitutes.

Although it is abundantly clear that electricity, especially DC currents, can induce osteogenesis, this fact by no means is sufficient to assume that its application in clinical situations is warranted. Hopefully, electrical therapy will prove to be an effective and efficient technique for modulating bone healing and repair. The clinical literature (Brighton, 1977a, 1984a; Connolly, 1981) *suggests* this as a strong possibility. However, these claims must be substantiated by rigorous evidence before they will be accepted by the wider scientific and medical community.

SUMMARY

This brief review, by no means exhaustive, has attempted to give an overview of what is known about the electrical nature of bone. The critical tone of this review is meant to compensate for more specious reviews that accept much of the faulty work at face value, and not to calumniate an entire area of investigation. It is amply apparent that bone can both produce and respond to electrical signals, and many aspects of these phenomena have become well established.

The stress-generated potentials that have been measured from bone in response to applied force are produced in dry bone by the piezoelectric nature of its collagen component, and in wet bone by streaming potentials. These SGPs have been demonstrated at the osteonal level and are thought to mediate Wolff's law. However, it is clear that bone cells can respond directly to pressure and stress without electrical signals acting as an intermediary (Rodan et al., 1975; Somjen et al., 1980; Yeh and Rodan, 1984; Binderman, et al., 1984). Also, SGPs, either by piezoelectric or streaming mechanisms, are a natural consequence of the "orderedness" of biological tissue and are not unique to bone. Thus it may be that SGPs are an epiphenomenon with no direct biological consequences and that the concept that SGPs must mediate Wolff's Law is a teleologic argument.

Bone has also been described as having steady metabolic potentials maintained by a "bone membrane." These potentials have been detected indirectly by ion flux measurements (Peterson et al., 1985; Trumbore et al., 1980) and by current measurements (Borgens, 1982), although the bone compartment was found to be positive with respect to the extracellular fluid in the first case and negative in the second. The measurement of bioelectric potentials, which presumably reflects some component of this potential, adds little to the understanding of bone physiology. Also, the common belief that the BEPs demonstrate a striking negativity at a fracture site is *not* supported by experimental evidence. Seemingly, the only consistent finding from the many BEP studies is that the proximal one-half of the rabbit tibia is negative to the distal half. The significance of this observation is not obvious. Therefore, it can be concluded that the nature of the steady metabolic potentials in bone is still unknown, although the accumulated evidence suggests that these potentials do exist.

The most well-established observation reported so far is the production of bone at a metallic cathode in response to small DC currents. This electrically induced osteogenesis has also been reported for other waveforms and at other locations. Although this phenomenon is well established, quantitation of the amount of induced osteogenesis has been difficult and has hampered efforts to define the parameters of the response. It is now incumbent upon researchers to move beyond the phenomenology of the response and increase efforts to standardize the effect and investigate the mechanisms involved. A significant step in this direction has been the in vitro studies of the effects of electricity on bone cells.

The application of electricity as an orthopedic therapy has preceded an understanding of the phenomenon and has resulted in an empirical approach to clinical applications. While this is understandable, and even acceptable, given the slow pace of the laboratory studies, the concern is that some investigators now consider accumulated case histories as adequate substitutes for experimental evidence. It must be remembered that, especially in the case of PEMF treatment, very rigid casting and strict non-weight-bearing are necessities for the success of electrical therapy. Only adequately controlled studies will be able to determine which component of the therapy, the conservative management or the electromagnetic fields, is most relevant to successful treatment. Only then will real progress be made in the orthopedic management of nonunions.

Finally, the view that the electrical phenomena of bone constitute a grand, unifying theory for bone growth, fracture healing, Wolff's law, and mineral metabolism is intriguing and quite exciting. The evidence against this theory is not so much from negative results, but from the inadequacy of the positive results so far proffered in its support.

As Santiago Ramón y Cajal (1937) once stated, "I insist upon these details because I wish to warn against the invincible attraction of theories which simplify and unify seductively."

REFERENCES

Anderson JC and Eriksson C (1968) Electrical properties of wet collagen. Nature 218:166–168.

Athenstaedt H (1970) Permanent longitudinal electric polarization and pyroelectric behavior of collaginous structures and nervous tissue in man and other vertebrates. Nature 228:830–834.

Baranowski TJ, Black J, Brighton CT, and Friedenberg ZB (1983) Electrical osteogenesis by low direct current. J. Orthop. Res. 1:120–128.

Barker AT (1980) Electromagnetic stimulation of bone healing, the need for a multicentre collaboration. J. Med. Eng. Technol. 4:271.

Barker AT, Jaffe LF, and Vanable JW Jr. (1982) The glabrous epidermis of cavies contains a powerful battery. Am. J. Physiol. 242:R358–366.

Barker AT, Dixon RA, Sharrard WJW, and Sutcliffe ML (1984) Pulsed magnetic field therapy for tibial non-union. Lancet 1:994–996.

Bassett CAL (1968) Biological significance of piezoelectricity. Calcif. Tissue Res. 1:252–272.

Bassett CAL (1971) Biophysical principles affecting bone structure. In Bourne GH (ed.): *The Biochemistry and Physiology of Bone,* ed. 2, Vol. III. Academic Press, New York, pp. 1–76.

Bassett CAL (1984a) The development and application of pulsed electromagnetic fields (PEMFs) for ununited fractures and arthrodeses. Orthop. Clin. 15:61–87.

Bassett CAL (1984b) The electrical management of ununited fractures. In Gossling HR and Pillsbury SL (eds.): *Complications of Fracture Management.* J.B. Lippincott Company, Philadelphia, pp. 9–39.

Bassett CAL and Becker RO (1962) Generation of electric potentials by bone in response to mechanical stress. Science 137:1063–1064.

Bassett CAL and Pawluk RJ (1975) Noninvasive methods for stimulating osteogenesis. J. Biomed. Mater. Res. 9:371–374.

Bassett CAL, Pawluk RJ, and Becker RO (1964) Effects of electric currents on bone *in vivo.* Nature 204:652–654.

Bassett CAL, Pawluk RJ, and Pilla AA (1974) Augmentation of bone repair by inductively coupled electromagnetic fields. Science 184:575–577.

Bassett LS, Tzitzikalakis G, Pawluk RJ, and Bassett CAL (1979) Prevention of disuse osteoporosis in the rat by means of pulsing electromagnetic fields. In Brighton CT, Black J, and Pollack SR (eds.): *Electrical Properties of Bone and Cartilage,* Grune & Stratton, New York, London, pp. 311–331.

Becker RO and Spadaro JA (1979) Experience with low-current/silver electrode treatment of nonunion. In Brighton CT, Black J, and Pollack SR (eds.): *Electrical Properties of Bone and Cartilage.* Grune and Stratton, New York, pp. 631–638.

Becker RO, Bassett CA, and Bachman CH (1964) Bioelectrical factors controlling bone structure. In Frost HM (ed.): *Bone Biodynamics.* Little, Brown Co., Boston, pp. 209–231.

Becker RO, Spadaro JA, and Marino AA (1977) Clinical experiences with low intensity direct current stimulation of bone growth. Clin. Orthop. 124:75–83.

Behari J and Singh S (1981) Bioelectric characteristics of unstressed *in vivo* bone. Med. Biol. Eng. Comput. 19:49–54.

Binderman I, Shimshoni Z, and Somjen D (1984) Biochemical pathways involved in the translation of physical stimulus into biological message. Calcif. Tissue Int. 36:S82–S85.

Borgens RB (1982) What is the role of naturally produced electric currents in vertebrate regeneration and healing? Int. Rev. Cytol. 76:245–298.

Borgens RB (1984) Endogenous ionic currents traverse intact and damaged bone. Science 225:478–482.

Borgens RB, Vanable JW Jr., and Jaffe LF (1977) Bioelectricity and regeneration: Large currents leave the stumps of regenerating newt limbs. Proc. Natl. Acad. Sci. U.S.A. 74:4528–4532.

Borgens RB, Jaffe LF, and Cohen MJ (1980) Large and persistent electrical currents enter the transected lamprey spinal cord. Proc. Natl. Acad. Sci. U.S.A. 77:1209–1213.

Borgens RB, Rouleau MF, and DeLanney LE (1983) A steady efflux of ionic current predicts hind limb development in the axolotl. J. Exp. Zool. 288:491–503.

Brighton CT (ed) (1977a) Symposium on bioelectrical effects on bone and cartilage. Clin. Orthop. 124.

Brighton CT (1977b) Editorial comment: Bioelectrical effects on bone and cartilage. 124:2–4.

Brighton CT (ed) (1984a) Symposium on electrically induced osteogenesis. Orthop. Clin. 15.

Brighton CT (1984b) Foreword: Symposium on electrically induced osteogenesis. Orthop. Clin. 15:1–2.

Brighton CT (1984c) The semi-invasive method of treating nonunion with direct current. Orthop. Clin. 15:33–45.

Brighton CT and Friedenberg ZB (1974) Electrical stimulation and oxygen tension. Ann. N.Y. Acad. Sci. 238:314–320.

Brighton CT and Pollack SR (1985) Treatment of recalcitrant non-union with a capacitively coupled electrical field. A preliminary report. J. Bone Joint Surg. 67-A:577–585.

Brighton CT and Hunt RM (1986) Ultrastructure of electrically induced osteogenesis in the rabbit medullary canal. J. Orthop. Res. 4:27–36.

Brighton CT, Black J, and Pollack SR (eds.) (1979a) *Electrical Properties of Bone and Cartilage.* Grune and Stratton, New York.

Brighton CT, Friedenberg ZB, and Black J (1979b) Evaluation of the use of constant direct current in the treatment of nonunion. In Brighton CT, Black J, and Pollack SR (eds): *Electrical Properties of Bone and Cartilage.* Grune and Stratton, New York, pp. 519–545.

Brighton CT, Friedenberg ZB, Black J, Esterhai JL, Mitchell JEI, and Montique F (1981) Electrically induced osteogenesis: Relationship between charge, current density, and the amount of bone formed: Introduction of a new cathode concept. Clin. Orthop. 161:122–132.

Brighton CT, Pfeffer GB, and Pollack SR (1983) *In vivo* growth plate stimulation in various capacitively coupled electrical fields. J. Orthop. Res. 1:42–49.

Brighton CT, Hozak WJ, Brager MD, Windsor RE, Pollack SR, Vreslovik EJ, and Kotwick JE (1985) Fracture healing in the rabbit fibula when subjected to various capacitively coupled electrical fields. J. Orthop. Res. 3:331–340.

Buch F, Albrektsson T, and Herbst E (1984) Direct current influence on bone formation in titanium implants. Biomaterials 5:341–346.

Buch F, Nannmark U, and Albrektsson T (1985) Bone tissue reactions during electrical stimulation. A vital microscopic study. 5th Annual BRAGS, Boston.

Buch F, Albrektsson T, and Herbst E (1986) The bone growth chamber for quantification of electrically induced osteogenesis. J. Orthop. Res. 4:194–203.

Burny F, Herbst E, and Hinsenkamp M (eds.) (1978) *Electric Stimulation of Bone Growth and Repair.* Springer-Verlag, Berlin, Heidelberg, New York.

Canas F, Terepka AR, and Neuman WF (1969) Potassium and milieu interieur of bone. Am. J. Physiol. 217:117–120.

Cerguiglini S, Cignitti M, Marchetti M, and Salleo A (1967) On the origin of electrical effects produced by stress in the hard tissues of living organisms. Life Sci. 6:2651–2660.

Cochran GVB (1972) Experimental methods for stimulation of bone healing by means of electrical energy. Bull. N.Y. Acad. Med. 48:899–911.

Cochran GVB, Pawluk RJ, and Bassett CAL (1968) Electromechanical characteristics of bone under physiologic moisture conditions. Clin. Orthop. 58:249–270.

Connolly JF (ed.) (1981a) Symposium on clinical applications of bioelectric effects. Clin. Orthop. 161.

Connolly JF (1981b) The orthopaedic-industrial complex, the clinical applications of bioelectric effects and a reminder from Ben Franklin. Clin. Orthop. 161:2–3.

Connolly JF, Hahn H, and Jardon OM (1977) The electrical enhancement of periosteal proliferation and delayed fracture healing. Clin. Orthop. 124:97–105.

Cruess RL (1984) Healing of bone, tendon, and ligaments. In Rockwood CA and Green DP (eds.): *Fractures in Adults,* Vol. 1. J.B. Lippincott Co., Philadelphia, pp. 147–172.

Cruess RL, Kan K, and Bassett CAL (1983) The effect of pulsing electromagnetic fields on bone metabolism in experimental disuse osteoporosis. Clin. Orthop. 173:245–250.

Dehass WG, Lazarovici MA, and Morrison DM (1979) The effect of low frequency magnetic fields on the healing of osteotomized rabbit radius. Clin. Orthop. 145:245–251.

Digby PSB (1966) Mechanism of calcification in mammalian bone. Nature 212:1250–1252.

Editorial (1980) Electricity and bones. Br. Med. J. Aug. 16:470–471.

Editorial (1981) Electromagnetism and bone. Lancet 1:815.

El Messiery MA (1981) Physical basis for piezoelectricity of bone matrix. I.E.E. Proc. 128:336–345.

El Messiery MA, Hastings GW, and Rakowski S (1979) Ferro-electricity of dry cortical bone. J. Biomed. Eng. 1:63–65.

Eriksson C (1976a) Bone morphogenesis and surface charge. Clin. Orthop. 121:295–302.

Eriksson C (1976b) Electrical properties of bone. In Bourne GH (ed.): *The Biochemistry and Physiology of Bone,* ed 2, Vol. IV. Academic Press, New York, pp. 329–384.

Eriksson C and Jones S (1977) Bone mineral and surface charge. Clin. Orthop. 128:351–353.

Esterhai JL, Friedenberg ZB, Brighton CT, and Black J (1985) Temporal course of bone formation in response to constant direct current stimulation. J. Orthop. Res. 3:137–139.

Ferrier J, Ross SM, Kanehisa J, and Aubin JE (1986) Ostoclasts and osteoblasts migrate in opposite directions in response to a constant electrical field. J. Cell. Physiol. 129:283–288.

Friedenberg ZB and Brighton CT (1966) Bioelectric potentials in bone. J. Bone Joint Surg. 48-A:915–923.

Friedenberg ZB and Kohanim M (1968) The effect of direct current on bone. Surg. Gynecol. Obstet. 127:97–102.

Friedenberg ZB and Smith HG (1969) Electrical potentials in intact and fractured tibia. Clin. Orthop. 63:222–225.

Friedenberg ZB, Andrews ET, Smolenski BI, Pearl BW, and Brighton CT (1970) Bone reactions to varying amounts of direct current. Surg. Gynecol. Obstet. 131:894–899.

Friedenberg ZB, Harlow MC, and Brighton CT (1971a) Healing of a nonunion of the medial malleolus by means of direct current—a case report. J. Trauma 11:883–885.

Friedenberg ZB, Roberts PG, Didizian NH, and Brighton CT (1971b) Stimulation of fracture healing by direct current in the rabbit fibula. J. Bone Joint Surg. 53-A: 1400–1408.

Friedenberg ZB, Harlow MC, Heppenstall RB, and Brighton CT (1973) The cellular origin of bioelectric potentials in bone. Calcif. Tissue Res. 13:53–62.

Friedenberg ZB, Zemsky LM, Pollis RP, and Brighton CT (1974) The response of non-traumatized bone to direct current. J. Bone Joint Surg. 56:1023–1030.

Frost HM (1973) *Bone Modeling and Skeletal Remodeling Errors.* C.C. Thomas, Springfield, IL, pp. 151–159.

Frost HM (1980) Skeletal physiology and bone remodeling. In Urist MR (ed.): *Fundamental and Clinical Bone Physiology.* Lippincott, Philadelphia, pp. 208–241.

Fukada E and Yasuda I (1957) On the piezoelectric effect on bone. J. Physiol. Soc. Jpn. 12:1158–1162.

Geddes LA and Baker LE (1975) *Principles of Applied Biomedical Instrumentation.* John Wiley and Sons, New York, pp. 304–308.

Geisler JZ and Neuman WF (1969) The membrane control of bone potassium. Proc. Soc. Exp. Biol. Med. 130:608–612.

Gerber H, Cordey J, and Perren SM (1978) Influence of magnetic fields on growth and regeneration in organ culture. In Burny F, Herbst E, and Hinsenkamp M (eds.): *Electric Stimulation of Bone Growth and Repair.* Springer-Verlag, Berlin, pp. 35–40.

Gross D and Williams WS (1982) Streaming potentials versus piezoelectricity in the electro-mechanical response of physiologically moist bone. J. Biomech. 15:277–295.

Hall BK (1983) Bioelectricity and cartilage. In Hall BK (ed.): *Cartilage, Vol. 3, Biomedical Aspects.* Academic Press, Inc., New York, pp. 309–338.

Hambury HJ, Watson J, Sivyer A, and Ashley DJB (1971) Effect of microamp electrical currents on bone *in vivo* and its measurement using strontium-85 uptake. Nature 231:190–191.

Herbst E (1978) Electrical stimulation of bone growth and repair: A review of different stimulation methods. In Burney F, Herbst E, and Hinsencamp M (eds.): *Electric Stimulation of Bone Growth and Repair.* Springer-Verlag, Berlin, Heidelberg, New York, pp. 1–13.

Inoue S, Ohashi T, Yasuda I, and Fukada E (1977) Electret induced callus formation in the rat. Clin. Orthop. 124:57–58.

Inoue S, Ohashi T, Fukada E, and Ashihara T (1979) Electric stimulation of osteogenesis in the rat: Amperage of three different stimulation methods. In Black J, Pollack S (eds.): *Electrical Properties of Bone and Cartilage.* Grune and Stratton, New York, pp. 199–213.

Jaffe LF (1979) Control of development by ionic currents. In Cone R and Dowlings J (eds.): *Membrane Transduction Mechanism.* Soc. Gen. Physiol. 33rd ed., pp. 199–231.

Jaffe LF and Nuccitelli R (1974) An ultrasensitive vibrating probe for measuring steady extracellular currents. J. Cell Biol. 63:614–628.

Jaffe LF and Nuccitelli R (1977) Electrical controls of development. Annu. Rev. Biophys. Bioeng. 6:445–476.

Jaffe LF and Vanable JW Jr. (1984) Electric fields and wound healing. Clin. Dermatol. 2:34–44.

Janssen LWM, Roelofs JMM, Visser WJ, and Wittebol P (1978) Hypothesis of bone remodeling and fracture healing by electrostimulation. In Burny F, Herbst E, and Hinsenkamp M (eds.): *Electric Stimulation of Bone Growth and Repair.* Spring-Verlag, Berlin, pp. 61–67.

Kenner GH, Gabrielson EW, Lovell JE, Marshall AE, and Williams WS (1975) Electrical modification of disuse osteoporosis. Calcif. Tissue Res. 18:111–117.

Klapper L and Stallard RE (1974) Mechanism of electrical stimulation of bone formation. Ann. N.Y. Acad. Sci. 238:530–539.

LaCroix P (1971) The internal remodeling of bones. In Bourne GH (ed.): *The Biochemistry and Physiology of Bone,* ed. 2, Vol. III. Academic Press, New York and London, pp. 119–144.

Lang SB (1966) Pyroelectric effect in bone and tendon. Nature 212:704.

Lanyon LE and Hartman W (1977) Strain related electrical potentials recorded *in vitro* and *in vivo*. Calcif. Tissue Res. 22:315–327.

Lavine LS and Grodzinsky AJ (1987) Electrical stimulation of repair of bone. J. Bone Joint Surg. 69-A:626–630.

Lavine LS, Lustrin I, and Shamos MH (1969) Experimental model for studying the effect of electric current on bone *in vivo*. Nature 224:1112–1113.

Lavine LS, Lustrin J, Shamos MH, and Moss ML (1971) The influence of electric current on bone regeneration *in vivo*. Acta Orthop. Scand. 42:305–314.

Levy D (1971) Induced osteogenesis by electrical stimulation. J. Electrochem. Soc. 118:1438–1442.

Liboff AR and Rinaldi RA (eds.) (1974) Electrically mediated growth mechanisms in living systems. Ann. N.Y. Acad. Sci. 238.

Lokietek W, Pawluk RJ, and Bassett CAL (1974) Muscle injury potentials: A source of voltage in the undeformed rabbit tibia. J. Bone Joint Surg. 56B:361–369.

López-Durán Stern L, and Yageya J (1980) Bioelectric potentials after fracture of the tibia in rats. Acta Orthop. Scand. 51:601–608.

Luben RA, Cain CD, Chen MC-Y, Rosen DM, and Adey WR (1982) Effects of electromagnetic stimuli on bone and bone cells *in vitro:* Inhibition of responses to parathyroid hormone by low-energy low-frequency fields. Proc. Natl. Acad. Sci. U.S.A. 79:4180–4184.

Lund EJ (1947) *Bioelectric Fields and Growth.* University of Texas Press, Austin.

McElhaney JH, Stalnaker R, and Bullard R (1968) Electric fields and bone loss of disuse. J. Biomech. 1:47–52.

McGinnis ME and Vanable JW Jr. (1986) Electrical fields in *Notophthalmus viridescens* limb stumps. Dev. Biol. 116:184–193.

McKibbin B (1978) The biology of fracture healing in long bones. J. Bone Joint Surg. 60-B: 150–162.

Nannmark U, Buch F, and Albrektsson T (1985) Vascular reactions during electrical stimulation. Acta Orthop. scand. 56:52–56.

Nelson DJ, Ehrenfeld J, and Lindemann B (1978) Volume changes and potential artifacts of epithelial cells of frog skin following impalement with microelectrodes filled with 3M KCl. J. Membr. Biol. Special issue, 91–119.

Noda M and Sato A (1985a) Calcification of cartilaginous matrix in culture by constant direct current stimulation. Clin. Orthop. 193:281–287.

Noda M and Sato A (1985b) Appearance of osteoclasts and osteoblasts in electrically stimulated bones cultured on chorioallantoic membranes. Clin. Orthop. 193:288–298.

Norimatsu H, Vander Wiel CJ, and Talmage RV (1979) Morphological support of a role for cells lining bone surfaces in maintenance of plasma calcium concentration. Clin. Orthop. 138:254–262.

Norton LA, Bourret LA, Majeska RJ, and Rodan GA (1979) Adherence and DNA synthesis changes in hard tissue cell culture produced by electric perturbation. In Brighton CT, Black J, Pollack SR (eds.): *Electrical Properties of Bone and Cartilage*. Grune and Stratton, New York, pp. 443–454.

Nuccitelli R (1983) Transcellular ion currents: Signals and effectors of cell polarity. Modern Cell Biol. 2:451–481.

Nuccitelli R and Jaffe L (1974) Spontaneous current pulses through developing fucoid eggs. Proc. Natl. Acad. Sci. U.S.A. 71:4855–4859.

Nunes MA and Lacaz Vieira F (1975) Negative potential level in the outer layer of the toad skin. J. Membr. Biol. 24:161–181.

O'Conner BT, Charlton HM, Currey JD, Kirby DRS, and Woods C (1969) Effects of electrical currents on bone *in vivo*. Nature 222:162–163.

Paterson DC (1984) Treatment of nonunion with a constant direct current: A totally implantable system. Orthop. Clin. 15:47–59.

Paterson DC, Hillier TM, Carter RF, Ludbrook J, Maxwell GM, and Savage JP (1977) Experimental delayed union of the dog tibia and its use in assessing the effect of an electrical bone growth stimulator. Clin. Orthop. 128:340–350.

Paterson DC, Lewis GN, and Cass CA (1980) Treatment of delayed union and nonunion with an implanted direct current stimulator. Clin. Orthop. 148:117–128.

Peltier LF (1981) A brief historical note on the use of electricity in the treatment of fractures. Clin. Orthop. 161:4–7.

Peterson DR, Heideger WJ, and Beach KW (1985) Calcium homeostasis: The effect of parathyroid hormone on bone membrane electrical potential difference. Calcif. Tissue Int. 37:307–311.

Petersson C, Holmer NG, and Johnell O (1982) The effect of transistor-regulated direct current on non-fractured rabbit femur. Acta Orthop. Scand. 53:161–165.

Pienkowski D and Pollack SR (1983) The origin of stress-generated potentials in fluid-saturated bone. J. Orthop. Res. 1:30–41.

Pollack SR (1984) Bioelectrical properties of bone: Endogenous electrical signals. Orthop. Clin. 15:3–14.

Pollack SR, Petrov N, Salzstein R, Brankov G, and Blagoeva R (1984) An anatomical model for streaming potentials in osteons. J. Biomech. 17:627–636.

Ramón y Cajal S (1937) *Recollections of My Life*. Craigie EH and Cano J, translators. MIT Press, Cambridge, MA, and London, p. 303.

Ro J, Sudmann E, and Marton PF (1976) Effect of indomethacin on fracture healing in rats. Acta. Orthop. Scand. 47:588–599.

Robinson KR (1983) Endogenous electrical current leaves the limb and prelimb region of the *Xenopus* embryo. Dev. Biol. 97:203–211.

Robinson KR (1985) The responses of cells to electrical fields: A review. J. Cell Biol. 101:2023.

Rodan GA, Bourret LA, Harvey BA, and Mensi T (1975) Cyclic AMP and cyclic GMP: Mediators of mechanical effects on bone remodeling. Science 189:467–468.

Rodan GA, Bourret LA, and Norton LA (1978) DNA synthesis in cartilage cells is stimulated by oscillating electric fields. Science 199:690–692.

Rosene HF (1947) Bibliography of continuous bioelectric currents and bioelectric fields in animals and plants. In Lund EJ (ed.): *Bioelectric Fields and Growth*. University of Texas Press, Austin, TX, pp. 301–391.

Rubinacci A and Tessari L (1983) A correlation analysis between bone formation rate and bioelectric potentials in rabbit tibia. Calcif. Tissue Int. 35:728–731.

Rubinacci A, Brigatti L, and Tessari L (1984) A reference curve for axial bioelectric potentials in rabbit tibia. Bioelectromagnetics 5:193–202.

Sevitt S (1981) *Bone Repair and Fracture Healing in Man*. Churchill Livingston, Edinburgh, pp. 1–179.

Shamos MH, Lavine LS, and Shamos MI (1963) Piezoelectric effect in bone. Nature 197:81.

Simmons OJ (1980) Fracture Healing. In Urist MR (ed.): *Fundamental and Clinical Bone Physiology*. J.B. Lippincott Co., Philadelphia, pp. 283–330.

Singh S and Saha S (1984) Electrical properties of bone: A review. Clin. Orthop. 186:249–271.

Somjen D, Binderman I, Berger E, and Harell A (1980) Bone modeling induced by physical stress is prostaglandin mediated. Biochem. Biophys. Acta 629:91–100.

Spadaro JA (1977) Electrically stimulated bone growth in animals and man. Clin. Orthop. 122:325–332.

Spadaro JA (1982) Electrically enhanced osteogenesis at various metal cathodes. J. Biomed. Mater. Res. 16:861–873.

Spadaro JA and Becker RO (1979) Function of implanted cathodes in electrode-induced bone growth. Med. Biol. Eng. Comput. 17:769–775.

Spadaro JA, Mino DE, Chase SE, Werner FW, and Murray DG (1986) Mechanical factors in electrode-induced osteogenesis. J. Orthop. Res. 4:37–44.

Starkebaum W, Pollack SR, and Korostoff E (1979) Microelectrode studies of stress-generated potentials in four-point bending of bone. J. Biomed. Mater. Res. 13:729–751.

Stern PH and Raisz LG (1979) Organ culture of bone. In Simmons DJ and Kunin AS (eds.): *Skeletal Research*. Academic Press, New York, pp. 22–60.

Sudmann E and Bang G (1979) Indomethacin-induced inhibition of Haversian remodelling in rabbits. Acta. Orthop. Scand. 50:621–627.

Talmage RV (1969) Calcium homeostasis-calcium transport-parathyroid action. Clin. Orthop. 67:210–224.

Talmage RV (1970) Morphological and physiological considerations in a new concept of calcium transport in bone. Am. J. Anat. 129:467–476.

Tasaki I and Singer I (1968) Some problems involved in electrical measurements of biological systems. Ann. N.Y. Acad. Sci. 148:36–53.

Tonna EA and Cronkite EP (1961) Cellular response to fracture studied with tritiated thymidine. J. Bone Joint Surg. 43A:352–362.

Treharne RW (1981) Review of Wolff's law and its proposed means of operation. Orthop. Rev. 10:35–47.

Treharne RW, Brighton CT, Korostoff E, and Pollack SR (1979) Application of direct, pulsed, and SGP-shaped currents to *in vitro* fetal rat tibia. In Brighton CT, Black J, and Pollack SR (eds.): *Electrical Properties of Bone and Cartilage*. Grune and Stratton, New York, pp. 169–180.

Trumbore DC, Heideger WJ, and Beach KW (1980) Electrical potential difference across bone membrane. Calcif. Tissue Int. 32:159–168.

Watson J. (1981) Electricity and bone healing. IEE Proc. 128:329–335.

Watson J, DeHass WG, and Hauser SS (1975) Effect of electric fields on growth rate of embryonic chick tibiae *in vitro*. Nature 254:331–332.

Watson J. (1981) Electricity and bone healing. IEE Proc. 128:329–335.

Watson J, DeHass WG, and Hauser SS (1975) Effect of electric fields on growth rate of embryonic chick tibiae *in vitro*. Nature 254:331–332.

Weigert M and Werhahn C (1977) The influence of electric potentials on plated bones. Clin. Orthop. 124:20–30.

Weinberg C, Ilfeld FW, Rosen V, August W, and Baddorf RL (1982) Electrical potentials in medullary bone. Clin. Orthop. 171:256–263.

Weisenseel MH, Nuccitelli R, and Jaffe LF (1975) Large electrical currents traverse growing pollen tubes. J. Cell Biol. 66:556–567.

Williams WS, and Breger L (1975) Piezoelectricity in tendon and bone. J. Biomech. 8:407–413.

Yasuda I (1953) Fundamental aspects of fracture treatment. J. Kyoto Med. Soc. 4:395. Translated and reprinted in Clin. Orthop. 124:5–8, 1977.

Yasuda I (1974) Mechanical and electrical callus. Ann. N.Y. Acad. Sci. 238:457–465.

Yasuda I (1977) Electrical callus and callus formation by electret. Clin. Orthop. 124:53–56.

Yeh C-K and Rodan GA (1984) Tensile forces enhance prostaglandin E synthesis in osteoblastic cells grown on collagen ribbons. Calcif. Tissue Int. 36:S67–S71.

Zichner L and Happel M (1979) Treatment of congenital and acquired nonunions by means of an invasive device. In Brighton CT, Black J, and Pollack SR (eds.): *Electrical Properties of Bone and Cartilage*. Grune and Stratton, New York, pp. 581–596.

Author Index

Subject Index